AF322676

Computational Simulation Tools in Engineering

Computational Simulation Tools in Engineering

V. Ramesh Kumar

Professor & Vice-Principal
University college of Technology, Osmania University,
Hyderabad, Telangana, India.

T. Bala Narsaiah

Professor,
Department of Chemical Engineering,
Jawaharlal Nehru Technological University College of Engineering,
Anantapur, Andhra Pradesh, India.

K. Ravichand

Senior Advisor
Aassaanedu Care Foundation,
Chennai, Tamilnadu, India.

 **BS Publications**
A unit of **BSP Books Pvt., Ltd.**
4-4-309/316, Giriraj Lane,
Sultan Bazar, Hyderabad - 500 095.

Computational Simulation Tools in Engineering
by V. Ramesh Kumar, T. Bala Narsaiah and K. Ravichand

Published by:

 **BS Publications**

A unit of **BSP Books Pvt., Ltd.**
4-4-309/316, Giriraj Lane,
Sultan Bazar, Hyderabad - 500 095.
Phone : 040 - 23445688, 23445600
e-mail : info@bspbooks.net
website: www.bspbooks.net

ISBN: 978-93-87593-40-4

Contents

Section - I

Heat Transfer

Section – II

Fluid Flow

Section - III
Thermodynamics

Section - IV
Mass Transfer

Section - V
Chemical Reaction Engineering

Section - VI
Equipment Design

Chapter 27

Chapter 28

Section - VII
Process Control

Chapter 29

Chapter 30

Chapter 31

Chapter 32

Section - VIII
Process Simulation

Chapter 33

Chapter 34

Chapter 35

Chapter 36

Preface

"I hear and I forget. I see and I remember. I do and I understand."
- Confucius

Chemical Engineering or any engineering student need to solve different problems of heat, mass and momentum transfer. These problems are formulated based on the physical description of the process or equipment. The conventional procedure to solve such problems is to write computer programs using C/C++ language. However, with the advancement of computers and availability of various software tools, it is very convenient to solve complex problems with simple code using MS-Excel, MATLAB, CHEMCAD and COMSOL.

In UG, PG curriculum of chemical and allied engineering programs, there are some subjects on "process modeling and simulation" or "computer applications in chemical engineering" followed by a laboratory sessions like "process simulation laboratory", Advanced Simulation Laboratory, Computational Fluid Dynamics laboratory, Computer Applications in Process Industries (CAPI) laboratory etc. The aim of this book is to illustrate different types of models from reputed text books and provide solutions to them using MS-Excel, MATLAB, CHEMCAD and COMSOL. This gives complete understanding and apply the concepts to solve the models in computational tools.

Purpose of the Book

Computer simulation is intensively used for data analysis, supporting design decisions, investigating the feasibility, sizing the equipment, and finally for studying process dynamics and control issues. This book gives hands-on experience procedures and solutions using various computational tools like MS-Excel, MATLAB, CHEMCAD and COMSOL for each model discussed.

The book aims to equip the young researchers and students with the skills developed in this field, and provide an insight into engineering modeling and numerical software used. The purpose is also to demonstrate how these tools can be used effectively in solving real-world problems.

The book extends over eight Sections. The first six sections deal with the fundamentals models of engineering majorly unit operations, while the last two sections deal with process control and complete simulation of processes consisting of various unit operations.

Section-I deals with Heat Transfer, consisting of the models related to conduction, convection and radiation. Case studies to calculate the insulation thickness, temperature profiles for a given process, simulate heat exchangers, estimate the flare stack parameters were included.

Section-II deals with Fluid Flow, emphasizing the case studies to estimate pressure drops/gains, velocity.

Section-III explains the underlying Thermodynamic principles. Case studies include the pure component property estimation, mixture property estimations and correlating the experimental data to the fundamental equations.

Section-IV deals with study of Distillation column dynamics which can be verified with fundamentals principles.

Section-V emphasizes the sizing of reactors and concentration profiles for various operating scenarios. Plug flow reactor, Batch Reactor, Continuous Stirred Tank reactor and fluidized bed reactor were included.

Section-VI elaborates the design concepts. Design of Oil-Water gas separator, process design of Column, Simple Shell & Tube Heat Exchanger, control valve, centrifuge, evaporator were included as case studies taken from reputed text books.

Section-VII deals with basic Process Control concepts implemented in computation tool, Temperature control of water heating tank, conncentration control of CSTR using conventional PID control technique.

Finally, Section-VIII deals with process simulation having various unit operations dealt in previous sections with recycle loops for handling the unconverted reactants. Case studies such as refrigeration where there was no feed and no product were also added. Other case studies primarily deals with throughput or energy optimization of the given process.

Appendix was included primarily for the benefit of beginners in using computational tools such as MATLAB and COMSOL where basics were explained such that they will cope up with the case studies discussed while solving models rather than concentrating to get familiarity on the tool usage.

Target Audience

This book is intended for students and faculty those are undergoing a practical course at both UG and PG levels of chemical and allied engineering programs.

-Authors

Acknowledgements

One of the authors is very much thankful to Dr. S. Sridhar for permitting and organizing training classes to his scholars at IICT and this training program motivated to write this book. Authors would like to acknowledge the help extended by faculty and students during various simulation sessions conducted in their respective institutions which lead to consolidation of various case studies.

Authors would like to thank for the services provided by Mr. K. Ravindranath, Manager, SciTech Patent Art Services, in designing the book cover page.

And last but not the least we express our gratitude and love to our families, for continuous support and understanding.

-Authors

About the Author

V. Ramesh Kumar is Professor & Vice-Principal at University College of Technology, Osmania University, Hyderabad. He has 21 years of teaching experience and specialized in computational simulations. He has numerous publications, held various positions such as Additional controller of examinations, BOS chairman etc. His interests include Chemical Engineering Three Phase Reactors, Modelling Simulation and Otimization of Chemical Processes.

T. Bala Narsaiah, is Professor in the Department of Chemical Engineering at Jawaharlal Nehru Technological University College of Engineering, Anantapur, Andhra Pradesh, India. He has 19 years of teaching experience. He has published 25 papers and 3 book chapters in international and national journals. He published 15 papers in international conference proceedings and presented 20 papers in national & international conferences. He has published a book as editor on "innovative Technologies for the Treatment of Industrial Wastewater: A Sustainable Approach", by Apple Academic Press in the year 2017. Presently, he is Chairman, UG Board of Studies, Chemical Engineering, JNTUA, Anantapur and Chairman, Research Review Committee, Food Technology at JNTUA, Anantapur. Presently, he is a member of Telangana Pollution Control Board (e-Waste committee). He is member of professional bodies - IIChE, ISTE and fellow of Institution of Engineers (India).

K. Ravichand is Senior Advisor at Aassaanedu Care Foundation and involved in the activity of training the faculty towards outcome-based education. He completed his master's degree in Chemical Engineering from IIT Bombay and has total 17 years experience consisting of 3.5 years as process engineer for the plant design construction and commissioning of 1350 MTPD Ammonia plant, 8+ years experience as developer and Manager for the Computational softwares PRO-II® and CATIA V4®, 5 years as Chemical, Petroleum Engineering faculty. He has 2 Journal and 9 Conference publications. He was resource person for 5 workshops and 4 invited lectures on computation tools. His accomplishments include reviving a sick unit of 5T/month Lead nitrate plant, process development for the pigment grade iron oxide from waste iron oxides produced in organic reduction processes. He is member of IEEE, IIChE, ISTE and NSC.

Introduction

"Computer users often quote the acronym "GIGO": 'Garbage In-Garbage out: in some cases, it is possible that the user has not thought sufficiently about the experimental conditions, or the sample preparation. In such a case, however good the computer or the software, the result is likely to be wrong!"

_ Thome B.

Computation is today regarded as an important tool needed for the advancement of scientific knowledge and engineering practice, in addition to the already existing tools of theoretical analysis and physical experimentation. Simulation techniques allow scientists to easily study complex natural phenomena and processes, which would otherwise be very difficult. The need for greater accuracy and detail in such simulations has created the need for faster and more efficient computer algorithms. It is because of these advancements that scientists and engineers can solve highly complex problems.

Computational Engineering refers to the design, development and application of computational systems for solving physical problems of higher complexity. These computational systems are used to obtain solutions of mathematical models that represent some physical process through the use of algorithms, software and other methods.

Computational simulation and modeling applied to engineering is a field that brings together the power of computers and the physical sciences.

To solve different problems in engineering using computer simulation, it is necessary to model the problem which represents a real-world phenomenon as a set of mathematical equations. This book deals with models involving simple algebraic equation, single & multiple non-linear equation, Single & multiple ordinary differential equations and partial differential equations of parabolic type.

The collection of problems from various reputed textbooks are aimed to utilize various numerical methods for various engineering subject areas such as Heat transfer, Fluid flow, Thermodynamics, Mass transfer, Reaction engineering, equipment design and process simulation. Most of the problems deals with chemical and pharmaceutical applications. Other applications such as Gas-Oil-Separator, Flare stack flow models and regression features are also included.

Purpose of the Book

This book gives hands-on exposure procedures and solutions using various computational tools such as MS-Excel®[1], MATLAB®[2], CHEMCAD®[3] and COMSOL®[4]for each problem discussed. Though the chapters were written such that the reader understands the problems with the description, Appendix was provided for beginners. Readers are strongly advised to go through the basic commands / programs mentioned in the Appendices before practising the book chapters.

The main purpose of this book is to give students, researchers and technicians broad coverage of the technological and scientific subjects in the modeling and simulation field required to underpin the success of engineering structures in a range of scientific and technical fields, providing in-depth study and training encompassing the principles and techniques in modelling and computational simulation. The book aims to equip the young researchers and technicians with the skills developed in this field, and provides an insight into engineering modeling and numerical software used. The purpose is also to demonstrate how these technologies can be used effectively in solving real-world problems.

Target Audience

- This book is intended for students and faculty those are undergoing a practical course at both UG and PG levels of chemical and allied engineering programs. Following table shows various models with the tools used in this book to solve them.

Description	Mathematic Model	Excel	MATLAB	CHEM CAD	COM SOL
Section – I – Heat Transfer					
Cylindrical pipe insulation	Algebraic equations	✓	✓		
Semi Infinite slab Unsteady state conduction	Algebraic equations	✓	✓		

---

[1]Ms-Excel is a registered trademark of Microsoft Corporation (http://www.microsoft.com)

[2]MATLAB is a registered trademark of The Math Works, Inc. (http://www.math- works.com)

[3]CHEMCAD is a registered trademark of Chemstations, Inc. (http://www.chemstations.com)

[4]COMSOL is a registered trademark of COMSOL AB (http://comsol.com)

Description	Mathematic Model	Excel	MATLAB	CHEM CAD	COM SOL
Counter current Heat Exchanger	Algebraic equations	✓	✓	✓	
Unsteady State Conduction from Flat Surface	Simultaneos Non linear equations	✓	✓		
Flare stack height simulation	Algebraic equations	✓	✓		
Heat Equation - Stability analysis using Implicit & Explicit Euler method	Simultaneos Non linear equations	✓	✓		
Section - II – Fluid Flow					
Simulation of multistage compressor	Simultaneos Non linear equations	✓	✓		
Flare sizing	Algebraic equations	✓	✓		
Pressure drop in Pipe lines	Algebraic equations	✓			
Flow in tube – PDE solver	Partial differential equations		✓		
Simulate the Gravity flow tank	Simultaneos Ordinary differential equations (ODEs)	✓	✓		
Section-III – Thermodynamics					
VLE calculation for multi-component system	Simultaneos Non linear equations	✓	✓		
Molar volume estimation using EOS methods	Single Non linear equation	✓	✓		
Bubble and Dew point estimation for multi component system	Simultaneos Non linear equations	✓	✓	✓	
CO_2 vapor pressure Regression	Linear, non-linear regression	✓	✓		
Regress feature – CO_2 emissions			✓		

Table *contd...*

Description	Mathematic Model	Excel	MATLAB	CHEM CAD	COM SOL
Section - IV – Mass Transfer					
Dynamics of Binary Distillation Column	Simulataneous ODEs		✓		
Section - V – Reaction Engineering					
Batch Reactor with series reaction	Simulataneous ODEs	✓	✓		
Design of Plug Flow Reactor	Simultaneos Non linear equations			✓	
Non-isothermal Plug Flow Reactor	Simulataneous ODEs	✓	✓		✓
3-zone circulating fluidized bed reactor	Simulataneous ODEs		✓		
Simulation of non-isothermal CSTR	Simulataneous ODEs		✓		
Section - VI – Equipment Design					
Three phase separator (Oil-Water-gas)	Polynomial fitting, Algebraic equations	✓	✓		
Column design using FUG method	Algebraic equations	✓	✓		
Shell & Tube Exchanger design using Kern method	Algebraic equations	✓			
Control valve sizing	Algebraic equations		✓		
Centrifuge design	Algebraic equations		✓		
Single Effect Evaporator	Algebraic equations	✓			
Section VII – Process Control					
Dynamics of 2nd order systems	Simultaneos Non linear equations, ODEs		✓		
Boiler efficiency	Simultaneos Non linear equations	✓	✓		
Predictive Control of waste water treatment process	Simulataneous ODEs		✓		

In addition to the models discussed in this book, there are open source web sites available with some more models. Michael B. Cutlip et.al., [1] solved the models in computational tools – POLYMATH, Excel and MATLAB. MATLAB Central [2] also continuously updates the models solved various professional with the File Exchange facility with search option by tag, Author etc.

References

1.   http://www.problemsolvingbook.com

2.   https://in.mathworks.com/matlabcentral/fileexchange

Heat Transfer

"The inside of a tube is, without a doubt, the poorest place in the world where a person could affect heat transfer, ... because it has a uniform cross section, it does not create randomness ... and it tends to give you the least for your money when you circulate a fluid through it."

Donald Q Kern., disparaged conventional tubes for heat transfer and emphasizing the need of extended surfaces.

Kern, D. Q., Speech delivered to Process Heat Exchanger Society, Houston, TX, January 20, 1958.

1

Thickness of a Cylindrical Pipe Insulation

Objective: Emphasize the following principle:

- Steady state conduction in composite wall (Temperature profiles with radius);

Problem Statement

The wall of a small cylindrical test chamber is made of stainless steel and glass, as shown in Fig.1.1. It is desired to reduce the heat loss from a chamber by adding an Insulation layer of 85% magnesia. Of interest is determining where to insert the insulation layer so that the rate of heat loss is lowest. One student recommended adding the 85% magnesia layer on the cold side (glass) while a second student suggested the hot side (steel). A third student claimed that it does not matter where to add the insulation. You are asked to carry out a study to determine the option that will result in the lowest rate of heat loss from the chamber. Also draw the temperature profile across the composite wall for best option.

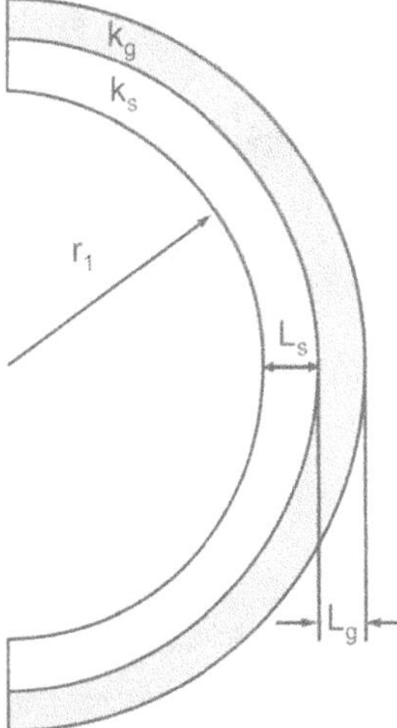

Fig. 1.1 Chamber Wall – Steel inside, glass outside

***Known quantities*:**

Chamber's inside radius, $r_1 = 8.5$ cm

Stainless steel thickess, $L_s = 1.5$ cm

Glass thickness, $L_g = 1$ cm

85% magnesia thickness, $L_m = 4$ cm

$T_{\infty, steel} = 100\ ^\circ C$

$h_{steel} = 8\ W/m^2 - ^\circ C$

$T_{\infty\ glass} = 10^\circ C$

$h_{glass} = 16\ W/m^2 - ^\circ C$

NAME	EXPRESSION	VALUE	DESCRIPTION
k_s	43 [W/(m*deg C)]	43[W/(m-K)]	1% Carbon Steel @ 20C
k_g	0.7[W/(m*deg C)]	0.7[W/m-K)]	Window Glass @ 20C
K_m	0.065[W/(m*deg C)]	0.065[W/(m-K)]	85% Magnesia @ 20C
T_inf1	100[deg C]	373. 15[K]	T Inside Chamber
T_inf4	10[deg C]	283. 15[K]	T Outside Chamber
h_1	8[W/(m^2*deg C)]	8[W/(m²-K)]	h inside Chamber
h_4	16[W/(m^2*deg C)]	16[W/(m²-K)]	h Outside Chamber

Theory

Conduction in Cylinder

Consider a hollow cylinder of length L (shown below), whose inner and outer surfaces are exposed to fluids at different temperatures. The system is analyzed by the standard method as follows:

- The heat transfer rate is obtained by using the temperature distribution with Fourier's law:

$$q_r = -kA\frac{dT}{dr} = -k(2\pi rL)\frac{dT}{dr}$$

$$= \frac{2\pi Lk\left(T_{s,1} - T_{s,2}\right)}{\ln\left(r_2/r_1\right)}$$

- Above Equation shows that the heat transfer rate q_r is a constant in the radial direction.

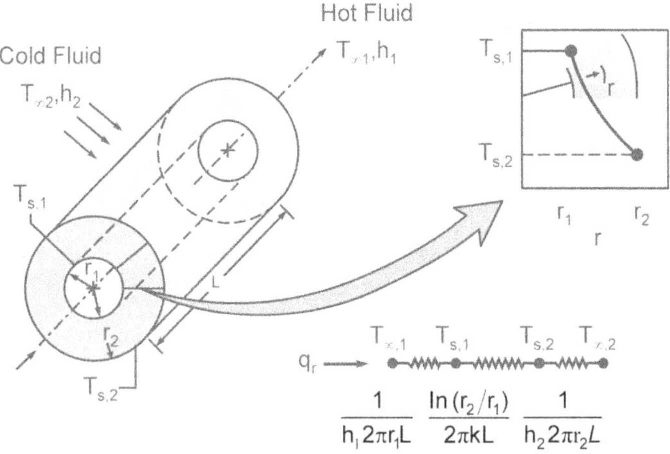

Fig. 1.2 Composite cylindrical wall resistance

- From equation, the thermal resistance for radial conduction in a cylindrical wall is

$$R_{t,cond} = \frac{\Delta T}{q_r} = \frac{\ln\left(r_2/r_1\right)}{2\pi L k}$$

Composite Cylindrical Wall

Consider a composite cylindrical wall of length L shown below.

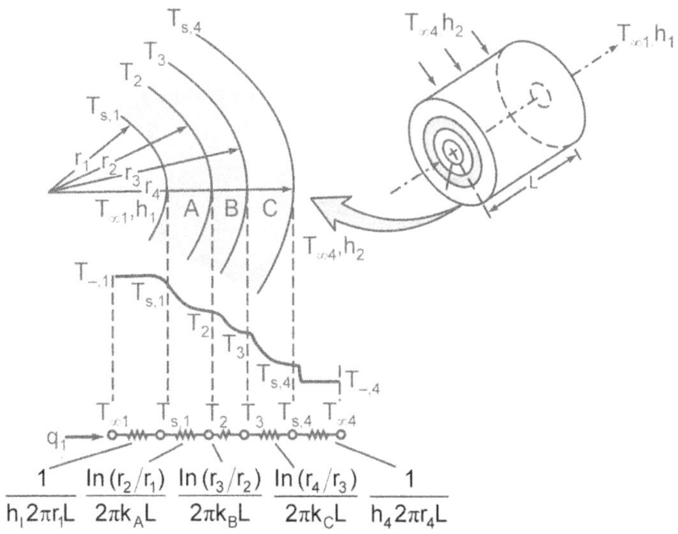

Fig. 1.3 Temperature profile across composite cylindrical wall

- Neglecting interfacial contact resistances, the heat transfer rate may be expressed as:

$$q_r = \frac{T_{\infty,1} - T_{\infty,4}}{\dfrac{1}{2\pi r_1 L h_1} + \dfrac{\ln(r_2/r_1)}{2\pi k_A L} + \dfrac{\ln(r_3/r_2)}{2\pi k_B L} + \dfrac{\ln(r_4/r_3)}{2\pi k_C L} + \dfrac{1}{2\pi r_4 L h_4}}$$

W M rite the following function in MATLAB editor

```matlab
function [q T]=compositewall(Ti,To,hi,ho,ri,x,k)
% function [q T]=compositewall(Ti,To,hi,ho,ri,x,k)
% Ti, To: Inside and outside Temperature, in K
% hi, ho: Convective heat transfer coefficients, W/mK
% ri: Inner radius, m
% x, k: thickness, thermal conductivities of composite wall
% q: heat loss rate per unit length, W
wallcount=length(x);
DT=Ti-To;
InnerConvRes=1/(2*pi*ri*hi);
OuterConvRes=1/(2*pi*ri*ho);
r(1)=ri;
for i=1:wallcount
    r(i+1)=r(i)+x(i);
end
CondRes=0;
for i=1:wallcount
    wallRes(i)=log(r(i-1)/r(i))/(2*pi*k(i));
    CondRes=CondRes+wallRes(i);
end

q=DT/(InnerConvRes+CondRes+OuterConvRes);
T(1)=Ti-q*InnerConvRes;
for i=1:wallcount
    T(i+1)=T(i)-q*wallRes(i);
end
T(i+2)=T(i)-q*OuterConvRes;
```

Execution Procedure: Execute the following commands from MATLAB command prompt & observe the results

```matlab
>> [q T]=compositewall(373, 273, 8, 16, 0.085, [0.015 0.01 0.04], [43 0.7 0.065])

q =

   88.2788


T =

   352.3382   352.2851   350.3721   283.3309   340.0412

>> [q T]=compositewall(373, 273, 8, 16, 0.085, [0.015 0.04 0.01], [43 0.865 0.7])

q =

   83.9468
```

```
T =

   353.3521   353.3016   284.1408   282.8239   274.3168

>> [q T]=compositewall(373, 273, 8, 16, 0.085, [0.04 0.015 0.01], [0.065 43 0.7])

q =

   76.2490

T =

   355.1538   283.1512   283.1192   281.9231   274.1961

>>
```

Draw Temperature profile with respect to radius once q and k are known:

Step 1: Write the following function and save the files as compwallode.m to evaluate ODE

```
function dtdr=compwallode(r,T)
global k;
q=76.249; % Watts
dtdr=-q/(k*2*pi*r);
```

Step 2: Write another function and saveit as compTprofile.m to solve ODE for the given composite wall with values specified for q, k and the radius range:

```
function compTprofile()
global k;
k=0.065; %W/mC
[r1 T1]=ode45('compwallode',[0.085 0.125],100);
len=length(T1);
k=43;
[r2 T2]=ode45('compwallode',[0.125 0.145],T1(len));
len=length(T2);
k=0.7;
[r3 T3]=ode45('compwallode',[0.14 0.15],T2(len));
r=[r1;r2;r3];
T=[T1;T2;T3];
plot (r*100, T);
title('Temperature profile of composite wall);
xlabel('radius, cm');
ylabel('Temperature, C');
```

Step 3: From Command prompt, execute the program:

Following graph will be the result:

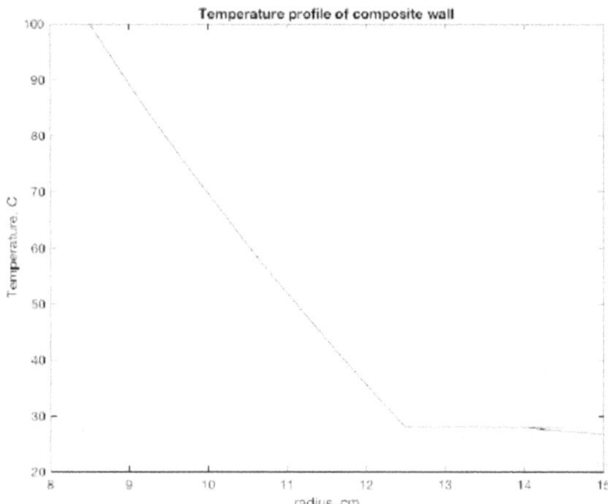

Fig. 1.4 Temperature profile across composite wall

Open an Excel sheet. Observe that the spreadsheet consists of multiple cells where you can enter the data (with columns numbered with Alphabets and the rows numbered with numerical digits).

Throughout this book, the cell numbers are used to explain the formulae used. For example A1 is a cell of Ath Column and 1st row. User is allowed to enter any text in the cells, however if it starts with '=' symbol, it will treat it as formula.

Excel Solution for the Problem:

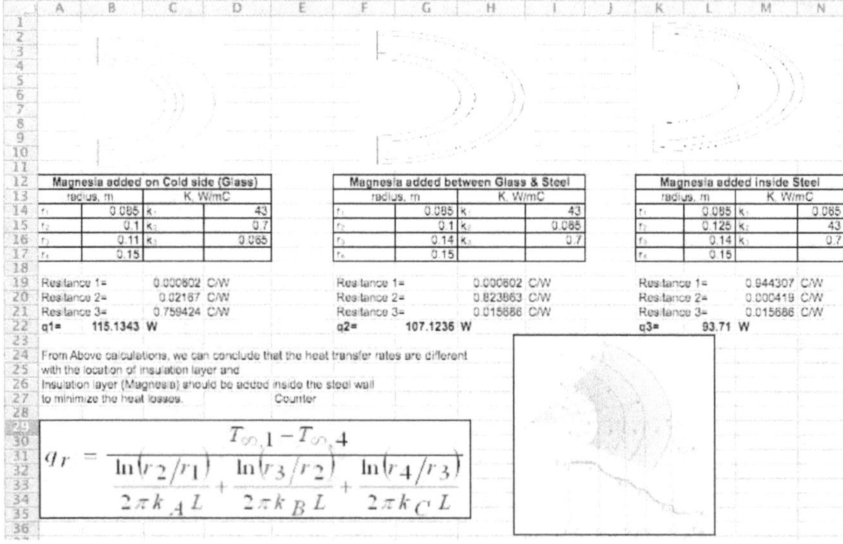

Magnesia added on Cold side (Glass)			
radius, m		K, W/mC	
r_1	0.085	k_1	43
r_2	0.1	k_2	0.7
r_3	0.11	k_3	0.065
r_4	0.15		

Magnesia added between Glass & Steel			
radius, m		K, W/mC	
r_1	0.085	k_1	43
r_2	0.1	k_2	0.065
r_3	0.14	k_3	0.7
r_4	0.15		

Magnesia added inside Steel			
radius, m		K, W/mC	
r_1	0.085	k_1	0.065
r_2	0.125	k_2	43
r_3	0.14	k_3	0.7
r_4	0.15		

Resitance 1= 0.000602 C/W
Resitance 2= 0.02167 C/W
Resitance 3= 0.759424 C/W
q1= 115.1343 W

Resitance 1= 0.000602 C/W
Resitance 2= 0.823863 C/W
Resitance 3= 0.015686 C/W
q2= 107.1236 W

Resitance 1= 0.944307 C/W
Resitance 2= 0.000419 C/W
Resitance 3= 0.015686 C/W
q3= 93.71 W

From Above calculations, we can conclude that the heat transfer rates are different with the location of insulation layer and Insulation layer (Magnesia) should be added inside the steel wall to minimize the heat losses.   Counter

$$q_r = \frac{T_{\infty,1} - T_{\infty,4}}{\dfrac{\ln(r_2/r_1)}{2\pi k_A L} + \dfrac{\ln(r_3/r_2)}{2\pi k_B L} + \dfrac{\ln(r_4/r_3)}{2\pi k_C L}}$$

Formulae Used for the Calculations:

Cell ID   Formula used

C19 = (LN(B15/B14))/(2*3.1416*D14*1)

C20 = (LN(B16/B15))/(2*3.1416*D15*1)

C21 = (LN(B17/B16))/(2*3.1416*D16*1)

Note that the formula can be entered for C19 and the formulae for C20, C21 can be obtained by holding the right bottom corner of C19 cell and drag to copy the formula for C20, C21

C19 = (LN(B15/B14))/(2*3.1416*D14*1)

C20 = (LN(B15/B14))/(2*3.1416*D14*1)

C21 = (LN(B17/B16))/(2*3.1416*D16*1)

Note that the formula can be entered for C19 and the formulae for C20, C21 can be obtained by holding the right bottom corner of C19 cell and drag to copy the formula for C20, C21

B22 = (100-10)/SUM(C19:C21)

H19 = (LN(G15/G14))/(2*3.1416*I14*1)

H20 = (LN(G16/G15))/(2*3.1416*I15*1)

H21 = (LN(G17/G16))/(2*3.1416*I16*1)

Note: Select cells C19 to C21, Copy them and past them in cell H19 also will give you the results same as per the mentioned formulae.

G22 = (100-10)/SUM(H19:H21)

M19 = (LN(L15/L14))/(2*3.1416*N14*1)

M20 = (LN(L16/L15))/(2*3.1416*N15*1)

M21 = (LN(L17/L16))/(2*3.1416*N16*1)

N22 = (100-10)/SUM(M19:M21)

References

Chapter 2 One dimensional steady state conduction, faculty.kfupm.edu.sa/CHE/Shammakh/files/myfiles/300-chapter-2.pdf.

COMSOL Multiphysics Instruction Manual – eportfolio.lib.ksu.edu.tw.

2

Unsteady State Conduction in Semi Infinite Slab

Objective: Emphasizes the following principle:

- Unsteady state conduction of Semi-infinite slab (Temperature profiles with time and distance);

Problem Statement

Example: A flat slab of plastic initially at 70^0F (21.1^0C) is placed between two platens at 250^0F (121.1^0C). The slab is 1.0.in. (2.54 cm) thick. (a) How long will it take to heat the slab to an average temperature of 210^0F (98.9^0C)?

Unsteady State Heat Conduction

- There are many heat transfer problems that are time dependent, i.e. *unsteady*, or *transient*.

- Transient problems typically arise when the boundary conditions are changed. The temperature at each point in the system will also begin to change. The changes will continue to occur until a *steady-state* temperature distribution is reached.

  Fig.2.1 represents a section through a large slab of material of thickness 2s, initially all at a uniform temperature T_a. At the start of heating one surface temperature is quickly increased to and subsequently held at temperature T_s. The temperature pattern shown in Fig.2.1 reflects conditions after a relatively short time t_r has elapsed since the start of heating.

Focus attention on the thin layer of thickness dx located at distance x from the left side of the slab. The two sides of the element are isothermal surfaces.

The rate of heat flow in at x is q_x. The rate of heat flow out at x+Δx is $q_{x+\Delta x}$.

Considering the heat balance for a time interval of Δt,

Heat in – heat out = accumulation of energy within the slab

$$(q_x - q_{x+\Delta x})\,\Delta t = \int (A\,\Delta x)\,Cp\,\Delta T$$

where ∫ is density of the slab material,

A Δx is the volume of the slab of thichness Δx

Cp is the specific heat of the slab material and ΔT is the temperature difference between points Dividing both sides with Δx Δt, and substituting the fourier's law equation for q, we get

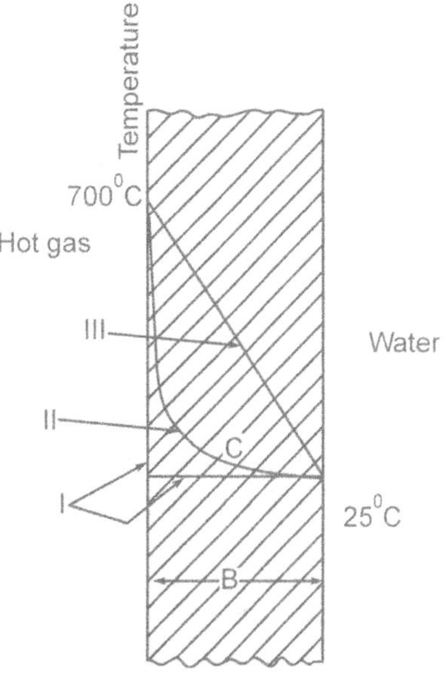

I.   at instant of exposure
II. during heating, at time t
III. at steady state

Fig. 2.1 Temperature distribution, unsteady state heating

$$\frac{\partial T}{\partial t} = \frac{k}{pc_p}\frac{\partial^2 T}{\partial x^2} = \alpha\frac{\partial^2 T}{\partial x^2} \qquad .....(2.1)$$

Where the term α is called as *thermal diffusivity* of the solid having units of area/time and is the property of the material.

Semi-Infinite Solid

Sometimes solids are heated or cooled in such a way that the temperature changes in the solid are confined to the region near one surface. Consider, for example, a very thick flat wall of a chimney, initially all at a uniform temperature T_a. Suppose that the inner surface of the wall is suddenly heated to, and held at, a high temperature T_s perhaps by suddenly admitting hot flue gas to the chimney. Temperatures inside the chimney wall will change with time, rapidly near the hot surface and more slowly farther away. If the wall is thick enough, there will be no measurable change in the temperature of the outer surface for a considerable time. Under these conditions the heat may be considered to be "penetrating" a solid of essentially infinite thickness. Fig.2.1 shows the temperature patterns in such a wall at various times after exposure to the hot gas, indicating the sharp discontinuity in temperature at the hot surface immediately after exposure and the progressive changes at interior points at later times.

For this situation integration of Eq.(2.1) with the appropriate boundary conditions, gives, for temperature T at any point a distance x from the hot surface, the equation

$$\frac{T_s - T}{T_s - T_a} = \frac{2}{\sqrt{\pi}} \int_o^z e^{-z^2}\, dZ \qquad .....(2.2)$$

where

$Z = x / 2\sqrt{\alpha t}$, dimensionless

α = thermal diffusivity

x = distance from surface

t = time after change in surface temperature, h

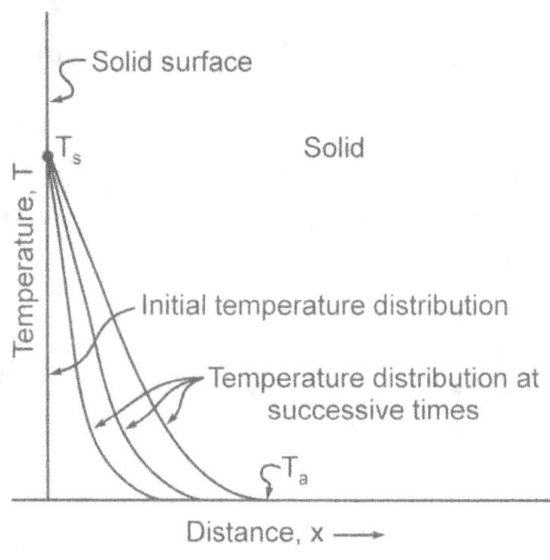

Fig. 2.2 Temperature distribution in unsteady state heating of semi-infinite solid

Analytical Solution:

(a) The properties of solid to be used are

$$k = 0.075\,\text{Btu} / \text{ft} - \text{h}\,^{\circ}\text{F} \qquad p = 56.2\,lb / \text{ft}^3 \qquad c_p = 0.40\,\text{Btu} / \text{lb} -^{\circ}\text{F}$$

$$s = \frac{0.5}{12} = 0.0417\,\text{ft} \quad T_s = 250^{\circ}\text{F} \quad T_a = 70^{\circ}\text{F} \quad \overline{T}_b = 210^{\circ}\text{F}$$

Then

$$\frac{T_s - \overline{T}_b}{T_s - T_a} = \frac{250 - 210}{250 - 70} = 0.222 \qquad \alpha = \frac{k}{\rho c_p} = \frac{0.075}{56.2 \times 0.40} = 0.00335$$

From eq.(2.2) which is the solution for eq.(2.1), time can be calculated.

$$N_{Fo} = 0.52\frac{0.00335t_T}{0.0417^2} \qquad t_r = 0.27 \text{ h} = 16 \text{ min}$$

$$\frac{T_s - \overline{T_b}}{T_s - T_a} = \frac{8}{\pi^2}\left(e^{-a_1 NFo} + \frac{1}{9}e^{-9a_1 NFa} + \frac{1}{25}e^{-25a_1 NFa} + ... \right) \qquad .....(2.3)$$

where

T_s = constant average temperature of surface of slab

T_a = initial temperature of slab

$\overline{T_b}$ = average temperature of slab at time t_T

N_{Fa} = Fourier number, defined as $\alpha t_r/s^2$

α = thermal diffusivity

t_T = time of heating or cooling

s = one-half slab thickness

$a_1 = (\pi/2)^2$

Solution for the Problem in MATLAB:

Write the following function in MATLAB editor (save it as semiinfiniteslab.m)

```
function [x T]=semiinfiniteslab(alpha,Tslabinitial, Tslabfinal, time)
% semiinfiniteslab(k,rho,cp,Tslabinitial, Tslabfinal)
% This function plots the temperature distribution in
%    unsteady state heating of semi-infinite solid.
% Reference: Unit Operations of Chemical Engineering, 5th Ed (Fig. 10.7)
% alpha=k/(rho*cp); %Computation of Thermal Diffusivity
% Tslabinitial: Slab steady state suface temperature before step change (K
% Tslabfinal: Slab suface temperature after step change (K)
% time: time at which Temperature profile in the slab is evaluated (hr)
% x: distance from slab surface (m)
% T: Slab temperature at distance x (K)
x=0:0.01:0.25; % distance from slab surface (m)
z=x/(2*sqrt(alpha)*time);
T=-1*((Tslabfinal-Tslabinitial)*erf(z)-Tslabfinal);
```

Write the following function in MATLAB editor (and save it as plotSISProfile.m) to plot the temperature profiles:

```
function plotSISProfile()
% This function plots the semi-infinite slab temperature profiles using
%  the results of semiinfiniteslab.m file
% Reference: Unit Operations of Chemical Engineering, 5th Ed (Ex. 10.4)
Tslabinitial=294.25; %Kelvin
Tslabfinal=394.25;
alpha=0.00335; %m2/h
```

```
for t=0.1:0.1:2
    [x T]=semiinfiniteslab(alpha,Tslabinitial, Tslabfinal, t);
    plot(x,T);
    hold on;
end
xlabel('Distance in m');
ylabel('Temperature in K');
```

Execution Procedure: Execute the following commands from MATLAB command prompt & observe the results

```
>> plotSISProfile
```

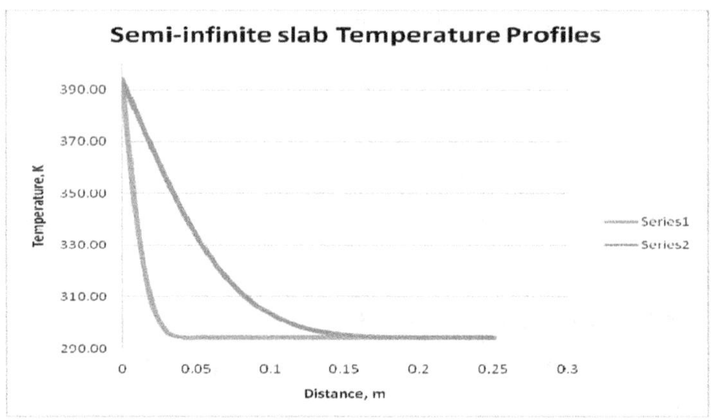

Fig. 2.3 Temperature profile from MATLAB Program

Excel Solution for the Problem

Open an excel sheet and provide the data in rows 1 to 7 and column A as shown in the screenshot (Fig.2.3) and then apply formulae as described and shown in Fig.2.4.

	A	B	C	D	E	F	G	H	I	J	K	L	M	N	O	P	Q		S	T	U	V
1																						
2	T initial, K	294.25																				
3	T final, K	394.25																				
4	Alpha, m²/h	0.0035																				
5																						
6	x Values estimation																					
7	x and t	0.1	0.2	0.3	0.4	0.5	0.6	0.7	0.8	0.9	1	1.1	1.2	1.3	1.4	1.5	1.6	1.7	1.8	1.9	2	
8	0	0	0	0	0	0	0	0	0	0	0	0	0	0	0	0	0	0	0	0	0	
9	0.01	0.535	0.378	0.309	0.267	0.239	0.218	0.202	0.189	0.178	0.169	0.161	0.154	0.148	0.143	0.138	0.134	0.130	0.126	0.123	0.120	
10	0.02	1.069	0.756	0.617	0.535	0.478	0.436	0.404	0.378	0.356	0.338	0.322	0.309	0.296	0.286	0.276	0.267	0.259	0.252	0.245	0.239	
11	0.03	1.604	1.134	0.926	0.802	0.717	0.655	0.606	0.567	0.535	0.507	0.483	0.463	0.445	0.429	0.414	0.401	0.389	0.378	0.368	0.359	
12	0.04	2.138	1.512	1.234	1.069	0.956	0.873	0.808	0.756	0.713	0.676	0.645	0.617	0.593	0.571	0.552	0.535	0.519	0.504	0.491	0.478	
13	0.05	2.673	1.890	1.543	1.336	1.195	1.091	1.010	0.945	0.891	0.845	0.806	0.772	0.741	0.714	0.690	0.668	0.648	0.630	0.613	0.598	
14	0.06	3.207	2.268	1.852	1.604	1.434	1.309	1.212	1.134	1.069	1.014	0.967	0.926	0.889	0.857	0.828	0.802	0.778	0.756	0.736	0.717	
15	0.07	3.742	2.646	2.160	1.871	1.673	1.528	1.414	1.323	1.247	1.183	1.128	1.080	1.038	1.000	0.966	0.935	0.907	0.882	0.858	0.837	
16	0.08	4.276	3.024	2.469	2.138	1.912	1.746	1.616	1.512	1.425	1.352	1.289	1.234	1.186	1.143	1.104	1.069	1.037	1.008	0.981	0.956	
17	0.09	4.811	3.402	2.777	2.405	2.151	1.964	1.818	1.701	1.604	1.521	1.450	1.389	1.334	1.286	1.242	1.203	1.167	1.134	1.104	1.076	
18	0.1	5.345	3.780	3.086	2.673	2.390	2.182	2.020	1.890	1.782	1.690	1.612	1.543	1.482	1.429	1.380	1.336	1.296	1.260	1.226	1.195	
19	0.11	5.880	4.158	3.395	2.940	2.630	2.400	2.222	2.079	1.960	1.859	1.773	1.697	1.631	1.571	1.518	1.470	1.426	1.386	1.349	1.315	
20	0.12	6.414	4.536	3.703	3.207	2.869	2.619	2.424	2.268	2.138	2.028	1.934	1.852	1.779	1.714	1.656	1.604	1.556	1.512	1.472	1.434	
21	0.13	6.949	4.914	4.012	3.474	3.108	2.837	2.626	2.457	2.316	2.197	2.095	2.006	1.927	1.857	1.794	1.737	1.685	1.638	1.594	1.554	
22	0.14	7.483	5.292	4.320	3.742	3.347	3.055	2.828	2.646	2.494	2.366	2.256	2.160	2.075	2.000	1.932	1.871	1.815	1.764	1.717	1.673	
23	0.15	8.018	5.669	4.629	4.009	3.586	3.273	3.030	2.835	2.673	2.535	2.417	2.315	2.224	2.143	2.070	2.004	1.945	1.890	1.839	1.793	
24	0.16	8.552	6.047	4.938	4.276	3.825	3.491	3.232	3.024	2.851	2.704	2.579	2.469	2.372	2.286	2.208	2.138	2.074	2.016	1.962	1.912	
25	0.17	9.087	6.425	5.246	4.543	4.064	3.710	3.435	3.213	3.029	2.874	2.740	2.623	2.520	2.429	2.346	2.272	2.204	2.142	2.085	2.032	
26	0.18	9.621	6.803	5.555	4.811	4.303	3.928	3.637	3.402	3.207	3.043	2.901	2.777	2.668	2.571	2.484	2.405	2.334	2.268	2.207	2.151	
27	0.19	10.156	7.181	5.864	5.078	4.542	4.146	3.839	3.591	3.385	3.212	3.062	2.932	2.817	2.714	2.622	2.539	2.463	2.394	2.330	2.271	
28	0.2	10.690	7.559	6.172	5.345	4.781	4.364	4.041	3.780	3.563	3.381	3.223	3.086	2.965	2.857	2.760	2.673	2.593	2.520	2.453	2.390	
29	0.21	11.225	7.937	6.481	5.612	5.020	4.583	4.243	3.969	3.742	3.550	3.384	3.240	3.113	3.000	2.898	2.806	2.722	2.646	2.575	2.510	
30	0.22	11.759	8.315	6.789	5.880	5.259	4.801	4.445	4.158	3.920	3.719	3.546	3.395	3.261	3.143	3.036	2.940	2.852	2.772	2.698	2.630	
31	0.23	12.294	8.693	7.098	6.147	5.498	5.019	4.647	4.347	4.098	3.888	3.707	3.549	3.410	3.286	3.174	3.074	2.982	2.898	2.820	2.748	

	0.1	0.2	0.3	0.4	0.5	0.6	0.7	0.8	0.9	1	1.1	1.2	1.3	1.4	1.5	1.6	1.7	1.8	1.9	2
0.24	12.829	9.071	7.407	6.414	5.737	5.237	4.849	4.536	4.276	4.057	3.868	3.703	3.558	3.429	3.312	3.207	3.111	3.024	2.943	2.869
0.25	13.363	9.449	7.715	6.682	5.976	5.455	5.051	4.725	4.454	4.226	4.029	3.858	3.706	3.571	3.450	3.341	3.241	3.150	3.066	2.988

Temprature estimation

x and t	0.1	0.2	0.3	0.4	0.5	0.6	0.7	0.8	0.9	1	1.1	1.2	1.3	1.4	1.5	1.6	1.7	1.8	1.9	2
0	394.25	394.25	394.25	394.25	394.25	394.25	394.25	394.25	394.25	394.25	394.25	394.25	394.25	394.25	394.25	394.25	394.25	394.25	394.25	394.25
0.01	339.22	353.55	360.50	364.80	367.78	370.01	371.76	373.18	374.36	375.36	376.22	376.98	377.64	378.24	378.78	379.26	379.70	380.11	380.48	380.83
0.02	307.31	322.75	332.52	339.22	344.15	347.96	351.02	353.55	355.68	357.51	359.10	360.50	361.75	362.87	363.88	364.80	365.64	366.41	367.12	367.78
0.03	296.58	305.13	313.29	319.93	325.30	329.70	333.39	336.52	339.22	341.58	343.66	345.52	347.19	348.70	350.07	351.33	352.48	353.55	354.54	355.46
0.04	294.50	297.50	302.34	307.31	311.88	315.95	319.56	322.75	325.60	328.15	330.44	332.52	334.42	336.15	337.75	339.22	340.58	341.85	343.04	344.15
0.05	294.27	295.00	297.16	300.13	303.35	306.53	309.56	312.39	315.02	317.45	319.70	321.77	323.70	325.49	327.16	328.72	330.18	331.55	332.84	334.05
0.06	294.25	294.38	295.13	296.58	298.50	300.66	302.90	305.13	307.31	309.40	311.40	313.29	315.09	316.79	318.41	319.93	321.38	322.75	324.06	325.30
0.07	294.25	294.27	294.48	295.07	296.05	297.33	298.80	300.39	302.03	303.68	305.31	306.91	308.47	309.98	311.44	312.84	314.19	315.48	316.73	317.92
0.08	294.25	294.25	294.30	294.50	294.93	295.61	296.48	297.50	298.63	299.83	301.07	302.34	303.60	304.85	306.09	307.31	308.50	309.65	310.78	311.88
0.09	294.25	294.25	294.26	294.32	294.48	294.80	295.26	295.87	296.58	297.39	298.27	299.20	300.17	301.15	302.15	303.15	304.14	305.13	306.11	307.07
0.1	294.25	294.25	294.25	294.27	294.32	294.45	294.68	295.00	295.42	295.93	296.52	297.16	297.85	298.59	299.35	300.13	300.92	301.73	302.54	303.35
0.11	294.25	294.25	294.25	294.25	294.27	294.32	294.42	294.58	294.81	295.11	295.47	295.89	296.36	296.88	297.43	298.01	298.62	299.25	299.89	300.55
0.12	294.25	294.25	294.25	294.25	294.25	294.27	294.31	294.38	294.50	294.66	294.87	295.13	295.44	295.78	296.17	296.58	297.03	297.50	297.99	298.50
0.13	294.25	294.25	294.25	294.25	294.25	294.26	294.27	294.30	294.36	294.44	294.55	294.71	294.89	295.11	295.37	295.65	295.97	296.30	296.67	297.05
0.14	294.25	294.25	294.25	294.25	294.25	294.25	294.26	294.27	294.29	294.33	294.39	294.48	294.58	294.72	294.88	295.07	295.28	295.51	295.77	296.05
0.15	294.25	294.25	294.25	294.25	294.25	294.25	294.25	294.26	294.27	294.28	294.31	294.36	294.42	294.49	294.59	294.71	294.85	295.00	295.18	295.37
0.16	294.25	294.25	294.25	294.25	294.25	294.25	294.25	294.25	294.26	294.26	294.28	294.30	294.33	294.37	294.43	294.50	294.59	294.69	294.80	294.93
0.17	294.25	294.25	294.25	294.25	294.25	294.25	294.25	294.25	294.25	294.25	294.26	294.27	294.29	294.31	294.34	294.38	294.43	294.50	294.57	294.66
0.18	294.25	294.25	294.25	294.25	294.25	294.25	294.25	294.25	294.25	294.25	294.25	294.26	294.27	294.28	294.29	294.32	294.35	294.38	294.43	294.48
0.19	294.25	294.25	294.25	294.25	294.25	294.25	294.25	294.25	294.25	294.25	294.25	294.25	294.26	294.26	294.27	294.28	294.30	294.32	294.35	294.38
0.2	294.25	294.25	294.25	294.25	294.25	294.25	294.25	294.25	294.25	294.25	294.25	294.25	294.25	294.26	294.26	294.27	294.27	294.29	294.30	294.32
0.21	294.25	294.25	294.25	294.25	294.25	294.25	294.25	294.25	294.25	294.25	294.25	294.25	294.25	294.25	294.25	294.26	294.26	294.27	294.28	294.29
0.22	294.25	294.25	294.25	294.25	294.25	294.25	294.25	294.25	294.25	294.25	294.25	294.25	294.25	294.25	294.25	294.25	294.26	294.26	294.26	294.27
0.23	294.25	294.25	294.25	294.25	294.25	294.25	294.25	294.25	294.25	294.25	294.25	294.25	294.25	294.25	294.25	294.25	294.25	294.25	294.26	294.26
0.24	294.25	294.25	294.25	294.25	294.25	294.25	294.25	294.25	294.25	294.25	294.25	294.25	294.25	294.25	294.25	294.25	294.25	294.25	294.25	294.25
0.25	294.25	294.25	294.25	294.25	294.25	294.25	294.25	294.25	294.25	294.25	294.25	294.25	294.25	294.25	294.25	294.25	294.25	294.25	294.25	294.25

Fig. 2.4 Excel sheet sholution (Screen captured image)

Formulae Used:

Note that When the formula copy/paste operation is performed in excel sheet, the cells are always taken based on the relative positions of the cells. If the user wish to take the value provided in a particular cell say B4 (Alpha value) in all the fomulae, then the same should be relative to cell position used as $B$4 instead of the normal cell B4. (Which will be changed relative to cell position when formulae is pasted).

	A	B	C	D		U
2	T initial, K	294.25				
3	T final, K	394.25				
4	Alpha, m²/h	0.0035				
5						
6	z Values estimation					
7	x and t	0.1	0.2	0.3	…	0.1
8	0	`=$A$8/SQRT($B$4*B7)`	`=$A$8/SQRT($B$4*C7)`	`=$A$8/SQRT($B$4*D7)`		`=$A$8/SQRT($B$4*U7)`
9	0.01	`=$A$9/SQRT($B$4*B7)`	==>			`=$A$9/SQRT($B$4*U7)`
10	0.02	`=$A$10/SQRT($B$4*B7)`				
11	0.03	`=$A$11/SQRT($B$4*B7)`				
12	0.04	`=$A$12/SQRT($B$4*B7)`				
13	0.05	`=$A$13/SQRT($B$4*B7)`				
14	0.06	`=$A$14/SQRT($B$4*B7)`				
15	0.07	`=$A$15/SQRT($B$4*B7)`				
16	0.08	`=$A$16/SQRT($B$4*B7)`				
17	0.09	`=$A$17/SQRT($B$4*B7)`				
18	0.1	`=$A$18/SQRT($B$4*B7)`				
19	0.11	`=$A$19/SQRT($B$4*B7)`				
20	0.12	`=$A$20/SQRT($B$4*B7)`				
21	0.13	`=$A$21/SQRT($B$4*B7)`				
22	0.14	`=$A$22/SQRT($B$4*B7)`				
23	0.15	`=$A$23/SQRT($B$4*B7)`				
24	0.16	`=$A$24/SQRT($B$4*B7)`				
25	0.17	`=$A$25/SQRT($B$4*B7)`				
26	0.18	`=$A$26/SQRT($B$4*B7)`				
27	0.19	`=$A$27/SQRT($B$4*B7)`				
28	0.2	`=$A$28/SQRT($B$4*B7)`				
29	0.21	`=$A$29/SQRT($B$4*B7)`				
30	0.22	`=$A$30/SQRT($B$4*B7)`				
31	0.23	`=$A$31/SQRT($B$4*B7)`				
32	0.24	`=$A$32/SQRT($B$4*B7)`				
33	0.25	`=$A$33/SQRT($B$4*B7)`				`=$A$33/SQRT($B$4*U7)`

Note the easier way to complete the writing the formulae for all the cell. The first formula in B8 can be typed and then copy / paste to the remaining cells B9 to B33; then change the cell numbers appropriately. Now, the entire cells can be selected i.e., B8:B33 and pasted to the remaining cells (OR simply drag to the remaining columns by holding the + symbol that appears in the right bottom corner of the selected cells).

22   Computational Simulation Tools in Engineering

	A	B	C	D	...	U
36	Temprature estimation					
37	x and t	0.1	0.2	0.3		2
38	0	=1*(((\$B\$3-\$B\$2)*ERF(B8))-\$B\$3)	=1*(((\$B\$3-\$B\$2)*ERF(C8))-\$B\$3)	=1*(((\$B\$3-\$B\$2)*ERF(D8))-\$B\$3)		=1*(((\$B\$3-\$B\$2)*ERF(U8))-\$B\$3)
39	0.01	=1*(((\$B\$3-\$B\$2)*ERF(B9))-\$B\$3)				
40	0.02					
41	0.03					
42	0.04					
43	0.05					
44	0.06					
45	0.07					
46	0.08					
47	0.09					
48	0.1					
49	0.11					
50	0.12					
51	0.13					
52	0.14					
53	0.15					
54	0.16					
55	0.17					
56	0.18					
57	0.19					
58	0.2					
59	0.21					
60	0.22					
61	0.23					
62	0.24					
63	0.25	=1*(((\$B\$3-\$B\$2)*ERF(B33))-\$B\$3)				=1*(((\$B\$3-\$B\$2)*ERF(U33))-\$B\$3)

In this example plot was drawn for 1st and last time steps by using Insert / Chart/Scattered chart (Smooth line scattered) option. Y-axis normally shown starting from zero. User need to select the axis & "format Axis" option should be used to change the scale.

Additional Practice

Solve Mass transfer and equation of motion for falling film

$$u_y \frac{\partial c_A}{\partial y} = D_{AB} \frac{\partial^2 c_A}{\partial z^2} \mu \frac{d^2 u_y}{dz^2} + pg = 0 \qquad .....(2.4)$$

Refer to the solution: Eqn 3.7 & 3.8 [Mass Transfer Operations, Robert E. Treybal, 3rd Edition].

References

McCabe, Smith and Hariott, Unit Operations of Chemical Engineering, McGraw Hill Publishers.

3

Counter Current Heat Exchanger

Objective: Emphasizes the following principle:

- Simulation of counter current heat exchanger for various scenarios

Problem Statement

Aniline is to be cooled from 200 to 150 F in a double pipe heat exchanger having a total outside area of 70 ft². For cooling, 8600 *lb*/h of toluene at a temperature of 100 F is available. Aniline flow rate is 10,000 *lb*/h.

(a) If flow is counter current, what are the toluene outlet temperature, the LMTD, overall heat transfer coefficient?

(b) What are they if the flow is parallel?

Theory

A counter current heat exchanger is an important example of a system described by equations that are usually solved. Fig.3.1 shows the system. The problem is to find the steady state outlet temperatures of the toluene, T_{C2}, and the heat transfer rate Q, given the inlet temperatures, flow rates, and area. The steady state equations for heat transfer are

$$Q = (10,000)(0.545)(200 - 100) = 2,72,500 \text{ Btu/h} \qquad \ldots (3.1)$$

$$Q = (8600)(0.44)(200 - T_{C2}) \qquad \ldots (3.2)$$

$$(\Delta T)_{LM} = \frac{(200 - T_{C2}) - (150 - 100)}{\ln\left(\dfrac{200 - T_{C2}}{150 - 100}\right)} \qquad \ldots (3.3)$$

$$Q = UA(\Delta T)_{LM} \qquad \ldots (3.4)$$

We have three eq.s(3.2 to 3.4) and four variables: $Q, (\Delta T)_{LM}$, T_{H2}, and T_{C2}. The procedure is:

1. Calculate Q_1 from Eq.(3.1).

2. Calculate T_{C2} from Eq.(3.2), using Q_1.

3. Calculate the log-mean-temperature driving force $(\Delta T)_{LM}$ from Eq.(3.3).

4. Calculate overall heat transfer rate U_2 from Eq.(3.4).

CHEMCAD Solution for the scenarios:

1. Launch CHEMCAD, create a new job.

2. Draw the flow sheet as shown in Fig. 3.1 with one counter current heat exchanger with water as coolant and Aniline on process side with the following data:
   Utility side: Toluene
   flow, *lb*/h   8600
   Temperature F   100
   Process side: Aniline
   Mass flow lb/h   10000
   Inlet Temperature, F   200
   Outlet Temperature, F   150
   Heat exchanger:
   Total Area/shell ft2   70

3. Solve the flow sheet and note the results.

4. Change the type of heat exchanger to concurrent type and note down the results.

5. Also, find the lowest temperature to which aniline can be cooled, by increasing toluene flow.

6. Solve the heat exchanger, if aniline is to cool in concurrent flow exchanger.

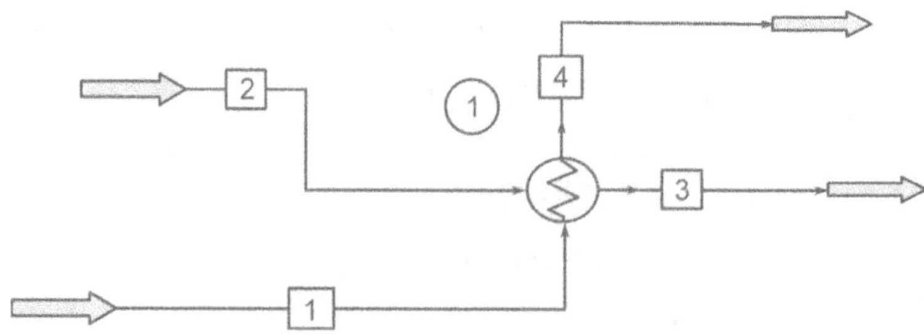

Fig. 3.1 Shell and Tube Heat Exchanger

Excel Solution for the Problem:

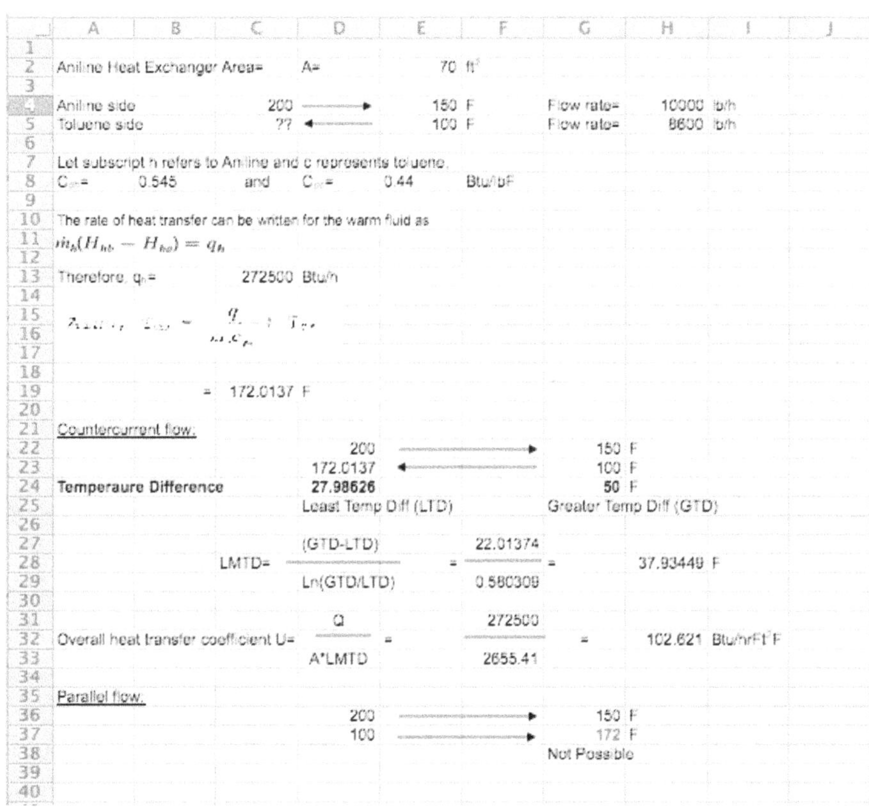

Formulae Used:

C13	= H4*B8*(C4-E4)	
C19	= (C13/(H5*E8))+E5	
D22	= C4	G22 =E4
D23	= C19	G23 =E5
D24	= D22-D23	G24 =G22-G23
F27	= G24-D24	
F29	= LN(G24/D24)	
G28	= F27/F29	
F31	= C13	
F33	= C13	

F33	= E2*H28
H32	= F31/F33

MATLAB Solution for the Problem:

Write simple MATLAB function to calculate the duty for any side of the heat exchanger and save it as calcduty.m

Content of the function is shown here.

```
function q=calcduty(tin,tout,flow,cp)
% this function calculates duty based on tube side/shell side data
% tin: Temperature at inlet, F
% tout: Temperature at outlet, F
% flow: mass flow rate, lb/h
% cp: Specific heat, BTU/lbF
q=flow*cp*(tin-tout);
```

Verify the function by executing once by providing hot side data:

```
>> q=calcduty(200,150,10000,0.545)

q =

      272500

>>
```

Write simple MATLAB function to calculate LMTD for the heat exchanger and save the file as calcLMTD.m

Content of the function is shown here.

```
function LMTD=calcLMTD(thin,thout,tcin,tcout,flowtype)
% This function calculates the LMTD based on the temperatures
% thin: Hot fluid inlet temperature, F
% thout: Hot fluid outlet temperature, F
% tcin: Cold fluid inlet temperature, F
% tcout: Cold fluid outlet temperature, F
% flowtype: 0, if Counter Current; 1, if Cocurrent
if(flowtype==1)
    TD1=thin-tcin;
    TD2=thout-tcout;
else
    TD1=thin-tcout;
    TD2=thout-tcin;
end
LMTD=(TD1-TD2)/log(TD1/TD2);
```

Verify the function by executing once by providing the data:

```
>> LMTD=calcLMTD(200,150,100,172,0)

LMTD =

   37.9429

>>
```

Write executive MATLAB function to calculate Overall heat transfer coefficient U for the heat exchanger and save it as calcCounterhxU.m Note that this function uses the other two functions mentioned earlier. Content of the function is shown here.

```
function U=calcCounterhxU(thin, thout,tcin, cph,cpc,flowh,flowc, area)
% This function calculates Overall Heat transfer coefficient U
% when complete information is provided on hot side and utility inlet is
% specified by the user
% thin: Hot fluid inlet temerature, F
% thout: Hot fluid outlet temerature, F
% tcin: Cold fluid inlet temerature, F
% cph: Hot fluid specific heat, BTU/lb.H
% cpc: Cold fluid specific heat, BTU/lb.H
% flowh: Hot fluid Flow rate, lb/H
% flowc: Cold fluid Flow rate, lb/H
% area: Heat transfer area, sq ft
% calculate the duty based on the hot side data
qh=calcduty(thin,thout,cph,flowh);
% Calculate the cold fluid outlet temperature based on energy conservation
tcout=qh/(flowc*cpc)+tcin;
% Calculate LMTD
flowtype=0; % 0 for counter flow and 1 for cocurrent flow
LMTD=calcLMTD(thin,thout,tcin,tcout,0);
%Calculate U
U=qh/(area*LMTD);
>> U=calcCounterhxU(200,150,100,0.545,0.44,10000,8600,70)

U =

   102.6205

>>
```

Repeat the similar procedure for the Cocurrent Heat exchanger and mention your observations.

References

Luyben W.L., Process modeling, simulation and control for chemical engineers, Mc-Graw Hill Publishers.

McCabe, Smith and Hariott, Unit Operations of Chemical Engineering, Mc-Graw Hill Publishers.

4

Unsteady State Conduction in a Slab

Objective: Emphasize the following principle:

Unsteady state conduction of Flat slab (*Use numerical methods to solve parabolic partial differential equations by explicit, implicit, and Crank-Nicolson methods* in MATLAB);

Problem Statement

A rod of steel is subjected to a temperature of $100°C$ on the left end and $25°C$ on the right end. If the rod is of length $0.05m$, use the explicit method to find the temperature distribution in the rod from $t = 0$ and $t = 9$ seconds. Use $\Delta x = 0.01\text{m}$, $\Delta t = 3s$. Given:

$$k = 54\frac{\text{W}}{\text{m}-\text{K}}, \ \rho = 7800\frac{\text{kg}}{\text{m}^3}, \ C = 490\frac{\text{J}}{\text{kg}-\text{K}}$$

The initial temperature of the rod is $20^{O}C$.

Theory

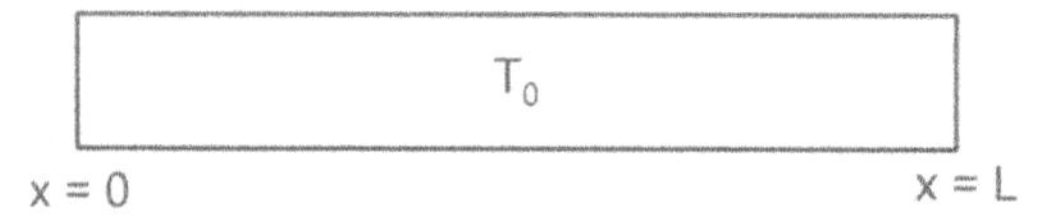

T_0

x = 0   x = L

Fig. 4.1 Two dimensional view of flat slab

T_0 is the initial rod temperature

T_1 is the new rod temperature at x = 0

Applying the microscopic energy balance to this system gives

$$\frac{\partial T}{\partial t} = \alpha\frac{\partial^2 T}{\partial x^2} \qquad\qquad .....(4.1)$$

with boundary conditions

$$T(x,0) = T_0 \qquad\qquad ..... (4.2)$$

$$T(0,t) = T_1 \qquad\qquad \text{..... (4.3)}$$

$$T(L,t) = T_0 \qquad\qquad \text{..... (4.4)}$$

This being a diffusive system or a parabolic partial differential equation, we apply the Crank-Nicolson finite-difference analogs to equation (4.1) and obtain

$$\frac{Ti+l,n - Ti,n}{\Delta t} = \alpha\left(\frac{T_{i+l,n+1} + T_{i,n+1} - 2T_{i+l,n} - 2T_{i,n} + T_{i+l,n-1} + T_{i,n-1}}{2(I_{n+1} - I_n)^2}\right) \text{.....(4.5)}$$

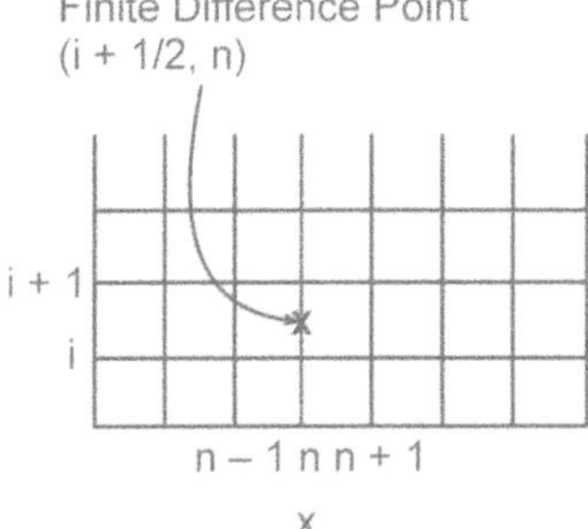

Fig. 4.2 Crank – Nicolson Grid

Fig.4.2 shows the finite-difference grid for this heat-transfer problem. We divide the distance grid into 20 elements. The Δt size in chosen so that $\Delta x^2/\Delta t = \alpha$. When the rod temperature is specified via the boundary condition, solid black dots are indicated in Fig.4.3.

Putting n = 1 and i = 0, we have two unknown temperatures $T_{1,1}$ and $T_{1,2}$ but just one algebraic relation. However, if we formulate the entire set of relations for this new time level, we get 19 equations in 19 unknowns. Science the equations are linear, we can express this set of equations in the matrix notation shown in Fig.8.16, where

$$a = -\frac{1}{2}\frac{a\Delta t}{\Delta x^2} \qquad\qquad \text{..... (4.6)}$$

$$b = 1 + \frac{a\Delta t}{\Delta x^2} \qquad\qquad \text{..... (4.7)}$$

$$c = -\frac{1}{2}\frac{a\Delta t}{\Delta x^2} \qquad\qquad \text{..... (4.8)}$$

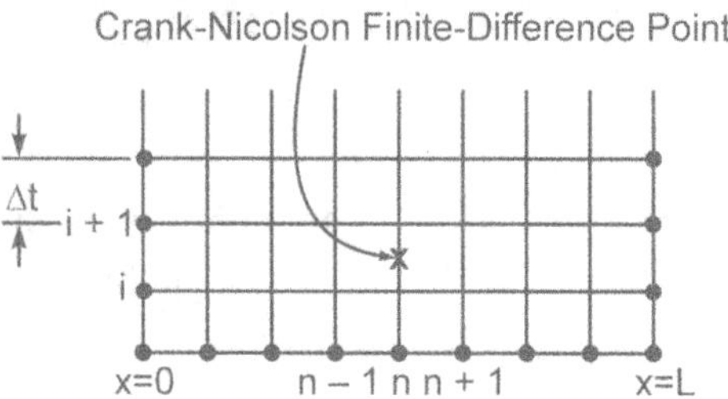

Fig. 4.3 Finite – Difference Grid.

$$d_n = \left(1 - \frac{a\Delta t}{\Delta x^2}\right)T_{i,\text{n}} - a(T_{i,\text{n}-1} + T_{i,\text{n}+1}) \qquad \ldots\ldots(4.9)$$

A MATLAB program can be implemented for getting solution to the problem. Fig. 4.4 shows the dynamic response curves for this heat conduction problem.

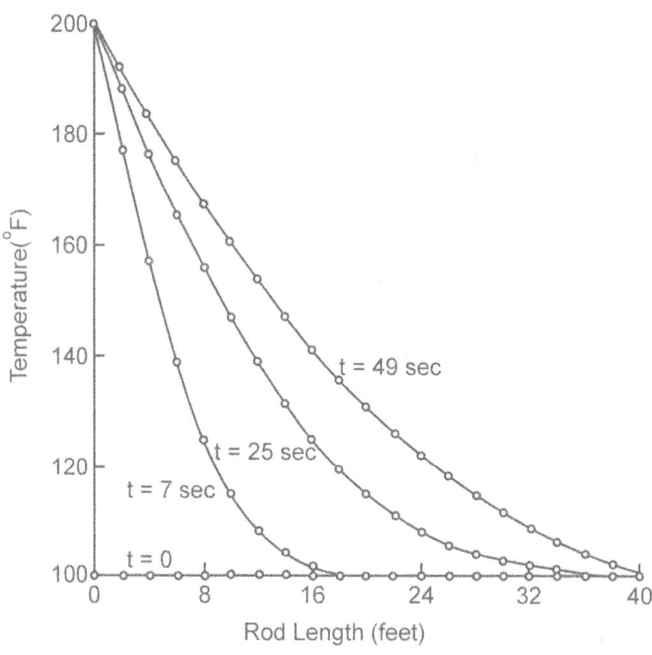

Fig. 4.4 Dynamic Response Curves for Unsteady – State Heat Conduction.

Conduction with Heat Source

Applying an energy balance to a non-dimensional solid, with the heat flux given by Fourier's law and with volumetric source term S (W/m^3) results in the following equations for steady state conduction in a flat plate of thickness 2R (b = 1), a cylinder of diameter 2R (b = 2), and a sphere of diameter 2R (b = 3). The parameter b is a measure of the curvature. The thermal conductivity is constant and there is convection at the surface, with heat transfer coefficient h and fluid temperature T_∞ .

$$\frac{d}{dr}\left(r^{b-1}\frac{dT}{dr}\right)+\frac{S}{K}r^{b-1}=0 \qquad\qquad .....(4.10)$$

$$\frac{dT(0)}{dr}=0\,(\text{symmetry condition})$$

$$-K\frac{dT}{dr}=h[T(R)-T_\infty]$$

The solutions to (1), for uniform S, are

$$\frac{T(r)-T_\infty}{SR^2/K}=\frac{1}{2b}\left[1-\left(\frac{r}{R}\right)^2\right]+\frac{1}{bBi}\begin{cases} b=1,\text{ plate, thicknes 2R}\\ b=2,\text{ cylinder, diameter 2R}\\ b=3,\text{ sphere, diameter 2R}\end{cases} .....(4.11)$$

where Bi = hR/K is the Biot number. For Bi<<1, the temperature in the solid is uniform. For Bi>>1, the surface temperature, $T(R)=T_\infty$.

The general second order linear PDE with two independent variables and one dependent variable is given by

$$A\frac{\partial^2 u}{\partial x^2}+B\frac{\partial^2 u}{\partial x\partial y}+C\frac{\partial^2 u}{\partial y^2}+D=0 \qquad\qquad ..... (4.12)$$

where A,B,C are functions of the independent variables, x , y , and D can be a function of $x,y,u,\dfrac{\partial u}{\partial x}$ and $\dfrac{\partial u}{\partial y}$. If $B^2-4AC=0$, Eq. (4.11) is called a parabolic partial differential equation. One of the simple examples of a parabolic PDE is the heat-conduction equation for a metal rod (Fig. 4.5)

$$\alpha\frac{\partial^2 T}{\partial x^2}=\frac{\partial T}{\partial t} \qquad\qquad ..... (4.13)$$

where

T = temperature as a function of location, x and time, t in which the thermal diffusivity, α is given by

$$\alpha=\frac{k}{\rho C}$$

where

k = thermal conductivity of rod material,

ρ = density of rod material,

C = specific heat of the rod material.

Fig. 4.5 A metal rod

Explicit Method of Solving Parabolic PDEs

To numerically solve parabolic PDEs such as Eq. (4.12), one can use finite difference approximations of the partial derivatives so that the dependent variable, T is now sought at particular nodes (x-location) and time (t) (Fig. 4.6). The left hand side second derivative is approximated by the central divided difference approximation as

$$\left.\frac{\partial^2 T}{\partial x^2}\right|_{i,j} \cong \frac{T_{i+1}^{j} - 2T_i^{j} + T_{i-1}^{j}}{\left(\Delta x\right)^2} \qquad \ldots\ldots(4.14)$$

where

i = node number along the $x-$ direction, $i = 0, 1, \ldots, n$,

j = node number along the time,

Δx = distance between nodes.

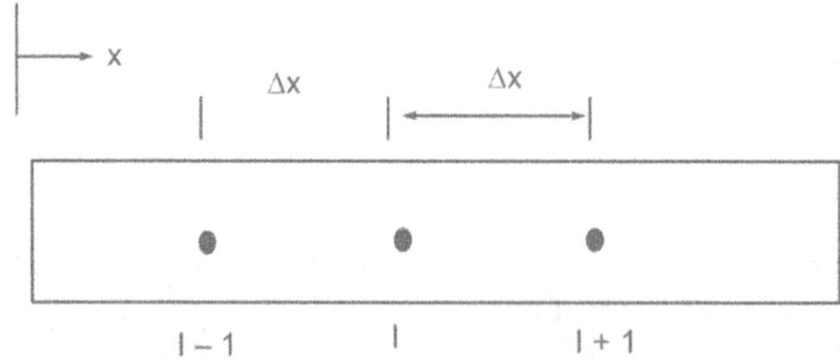

Fig. 4.6 Schematic diagram showing the node representation in the model

For a rod of length L which is divided into $n+1$ nodes,

$$\Delta x = \frac{L}{n} \qquad\qquad \ldots\ldots (4.15)$$

The time is similarly broken into time steps of Δt. Hence T_i^j corresponds to the temperature at node i, that is,

$$x = (i)(\Delta x)$$

and time,

$$t = (j)(\Delta t),$$

where

$$\Delta t = \text{time step.}$$

The time derivative of the right hand side of Eq. (4.12) is approximated by the forward divided difference approximation

$$\left. \frac{\partial T}{\partial t} \right|_{i,j} \cong \frac{T_i^{j+1} - T_i^j}{\Delta t} \qquad\qquad \ldots\ldots(4.16)$$

Substituting the finite difference approximations given by Eq.(4.13) and (4.15) in Eq. (4.12) gives

$$\alpha \frac{T_{i+1}^j - 2T_i^j + T_{i-1}^j}{(\Delta x)^2} = \frac{T_i^{j+1} - T_i^j}{\Delta t}$$

Solving for the temperature at the time node $j+1$, gives

$$T_i^{j+1} = T_i^j + \alpha \frac{\Delta t}{(\Delta x)^2} \left(T_{i+1}^j - 2T_i^j + T_{i-1}^j \right)$$

Choosing

$$\lambda = \alpha \frac{\Delta t}{(\Delta x)^2} \qquad\qquad \ldots\ldots (4.17)$$

$$T_i^{j+1} = T_i^j + \lambda \left(T_{i+1}^j - 2T_i^j + T_{i-1}^j \right) \qquad\qquad \ldots\ldots (4.18)$$

Eq.(4.17) can be solved explicitly because it can be written for each internal location node of the rod for time node $j + 1$ in terms of the temperature at time node j. In other words, if we know the temperature at node $j = 0$, and knowing the boundary temperatures, which is the temperature at the external nodes, we can find the temperature at the next time step. We continue the process by first finding the temperature at all nodes $j = 1$, and using these to find the temperature at the next time node, $j = 2$. This process continues till we reach the time at which we are interested in finding the temperature.

Solution for the Problem Statement:

$$\alpha = \frac{k}{\rho C}$$

$$= \frac{54}{7800 \times 490}$$

$$= 1.4129 \times 10^{-5} \text{ m}^2 / s$$

Then

$$\lambda = \alpha \frac{\Delta t}{(\Delta x)^2}$$

$$= 1.4129 \times 10^{-5} \frac{3}{(0.01)^2}$$

$$= 0.4239$$

Number of time steps $= \dfrac{t_{final} - t_{initial}}{\Delta t}$

$$= \frac{9 - 0}{3}$$

$$= 3$$

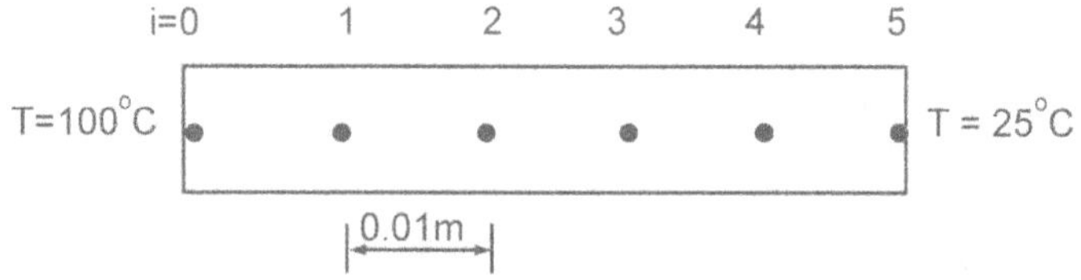

Fig. 4.7 Schematic diagram showing the node distribution in the rod

The boundary conditions

$$\left. \begin{array}{l} T_0^j = 100°C \\ T_5^j = 25°C \end{array} \right\} \quad \text{for all } j = 0, 1, 2, 3 \qquad \qquad .....(4.19)$$

The initial temperature of the rod is $20°C$, that is, all the temperatures of the nodes inside the rod are at $20°C$ when time, $t = 0$ sec except for the boundary nodes as given by Eq. (4.19). This could be represented as

$$T_i^0 = 20°C, \text{ for all } i = 1, 2, 3, 4 \qquad \qquad .....(4.20)$$

Initial temperature at the nodes inside the rod (when $t = 0$ sec)

$$T_0^0 = 100°C \quad \text{ from Equation} \qquad \qquad .....(4.21)$$

$$\left.\begin{array}{l} T_1^0 = 20°C \\ T_2^0 = 20°C \\ T_3^0 = 20°C \\ T_4^0 = 20°C \end{array}\right\} \quad \text{from Equation} \qquad \qquad .....(4.22)$$

$$T_5^0 = 25°C \quad \text{ from Equation} \qquad \qquad .....(4.23)$$

Temperature at the nodes inside the rod when $t=3$ sec

Setting $j = 0$ and $i = 0, 1, 2, 3, 4, 5$ in Eq. (4.17) gives the temperature of the nodes inside the rod when time, $t = 3\,\text{sec}$.

$$T_0^1 = 100°C \quad \text{ Boundary Condition} \qquad \qquad .....(4.24)$$

$$T_1^1 = T_1^0 + \lambda\left(T_2^0 - 2T_1^0 + T_0^0\right)$$
$$= 20 + 0.4239\left(20 - 2(20) + 100\right)$$
$$= 53.912°C$$

$$T_2^1 = T_2^0 + \lambda\left(T_3^0 - 2T_2^0 + T_1^0\right)$$
$$= 20 + 0.4239\left(20 - 2(20) + 20\right)$$
$$= 20°C$$

$$T_3^1 = T_3^0 + \lambda\left(T_4^0 - 2T_3^0 + T_2^0\right)$$
$$= 20 + 0.4239\left(20 - 2(20) + 20\right)$$
$$= 20°C$$

$$T_4^1 = T_4^0 + \lambda\left(T_5^0 - 2T_4^0 + T_3^0\right)$$
$$= 20 + 0.4239\left(25 - 2(20) + 20\right)$$
$$= 22.120°C$$

$$T_5^1 = 25°C \quad \text{Boundary Condition} \quad (4.18)$$

Time, t, sec	T0	T1	T2	T3	T4	T5
3	100	53.912	20	20	22.12	25
6*	100	59.073	34.375	20.899	22.442	25
9#	100	65.953	39.132	27.266	22.872	25

***Temperature at the nodes inside the rod when t = 6 sec**

Setting $j = 1$ and $i = 0, 1, 2, 3, 4, 5$ in Equation gives the temperature of the nodes inside the rod when time, $t = 6\,\text{sec}$

#Temperature at the nodes inside the rod when $t = 9$ sec

Setting $j = 2$ and $i = 0, 1, 2, 3, 4, 5$ in Eq.(4.17) gives the temperature of the nodes inside the rod when time, $t = 9\,\text{sec}$

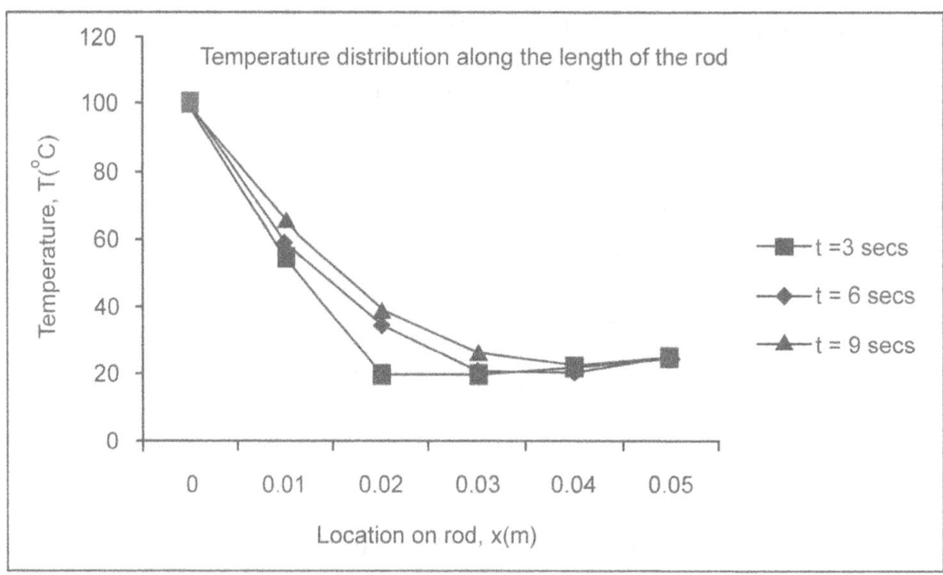

Fig. 4.8 Temperature distribution from explicit method

Implicit Method for Solving Parabolic PDEs

In the explicit method, one is able to find the solution at each node, one equation at a time. However, the solution at a particular node is dependent only on temperature from neighbouring nodes from the previous time step. For example, in the solution of Example 1, the temperatures at node 2 and 3 artificially stay at the initial temperature at $t = 3$ seconds. This is contrary to what we would expect physically from the problem.

Also the explicit method does not guarantee stability which depends on the value of the time step, location step and the parameters of the elliptic equation. For the PDE

$$\alpha \frac{\partial^2 T}{\partial x^2} = \frac{\partial T}{\partial t},$$

the explicit method is convergent and stable for

$$\frac{\alpha \Delta t}{(\Delta x)^2} \leq \frac{1}{2} \qquad \qquad .....(4.25)$$

These issues are addressed by using the implicit method. Instead of the temperature being found one node at a time, the implicit method results in simultaneous linear equations for the temperature at all interior nodes for a particular time.

The implicit method to solve the parabolic PDE given by equation (4.13) is as follows.

The second derivative on the left hand side of the equation is approximated by the central divided difference scheme at time level $j+1$ at node i as

$$\left. \frac{\partial^2 T}{\partial x^2} \right|_{i,j+1} \approx \frac{T_{i+1}^{j+1} - 2T_i^{j+1} + T_{i-1}^{j+1}}{(\Delta x)^2} \qquad \qquad .....(4.26)$$

The first derivative on the right hand side of the equation is approximated by backward divided difference approximation at time level $j+1$ and node i as

$$\left. \frac{\partial T}{\partial t} \right|_{i,j+1} \approx \frac{T_i^{j+1} - T_i^j}{\Delta t} \qquad \qquad .....(4.27)$$

Substituting Eq. (4.25) and (4.26) in Eq. (4.12) gives

$$\alpha \frac{T_{i+1}^{j+1} - 2T_i^{j+1} + T_{i-1}^{j+1}}{(\Delta x)^2} = \frac{T_i^{j+1} - T_i^j}{\Delta t}$$

giving

$$-\lambda T_{i-1}^{j+1} + (1+2\lambda)T_i^{j+1} - \lambda T_{i+1}^{j+1} = T_i^j \qquad \dots\dots (4.28)$$

where

$$\lambda = \alpha \frac{\Delta t}{(\Delta x)^2}$$

Now Eq. (4.11) can be written for all nodes (except the external nodes), at a particular time level. This results in simultaneous linear equations which can be solved to find the nodal temperature at a particular time.

Solution using Implicit Method:

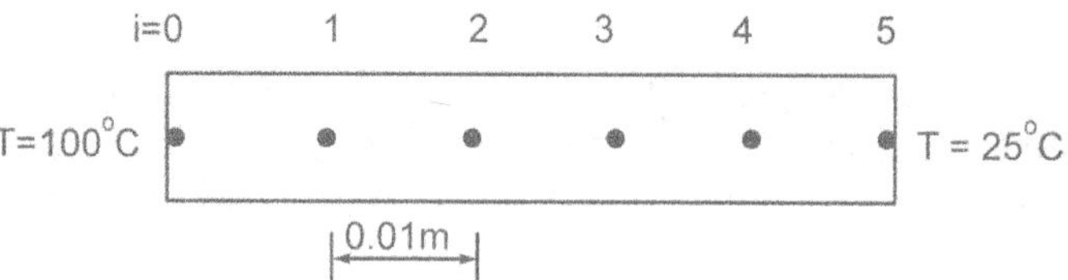

Fig. 4.9 Schematic diagram showing the node representation in the model

The boundary conditions

$$\left.\begin{array}{l} T_0^j = 100°C \\ T_5^j = 25°C \end{array}\right\} \quad \text{for } j = 0,1,2,3 \qquad \dots\dots(4.29)$$

The initial temperature of the rod is $20°C$, that is, the temperatures of all the nodes inside the rod are at $20°C$ when time, $t = 0$ except for the boundary nodes where the temperatures are given by satisfying the Eq.(4.27). This could be represented as

$$T_i^0 = 20°C, \text{ for } i = 1,2,3,4 \qquad \dots\dots(4.30)$$

Initial temperature at the nodes inside the rod (when $t = 0$ sec)

$$T_0^0 = 100°C \quad \text{from Equation} \qquad \dots\dots(4.31)$$

$$\left.\begin{array}{l} T_1^0 = 20°\text{C} \\[4pt] T_2^0 = 20°\text{C} \\[4pt] T_3^0 = 20°\text{C} \\[4pt] T_4^0 = 20°\text{C} \end{array}\right\} \quad \text{from Equation} \qquad\qquad .....(4.32)$$

$$T_5^0 = 25°\text{C} \qquad \text{from Equation} \qquad\qquad .....(4.33)$$

Temperature at the nodes inside the rod when t = 3 sec

$$\left.\begin{array}{l} T_0^1 = 100°\text{C} \\[4pt] T_5^1 = 25°\text{C} \end{array}\right\} \text{Boundary Condition} \qquad\qquad .....(4.34)$$

For all the interior nodes, putting $j = 0$ and $i = 1,2,3,4$ in Eq. (4.28) gives the following equations

$i = 1$

$$-\lambda T_0^1 + (1+2\lambda)T_1^1 - \lambda T_2^1 = T_1^0$$

$$(-0.4239 \times 100) + (1 + 2 \times 0.4239)T_1^1 - (0.4239 T_2^1) = 20$$

$$-42.39 + 1.8478 T_1^1 - 0.4239 T_2^1 = 20$$

$$1.8478 T_1^1 - 0.4239 T_2^1 = 62.390 \qquad\qquad .....(4.35)$$

$i = 2$

$$-\lambda T_1^1 + (1+2\lambda)T_2^1 - \lambda T_3^1 = T_2^0$$

$$-0.4239 T_1^1 + 1.8478 T_2^1 - 0.4239 T_3^1 = 20 \qquad\qquad .....(4.36)$$

$i = 3$

$$-\lambda T_2^1 + (1+2\lambda)T_3^1 - \lambda T_4^1 = T_3^0$$

$$-0.4239 T_2^1 + 1.8478 T_3^1 - 0.4239 T_4^1 = 20 \qquad\qquad .....(4.37)$$

$i = 4$

$$-\lambda T_3^1 + (1+2\lambda)T_4^1 - \lambda T_5^1 = T_4^0$$

$$-0.4239 T_3^1 + 1.8478 T_4^1 - (0.4239 \times 25) = 20$$

$$-0.4239 T_3^1 + 1.8478 T_4^1 - 10.598 = 20$$

$$-0.4239 T_3^1 + 1.8478 T_4^1 = 30.598 \qquad\qquad .....(4.38)$$

The simultaneous linear equations (4.35) – (4.38) can be written in matrix form as:

$$\begin{bmatrix} 1.8478 & -0.4239 & 0 & 0 \\ -0.4239 & 1.8478 & -0.4239 & 0 \\ 0 & -0.4239 & 1.8478 & -0.4239 \\ 0 & 0 & -0.4239 & 1.8478 \end{bmatrix} \begin{bmatrix} T_1^1 \\ T_2^1 \\ T_3^1 \\ T_4^1 \end{bmatrix} = \begin{bmatrix} 62.390 \\ 20 \\ 20 \\ 30.598 \end{bmatrix}$$

The above coefficient matrix is tri-diagonal. Special algorithms such as Thomas' algorithm can be used to solve simultaneous linear equation with tri-diagonal coefficient matrices. The solution is given by

$$\begin{bmatrix} T_1^1 \\ T_2^1 \\ T_3^1 \\ T_4^1 \end{bmatrix} = \begin{bmatrix} 39.451 \\ 24.792 \\ 21.438 \\ 21.477 \end{bmatrix}$$

Hence, the temperature at all the nodes at time, $t = 3\,\text{sec}$ is

$$\begin{bmatrix} T_0^1 \\ T_1^1 \\ T_2^1 \\ T_3^1 \\ T_4^1 \\ T_5^1 \end{bmatrix} = \begin{bmatrix} 100 \\ 39.451 \\ 24.792 \\ 21.438 \\ 21.477 \\ 25 \end{bmatrix}$$

Temperature at the nodes inside the rod when *t*=6 sec

$$\left.\begin{array}{l} T_0^2 = 100°C \\ T_5^2 = 25°C \end{array}\right\} \text{Boundary Condition} \qquad .....(4.39)$$

For all the interior nodes, putting $j = 1$ and $i = 1, 2, 3, 4$ in Eq. (4.27) gives the following equations

$i = 1$

$$-\lambda T_0^2 + (1 + 2\lambda)T_1^2 - \lambda T_2^2 = T_1^1$$

$$(-0.4239 \times 100) + (1 + 2 \times 0.4239)T_1^2 - 0.4239 T_2^2 = 39.451$$

$$-42.39 + 1.8478T_1^2 - 0.4239T_2^2 = 39.451$$

$$1.8478T_1^2 - 0.4239T_2^2 = 81.841 \qquad \qquad .....(4.40)$$

$i = 2$

$$-\lambda T_1^2 + (1 + 2\lambda)T_2^2 - \lambda T_3^2 = T_2^1$$

$$-0.4239T_1^2 + 1.8478T_2^2 - 0.4239T_3^2 = 24.792 \qquad \qquad .....(4.41)$$

$i = 3$

$$-\lambda T_2^2 + (1 + 2\lambda)T_3^2 - \lambda T_4^2 = T_3^1$$

$$-0.4239T_2^2 + 1.8478T_3^2 - 0.4239T_4^2 = 21.438$$

$i = 4$

$$-\lambda T_3^2 + (1 + 2\lambda)T_4^2 - \lambda T_5^2 = T_4^1$$

$$-0.4239T_3^2 + 1.8478T_4^2 - (0.4239 \times 25) = 21.477$$

$$-0.4239T_3^2 + 1.8478T_4^2 - 10.598 = 21.477$$

$$-0.4239T_3^2 + 1.8478T_4^2 = 32.075 \qquad \qquad .....(4.42)$$

The simultaneous linear eq. (4.40) – (4.42) can be written in matrix form as:

$$\begin{bmatrix} 1.8478 & -0.4239 & 0 & 0 \\ -0.4239 & 1.8478 & -0.4239 & 0 \\ 0 & -0.4239 & 1.8478 & -0.4239 \\ 0 & 0 & -0.4239 & 1.8478 \end{bmatrix} \begin{bmatrix} T_1^2 \\ T_2^2 \\ T_3^2 \\ T_4^2 \end{bmatrix} = \begin{bmatrix} 81.841 \\ 24.792 \\ 21.438 \\ 32.075 \end{bmatrix}$$

The solution of the above set of simultaneous linear equation is

$$\begin{bmatrix} T_1^2 \\ T_2^2 \\ T_3^2 \\ T_4^2 \end{bmatrix} = \begin{bmatrix} 51.326 \\ 30.669 \\ 23.876 \\ 22.836 \end{bmatrix}$$

Hence, the temperature at all the nodes at time, $t = 6 \sec$ is

$$\begin{bmatrix} T_0^2 \\ T_1^2 \\ T_2^2 \\ T_3^2 \\ T_4^2 \\ T_5^2 \end{bmatrix} = \begin{bmatrix} 100 \\ 51.326 \\ 30.669 \\ 23.876 \\ 22.836 \\ 25 \end{bmatrix}$$

Temperature at the nodes inside the rod when $t = 9$ sec

$$\left. \begin{array}{l} T_0^3 = 100°C \\ T_5^3 = 25°C \end{array} \right\} \text{Boundary Condition} \qquad \qquad .....(4.43)$$

For all the interior nodes, setting $j = 2$ and $i = 1, 2, 3, 4$ in Eq. (4.27) gives the following equations

$i = 1$

$$-\lambda T_0^3 + (1 + 2\lambda)T_1^3 - \lambda T_2^3 = T_1^2$$

$$(-0.4239 \times 100) + (1 + 2 \times 0.4239)T_1^3 - (0.4239T_2^3) = 51.326$$

$$-42.39 + 1.8478T_1^3 - 0.4239T_2^3 = 51.326$$

$$1.8478T_1^3 - 0.4239T_2^3 = 93.716 \qquad \qquad .....(4.44)$$

$i = 2$

$$-\lambda T_1^3 + (1 + 2\lambda)T_2^3 - \lambda T_3^3 = T_2^2$$

$$-0.4239T_1^3 + 1.8478T_2^3 - 0.4239T_3^3 = 30.669 \qquad \qquad .....(4.45)$$

$i = 3$

$$-\lambda T_2^3 + (1 + 2\lambda)T_3^3 - \lambda T_4^3 = T_3^2$$

$$-0.4239T_2^3 + 1.8478T_3^3 - 0.4239T_4^3 = 23.876 \qquad \qquad .....(4.46)$$

$i = 4$

$$-\lambda T_3^3 + (1 + 2\lambda)T_4^3 - \lambda T_5^3 = T_4^2$$

$$-0.4239T_3^3 + 1.8478T_4^3 - (0.4239 \times 25) = 22.836$$

$$-0.4239T_3^3 + 1.8478T_4^3 - 10.598 = 22.836$$

$$-0.4239T_3^3 + 1.8478T_4^3 = 33.434 \qquad \qquad .....(4.47)$$

The simultaneous linear eq. (4.44) – (4.47) can be written in matrix form as

$$\begin{bmatrix} 1.8478 & -0.4239 & 0 & 0 \\ -0.4239 & 1.8478 & -0.4239 & 0 \\ 0 & -0.4239 & 1.8478 & -0.4239 \\ 0 & 0 & -0.4239 & 1.8478 \end{bmatrix} \begin{bmatrix} T_1^3 \\ T_2^3 \\ T_3^3 \\ T_4^3 \end{bmatrix} = \begin{bmatrix} 93.716 \\ 30.669 \\ 23.876 \\ 33.434 \end{bmatrix}$$

The solution of the above set of simultaneous linear equation is

$$\begin{bmatrix} T_1^3 \\ T_2^3 \\ T_3^3 \\ T_4^3 \end{bmatrix} = \begin{bmatrix} 59.043 \\ 36.292 \\ 26.809 \\ 24.243 \end{bmatrix}$$

Hence, the temperature at all the nodes at time, $t = 9 \sec$ is

$$\begin{bmatrix} T_0^3 \\ T_1^3 \\ T_2^3 \\ T_3^3 \\ T_4^3 \\ T_5^3 \end{bmatrix} = \begin{bmatrix} 100 \\ 59.043 \\ 36.292 \\ 26.809 \\ 24.243 \\ 25 \end{bmatrix}$$

To better visualize the temperature variation at different locations at different times, the temperature distribution along the length of the rod at different times is plotted in Fig. 4.10.

Crank-Nicolson Method

The Crank-Nicolson method provides an alternative scheme to implicit method. The accuracy of Crank-Nicolson method is same in both space and time. In the implicit method, the approximation of $\dfrac{\partial^2 T}{\partial x^2}$ is of $O(\Delta x)^2$ accuracy, while the approximation for $\dfrac{\partial T}{\partial t}$ is of (Δt) accuracy. The accuracy in the Crank-Nicolson method is achieved by approximating the derivative at the midpoint of time step. To numerically solve PDEs such as Eq. (4.12), one can use finite difference approximations of the partial derivatives. The left hand side of the second derivative is approximated at node i as the average value of the central divided difference approximation at time level $j+1$ and time level j.

$$\left.\frac{\partial^2 T}{\partial x^2}\right|_{i,j} \approx \frac{1}{2}\left[\frac{T_{i+1}^j - 2T_i^j + T_{i-1}^j}{(\Delta x)^2} + \frac{T_{i+1}^{j+1} - 2T_i^{j+1} + T_{i-1}^{j+1}}{(\Delta x)^2}\right] \qquad \dots\dots (4.48)$$

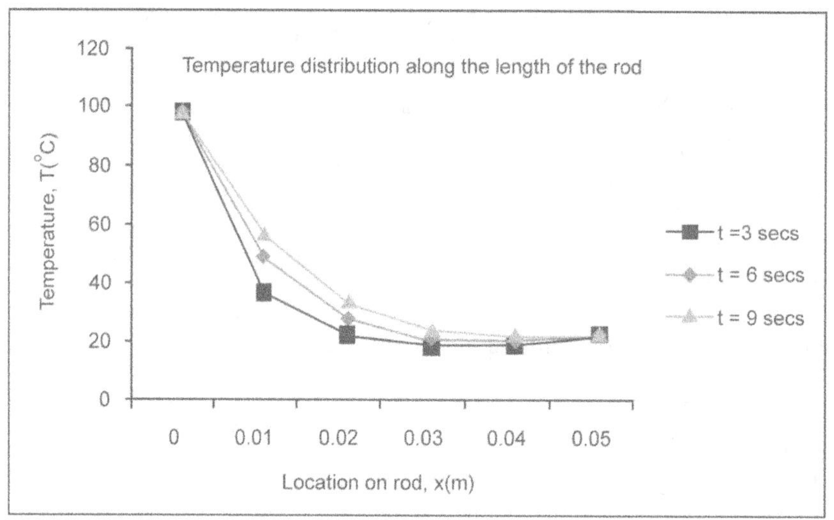

Fig. 4.10 Temperature distribution in rod from implicit method

The first derivative on the right side of Eq.(4.12) is approximated using forward divided difference approximation at time level $j+1$ and node i as

$$\left.\frac{\partial T}{\partial t}\right|_{i,j} \approx \frac{T_i^{j+1} - T_i^j}{\Delta t} \qquad \dots\dots (4.49)$$

Substituting Eq. (4.48) and (4.49) in Eq. (4.12) gives

$$\alpha \bullet \frac{1}{2}\left[\frac{T_{i+1}^j - 2T_i^j + T_{i-1}^j}{(\Delta x)^2} + \frac{T_{i+1}^{j+1} - 2T_i^{j+1} + T_{i-1}^{j+1}}{(\Delta x)^2}\right] = \frac{T_i^{j+1} - T_i^j}{\Delta t} \qquad \dots\dots (4.50)$$

giving

$$-\lambda T_{i-1}^{j+1} + 2(1+\lambda)T_i^{j+1} - \lambda T_{i+1}^{j+1} = \lambda T_{i-1}^j + 2(1-\lambda)T_i^j + \lambda T_{i+1}^j \qquad \dots\dots (4.51)$$

where

$$\lambda = \alpha \frac{\Delta t}{(\Delta x)^2}$$

Now Eq.(4.51) is written for all nodes (except the external nodes). This will result in simultaneous linear equations that can be solved to find the temperature at a particular time.

Solution using Crank-Nicolson Method:

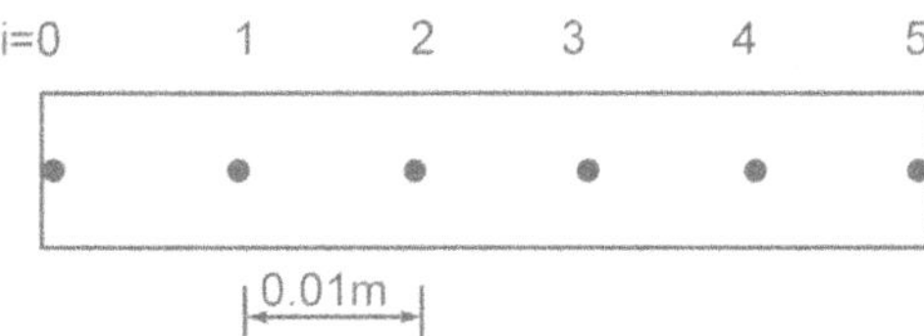

Fig. 4.11 Schematic diagram showing the node representation in the model

The boundary conditions are:

$$\left.\begin{array}{l} T_0^j = 100°C \\ T_5^j = 25°C \end{array}\right\} \quad \text{for } j = 0,1,2,3 \qquad .....(4.52)$$

The initial temperature of the rod is $20°C$, that is, all the temperatures of the nodes inside the rod are at $20°C$ at, $t = 0$ except for the boundary nodes given by Eq.(4.52). This could be represented as

$$T_i^0 = 20°C, \text{ for } i = 1,2,3,4$$

Initial temperature at the nodes inside the rod (when $t = 0$ sec)

$$T_0^0 = 100°C \quad \text{from Equation} \qquad (4.53)$$

$$\left.\begin{array}{l} T_1^0 = 20°C \\ T_2^0 = 20°C \\ T_3^0 = 20°C \\ T_4^0 = 20°C \end{array}\right\} \quad \text{from Equation} \qquad (4.54)$$

$$T_5^0 = 25°C \quad \text{from Equation} \qquad (4.55)$$

Temperature at the nodes inside the rod when $t=3$ sec

$$\left.\begin{array}{l} T_0^1 = 100°C \\ T_5^1 = 25°C \end{array}\right\} \text{Boundary Condition} \qquad (4.56)$$

For all the interior nodes, setting $j = 0$ and $i = 1,2,3,4$ in Eq. (4.51) gives the following equations

$i = 1$

$$-\lambda T_0^1 + 2(1+\lambda)T_1^1 - \lambda T_2^1 = \lambda T_0^0 + 2(1-\lambda)T_1^0 + \lambda T_2^0$$

$$(-0.4239 \times 100) + 2(1+0.4239)T_1^1 - 0.4239 T_2^1 = (0.4239)100 + 2(1-0.4239)$$

$$-42.39 + 2.8478 T_1^1 - 0.4239 T_2^1 = 42.39 + 23.044 + 8.478$$

$$2.8478 T_1^1 - 0.4239 T_2^1 = 116.30$$

$$2.8478 T_1^1 - 0.4239 T_2^1 = 116.30 \qquad .....(4.57)$$

$i = 2$

$$-\lambda T_1^1 + 2(1+\lambda)T_2^1 - \lambda T_3^1 = \lambda T_1^0 + 2(1-\lambda)T_2^0 + \lambda T_3^0$$

$$-0.4239 T_1^1 + 2(1+0.4239)T_2^1 - 0.4239 T_3^1$$
$$= (0.4239)20 + 2(1-0.4239)20 + (0.4239)20$$

$$-0.4239 T_1^1 + 2.8478 T_2^1 - 0.4239 T_3^1 = 40.000 \qquad .....(4.58)$$

$i = 3$

$$-\lambda T_2^1 + 2(1+\lambda)T_3^1 - \lambda T_4^1 = \lambda T_2^0 + 2(1-\lambda)T_3^0 + \lambda T_4^0$$

$$-0.4239 T_2^1 + 2(1+0.4239)T_3^1 - 0.4239 T_4^1$$
$$= (0.4239)20 + 2(1-0.4239)20 + (0.4239)20$$
$$-0.4239 T_2^1 + 2.8478 T_3^1 - 0.4239 T_4^1 = 40.000 \qquad .....(4.59)$$

$i = 4$

$$-\lambda T_3^1 + 2(1+\lambda)T_4^1 - \lambda T_5^1 = \lambda T_3^0 + 2(1-\lambda)T_4^0 + \lambda T_5^0$$

$$-0.4239 T_3^1 + 2(1+0.4239)T_4^1 - (0.4239)25$$
$$= (0.4239)20 + 2(1-0.4239)20 + (0.4239)25$$

$$-0.4239 T_3^1 + 2.8478 T_4^1 - 10.598 = 8.478 + 23.044 + 10.598$$

$$-0.4239 T_3^1 + 2.8478 T_4^1 = 52.718$$

The coefficient matrix in the above set of equations is tridiagonal. Special algorithms such as Thomas' algorithm are used to solve equation with tridiagonal coefficient matrices

$$\begin{bmatrix} 2.8478 & -0.4239 & 0 & 0 \\ -0.4239 & 2.8478 & -0.4239 & 0 \\ 0 & -0.4239 & 2.8478 & -0.4239 \\ 0 & 0 & -0.4239 & 2.8478 \end{bmatrix} \begin{bmatrix} T_1^1 \\ T_2^1 \\ T_3^1 \\ T_4^1 \end{bmatrix} = \begin{bmatrix} 116.30 \\ 40.000 \\ 40.000 \\ 52.718 \end{bmatrix}$$

The above matrix is tridiagonal. Solving the above matrix we get

$$\begin{bmatrix} T_1^1 \\ T_2^1 \\ T_3^1 \\ T_4^1 \end{bmatrix} = \begin{bmatrix} 44.372 \\ 23.746 \\ 20.797 \\ 21.607 \end{bmatrix}$$

Hence, the temperature at all the nodes at time, $t = 3$ sec is

$$\begin{bmatrix} T_0^1 \\ T_1^1 \\ T_2^1 \\ T_3^1 \\ T_4^1 \\ T_5^1 \end{bmatrix} = \begin{bmatrix} 100 \\ 44.372 \\ 23.746 \\ 20.797 \\ 21.607 \\ 25 \end{bmatrix}$$

Temperature at the nodes inside the rod when $t = 6$ sec

$$\left. \begin{array}{l} T_0^2 = 100°C \\ T_5^2 = 25°C \end{array} \right\} \text{Boundary Condition} \tag{4.60}$$

For all the interior nodes, putting $j = 1$ and $i = 1, 2, 3, 4$ in Eq. (4.51) gives the following equations.

$i = 1$

$$-\lambda T_0^2 + 2(1+\lambda)T_1^2 - \lambda T_2^2 = \lambda T_0^1 + 2(1-\lambda)T_1^1 + \lambda T_2^1$$

$$(-0.4239 \times 100) + 2(1+0.4239)T_1^2 - 0.4239T_2^2 =$$
$$(0.4239)100 + 2(1-0.4239)44.372 + (0.4239)23.746$$

$$-42.39 + 2.8478T_1^2 - 0.4239T_2^2 = 42.39 + 51.125 + 10.066$$

$$2.8478T_1^2 - 0.4239T_2^2 = 145.971 \tag{.....(4.61)}$$

$i = 2$

$$-\lambda T_1^2 + 2(1+\lambda)T_2^2 - \lambda T_3^2 = \lambda T_1^1 + 2(1-\lambda)T_2^1 + \lambda T_3^1$$

$$-0.4239T_1^2 + 2(1+0.4239)T_2^2 - 0.4239T_3^2 =$$
$$(0.4239)44.372 + 2(1-0.4239)23.746 + (0.4239)20.797$$

$$-0.4239T_1^2 + 2.8478T_2^2 - 0.4239T_3^2 = 18.809 + 27.360 + 8.8158$$

$$-0.4239T_1^2 + 2.8478T_2^2 - 0.4239T_3^2 = 54.985 \qquad .....(4.62)$$

$i = 3$

$$-\lambda T_2^2 + 2(1+\lambda)T_3^2 - \lambda T_4^2 = \lambda T_2^1 + 2(1-\lambda)T_3^1 + \lambda T_4^1$$

$$-0.4239T_2^2 + 2(1+0.4239)T_3^2 - 0.4239T_4^2 =$$
$$(0.4239)23.746 + 2(1-0.4239)20.797 + (0.4239)21.607$$

$$-0.4239T_2^2 + 2.8478T_3^2 - 0.4239T_4^2 = 10.066 + 23.962 + 9.1592$$

$$-0.4239T_2^2 + 2.8478T_3^2 - 0.4239T_4^2 = 43.187 \qquad .....(4.63)$$

$i = 4$

$$-\lambda T_3^2 + 2(1+\lambda)T_4^2 - \lambda T_5^2 = \lambda T_3^1 + 2(1-\lambda)T_4^1 + \lambda T_5^1$$

$$-0.4239T_3^2 + 2(1+0.4239)T_4^2 - (0.4239)25 =$$
$$(0.4239)20.797 + 2(1-0.4239)21.607 + (0.4239)25$$

$$-0.4239T_3^2 + 2.8478T_4^2 - 10.598 = 8.8158 + 24.896 + 10.598$$

$$-0.4239T_3^2 + 2.8478T_4^2 = 54.908 \qquad .....(4.64)$$

The simultaneous linear equations $(4.61) - (4.64)$ can be written in matrix form as

$$\begin{bmatrix} 2.8478 & -0.4239 & 0 & 0 \\ -0.4239 & 2.8478 & -0.4239 & 0 \\ 0 & -0.4239 & 2.8478 & -0.4239 \\ 0 & 0 & -0.4239 & 2.8478 \end{bmatrix} \begin{bmatrix} T_1^2 \\ T_2^2 \\ T_3^2 \\ T_4^2 \end{bmatrix} = \begin{bmatrix} 145.971 \\ 54.985 \\ 43.187 \\ 54.908 \end{bmatrix}$$

Solving the above set of equations, we get

$$\begin{bmatrix} T_1^2 \\ T_2^2 \\ T_3^2 \\ T_4^2 \end{bmatrix} = \begin{bmatrix} 55.883 \\ 31.075 \\ 23.174 \\ 22.730 \end{bmatrix}$$

Hence, the temperature at all the nodes at time, $t = 6\,\text{sec}$ is

$$\begin{bmatrix} T_0^2 \\ T_1^2 \\ T_2^2 \\ T_3^2 \\ T_4^2 \\ T_5^2 \end{bmatrix} = \begin{bmatrix} 100 \\ 55.883 \\ 31.075 \\ 23.174 \\ 22.730 \\ 25 \end{bmatrix}$$

Temperature at the nodes inside the rod when t=9 sec

$$\left. \begin{array}{l} T_0^3 = 100°C \\[2mm] T_5^3 = 25°C \end{array} \right\} \text{Boundary Condition (E3.1)}$$

For all the interior nodes, setting $j = 2$ and $i = 1, 2, 3, 4$ in Eq.(4.51) gives the following equations

$i = 1$

$$-\lambda T_0^3 + 2(1+\lambda)T_1^3 - \lambda T_2^3 = \lambda T_0^2 + 2(1-\lambda)T_1^2 + \lambda T_2^2$$

$$(-0.4239 \times 100) + 2(1+0.4239)T_1^3 - 0.4239 T_2^3 =$$
$$(0.4239)100 + 2(1-0.4239)55.883 + (0.4239)31.075$$

$$-42.39 + 2.8478 T_1^3 - 0.4239 T_2^3 = 42.39 + 64.388 + 13.173$$

$$2.8478 T_1^3 - 0.4239 T_2^3 = 162.34 \qquad\qquad .....(4.65)$$

$i = 2$

$$-\lambda T_1^3 + 2(1+\lambda)T_2^3 - \lambda T_3^3 = \lambda T_1^2 + 2(1-\lambda)T_2^2 + \lambda T_3^2$$

$$-0.4239 T_1^3 + 2(1+0.4239)T_2^3 - 0.4239 T_3^3 =$$
$$(0.4239)55.883 + 2(1-0.4239)31.075 + (0.4239)23.174$$

$$-0.4239 T_1^3 + 2.8478 T_2^3 - 0.4239 T_3^3 = 23.689 + 35.805 + 9.8235$$

$$-0.4239 T_1^3 + 2.8478 T_2^3 - 0.4239 T_3^3 = 69.318 \qquad\qquad .....(4.66)$$

$i = 3$

$$- \lambda T_2^3 + 2(1+\lambda)T_3^3 - \lambda T_4^3 = \lambda T_2^2 + 2(1-\lambda)T_3^2 + \lambda T_4^2$$

$$-0.4239T_2^3 + 2(1+0.4239)T_3^3 - 0.4239T_4^3 =$$

$$(0.4239)31.075 + 2(1-0.4239)23.174 + (0.4239)22.730$$

$$-0.4239T_2^3 + 2.8478T_3^3 - 0.4239T_4^3 = 13.173 + 26.701 + 9.635$$

$$-0.4239T_2^3 + 2.8478T_3^3 - 0.4239T_4^3 = 49.509 \qquad .....(4.67)$$

$i = 4$

$$- \lambda T_3^3 + 2(1+\lambda)T_4^3 - \lambda T_5^3 = \lambda T_3^2 + 2(1-\lambda)T_4^2 + \lambda T_5^2$$

$$-0.4239T_3^3 + 2(1+0.4239)T_4^3 - (0.4239)25 =$$

$$(0.4239)23.174 + 2(1-0.4239)22.730 + (0.4239)25$$

$$-0.4239T_3^3 + 2.8478T_4^3 - 10.598 = 9.8235 + 26.190 + 10.598$$

$$-0.4239T_3^3 + 2.8478T_4^3 = 57.210 \qquad .....(4.68)$$

The simultaneous linear eq.s(4.65) – (4.68) can be written in matrix form as

$$
\begin{bmatrix}
2.8478 & -0.4239 & 0 & 0 \\
-0.4239 & 2.8478 & -0.4239 & 0 \\
0 & -0.4239 & 2.8478 & -0.4239 \\
0 & 0 & -0.4239 & 2.8478
\end{bmatrix}
\begin{bmatrix}
T_1^3 \\ T_2^3 \\ T_3^3 \\ T_4^3
\end{bmatrix}
=
\begin{bmatrix}
162.34 \\ 69.318 \\ 49.509 \\ 57.210
\end{bmatrix}
$$

Solving the above set of equations, we get

$$
\begin{bmatrix}
T_1^3 \\ T_2^3 \\ T_3^3 \\ T_4^3
\end{bmatrix}
=
\begin{bmatrix}
62.604 \\ 37.613 \\ 26.562 \\ 24.042
\end{bmatrix}
$$

Hence, the temperature at all the nodes at time, $t = 9 \sec$ is

$$
\begin{bmatrix} T_0^3 \\ T_1^3 \\ T_2^3 \\ T_3^3 \\ T_4^3 \\ T_5^3 \end{bmatrix} = \begin{bmatrix} 100 \\ 62.604 \\ 37.613 \\ 26.562 \\ 24.042 \\ 25 \end{bmatrix}
$$

To better visualize the temperature variation at different locations at different times, the temperature distribution along the length of the rod at different times is plotted in Fig. 4.12.

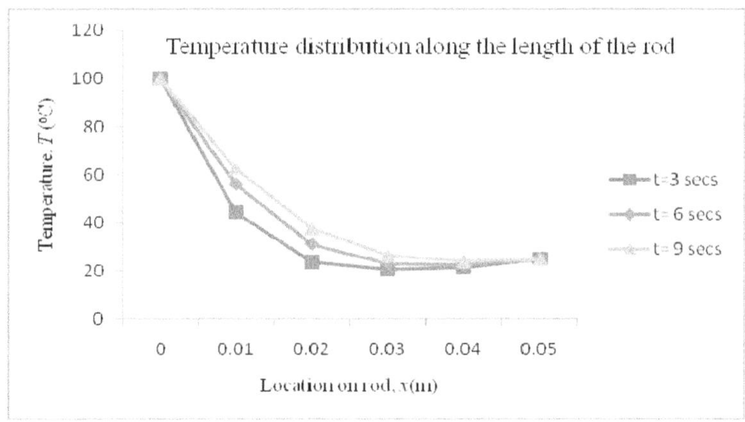

Fig. 4.12 Temperature distribution in rod from Crank-Nicolson method

Analytical Method

The parabolic heat conduction equation given by Equation (2) is formulated as:

$$
\alpha \frac{\partial^2 T}{\partial x^2} = \frac{\partial T}{\partial t} \qquad 0 < x < 0.05, \ t > 0
$$

with boundary conditions

$$T = 100°C \ \text{ at } \ x = 0, \, t > 0 \qquad\qquad ....(4.69)$$

$$T = 25°C \ \text{ at } \ x = 0.05, \, t > 0 \qquad\qquad .....(4.70)$$

and initial conditions

$$T = 20°C \ \text{ at } t = 0, \ 0 < x < 0.05 \qquad\qquad .....(4.71)$$

We split the problem into a steady state problem and a transient (homogeneous) problem. The solutions of the steady state problem and transient problem are found separately and by applying the principle of superposition, the final solution would be obtained. This formulation can be represented as

$$T(x,t) = T_s(x) + T_h(x,t) \qquad\qquad .....(4.72)$$

where

T_s = solution for steady state problem,

T_h = solution for transient problem.

Steady State Solution

$$T_h(x,t) = \sum_{m=1}^{\infty} \left\{ \frac{10(-1)^m - 160}{m\pi} \right\} e^{-\alpha[2.5m\pi]^2 t} \sin(2.5m\pi x) \qquad ....(4.73)$$

Substituting Eq. (4.73) in Eq. (4.72) we have

$$T(x,t) = -1500x + 100$$

$$+ \sum_{m=1}^{\infty} \left\{ \frac{10(-1)^m - 160}{m\pi} \right\} e^{-\alpha[20m\pi]^2 t} \sin(20m\pi x) \qquad .....(4.74)$$

Now

$$\alpha = \frac{k}{\rho C}$$

$$= \frac{54}{7800 \times 490}$$

$$= 1.4129 \times 10^{-5} \ m^2 / s$$

and substituting the value of α in Eq. (4.74) gives

$$T(x,t) = -1500x + 100$$

$$+ \sum_{m=1}^{\infty} \left\{ \frac{10(-1)^m - 160}{m\pi} \right\} e^{-1.4129 \times 10^{-5}[20m\pi]^2 t} \sin(20m\pi x) \qquad .....(4.75)$$

Eq.(4.75) is the analytical solution of the problem. Substituting the values of x and t gives the temperature inside the rod at a particular location and time. For example using the analytical solution, we will find the temperature of the rod at the first node, that is, $x = 0.01\,m$ when $t = 9$ secs.

$$T(0.01,9) = -1500(0.01) + 100$$

$$+ \sum_{m=1}^{\infty} \left\{ \frac{10(-1)^m - 160}{m\pi} \right\} e^{-1.4129\times10^{-5}[20m\pi]^2\times9} \sin(0.2m\pi = 62.510°C$$

Similarly using Eq. (4.81), the temperature of the rod at any location at any time can be found by substituting the corresponding values of x and t.

Comparison of the Three Numerical Methods

To compare all three numerical methods with the analytical solution, the temperature values obtained at all the interior nodes at time, $t = 9$ sec are presented in the Table 4.1. From Table 4.1, it is clear that among the numerical methods used to solve partial differential equations, Crank-Nicolson method provides better accuracy compared to the other two numerical methods (Explicit Method and Implicit Method) explained in this chapter.

Table 4.1 Comparison of temperature obtained at interior nodes using different methods discussed in this chapter (absolute true error is given in parenthesis)

Temperature at Nodes	Explicit Method ($°C$)	Implicit Method ($°C$)	Crank-Nicolson Method ($°C$)	Analytical Solution ($°C$)
T_1^3	65.953(3.443)	59.043(3.467)	62.604(0.094)	62.510
T_2^3	39.132(2.048)	36.292(0.792)	37.613(0.529)	37.084
T_3^3	27.266(1.422)	26.809(0.965)	26.562(0.282)	25.844
T_4^3	22.872(0.738)	24.243(0.633)	24.042(0.432)	23.610

Solution provided in CD/Publisher Website

References

Ramirez. W.F., Computational Methods for Process simulation, Butterworth Heinemann Publishers.

http://nm.mathforcollege.com/topics/pde_parabolic.html

5

Flare Slack Height Estimation

Objective: Emphasize the following principle.

Use API 521 to design flare stack for various scenarios

Problem Statement

Estimate flare stack height for plant with various scenario venting.

  (i)   Natural gas, 56,100 *lb*/hr
 (ii) Propane, 6700 *lb*/hr
(iii) Ammonia, 3400 *lb*/hr

with continuous exposure to operating

Flare Stack Height and Distance

Flare height and distance are dictated by radiation from the flame. In oilfield emergency flaring, distance is often used to provide protection from radiation. Height can be used in combination with or instead of distance to provide protection. Prevailing wind speed and direction are important factors. On windless days the flame is vertical, whereas maximum flare radiation occurs for a windblown flame in the direction of the processing facilities.

The following equation is given by API RP 521 (1990) for the distance required between a flame and a point of exposure where thermal radiation must be limited.

$$D = \sqrt{FQ / (4\pi K)} \qquad\qquad .....(5.1)$$

where

$D =$ minimum distance from the midpoint of the flame to the object being considered, ft

$F =$ fraction of heat that is radiated

$Q =$ heat release (lower heating value), Btu/hr

$K =$ allowable radiation, Btu/(hr-ft^2)

Flame radiant-heat fraction data are approximate. Read lists the following data for various flame types.

Flame Type	F
Clear, blue	0.07
Bright yellow	0.15
Slight smoking	0.13
Heavy smoking	0.10

Data from API RP 521 indicate the influence of type of gas and burner diameter, as follows.

Gas	Burner Dian., in	F
Methane	0.36	0.116
	0.75	0.160
	1.61	0.161
	3.31	0.147
Natural Gas	8.0	0.192
	16.0	0.232

Oenbring and Sifferman recommend a value of 0.25 for F for a 16.8 mol wt natural gas and 0.50 for a 40 mol wt gas. Chamberlain states that F is function of gas velocity. Obviously, the phenomenon is complex.

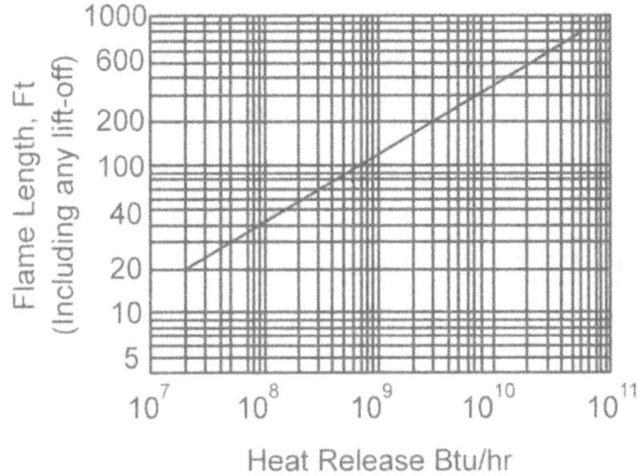

Fig. 5.1 Flame length vs. Heat Release (API RP 521, 1990)

Endurable radiation in the flare area is a total of solar plus flare radiation. At 40° north latitude, maximum solar radiation to earth is 246.

Btu/(hr-fr^2). According to Reed if work is to be carried out with only minor discomfort total radiation maybe 1000Btu/hft2.If greater discomfort is tolerable, radiation can be increased to 1,500 Btu/(hr-ft^2), if those who are thus exposed move about regularly. Radiant density at 2,000 Btu/(hr-ft^2) may be endured without injury if skin is well covered and if there is brisk movement.

The recommendations in Table 5.1 are given in API RP 521 (1990).

Flame length L. (ft) may be estimated by

$$L = Q / (A \times 8,000,000) \qquad\qquad .....(5.2)$$

where A = area of flare tip, ft^2

Reed cautions that such formulas are approximate at best. There is no substitute for pilot plant data under similar conditions and at as large a scale as possible.

The recommendation of API RP 521 (1990) for flame length is shown in Fig.5.1. In addition to length, the effect of wind on the flame angle must be considered. Fig.5.2 shows the API RP 521 recommendation for wind effect.

Table 5.1 Maximum design flare radiation levels during solar radiation

Condition	Allowable Radiation, K Btu/(hr–ft^2)
Any location where personnel are continuously exposed	500
Where emergency actions lasting several minutes may be required by personnel without shielding but appropriately clothed	1,500
Where emergency actions up to 1 minute may be required by personnel without shielding but appropriately clothed	2,000
Where exposure to personnel should be limited to escape only limited to a few seconds	3,000
Where operators are not likely to be performing duties and shelter from radiant heat is available (for example, behind equipment)	5,000

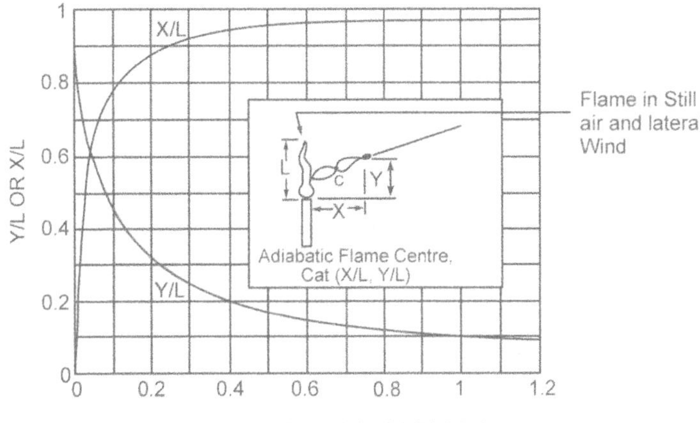

Fig. 5.2 Approximate flame distortion due to lateral wind for Flare stacks
(API RP 521, 1990)

Excel Solution for the Problem:

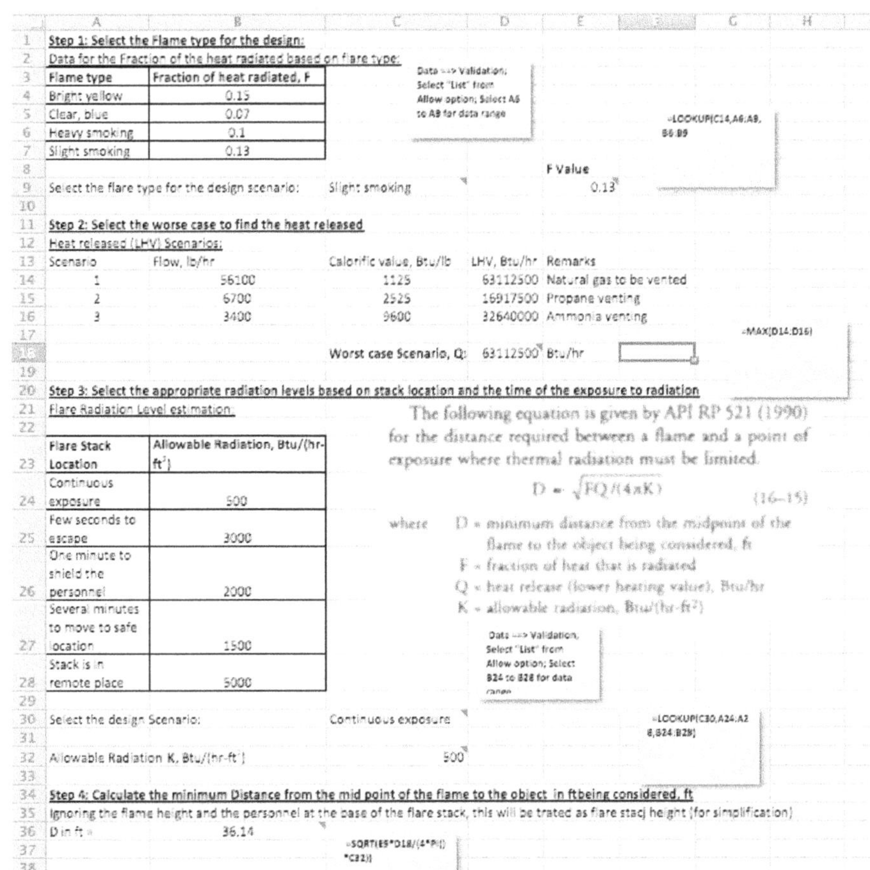

	A	B	C	D	E	F	G	H
1	Step 1: Select the Flame type for the design:							
2	Data for the Fraction of the heat radiated based on flare type:							
3	Flame type	Fraction of heat radiated, F		Data ---> Validation; Select "List" from Allow option; Select A5 to A8 for data range				
4	Bright yellow	0.15						
5	Clear, blue	0.07				=LOOKUP(C14,A6:A9, B6:B9)		
6	Heavy smoking	0.1						
7	Slight smoking	0.13						
8				F Value				
9	Select the flare type for the design scenario:		Slight smoking	0.13				
10								
11	Step 2: Select the worse case to find the heat released							
12	Heat released (LHV) Scenarios:							
13	Scenario	Flow, lb/hr	Calorific value, Btu/lb	LHV, Btu/hr	Remarks			
14	1	56100	1125	63112500	Natural gas to be vented			
15	2	6700	2525	16917500	Propane venting			
16	3	3400	9600	32640000	Ammonia venting			
17						=MAX(D14:D16)		
18			Worst case Scenario, Q:	63112500	Btu/hr			
19								
20	Step 3: Select the appropriate radiation levels based on stack location and the time of the exposure to radiation							
21	Flare Radiation Level estimation:							
22								
23	Flare Stack Location	Allowable Radiation, Btu/(hr-ft^2)						
24	Continuous exposure	500						
25	Few seconds to escape	3000						
26	One minute to shield the personnel	2000						
27	Several minutes to move to safe location	1500						
28	Stack is in remote place	5000						
29								
30	Select the design Scenario:		Continuous exposure			=LOOKUP(C30,A24:A2 8,B24:B28)		
31								
32	Allowable Radiation K, Btu/(hr-ft^2)		500					
33								
34	Step 4: Calculate the minimum Distance from the mid point of the flame to the object in ftbeing considered, ft							
35	Ignoring the flame height and the personnel at the base of the flare stack, this will be trated as flare stacj height (for simplification)							
36	D in ft =	36.14						
37			=SQRT(E5*D18/(4*PI() *C32))					
38								

The following equation is given by API RP 521 (1990) for the distance required between a flame and a point of exposure where thermal radiation must be limited.

$$D = \sqrt{FQ/(4\pi K)} \qquad (16\text{--}15)$$

where

D = minimum distance from the midpoint of the flame to the object being considered, ft

F = fraction of heat that is radiated

Q = heat release (lower heating value), Btu/hr

K = allowable radiation, Btu/(hr-ft^2)

Note:

1. Key formulae to be used are mentioned as comments in the excel sheet itself.

2. Users will learn list boxes in Excel Sheet.

MATLAB Solution for the Problem:

Write simple MATLAB function to calculate stack height and save the file as flarestackheight.m content of the function is shown below:

```
function d=flarestackheight(q)
% This function calculates the flare stack height
% Reference: Francis S. Manning, Richard E. Thompson, Oilfield Processing
%               of Petroleum: Crude oil
% q: Heat release in worst case scenario of the plant, Btu/hr
% d: estimated flare stack height, ft
% This function asks data from user during the program execution
disp('1-Bright Yellow; 2-Clear Blue; 3-Heavy smoking;4-Slight smoking');
flametype=input('Enter the Flame type=');
Foption=[0.15 0.07 0.1 0.13];
F=Foption(flametype);
disp('1-Continuous exposure; 2-Few Seconds to escape;');
disp('3-Several minutes to move to safe place;');
disp('4-One minute to move to safe place; 5-Stack in remote location;');
raditiontype=input('\nEnter the radiation type=');
AllowableR=[500 3000 2000 1500 5000];%Btu/(hr-ft2)
k=AllowableR(raditiontype);
d=sqrt(F*q/(4*pi*k));
```

Execution Procedure:

```
>> flarestackheight(63112500)
1-Bright Yellow; 2-Clear Blue; 3-Heavy smoking;4-Slight smoking
Enter the Flame type=4
1-Continuous exposure; 2-Few Seconds to escape;
3-Several minutes to move to safe place;
4-One minute to move to safe place; 5-Stack in remote location;

Enter the radiation type=1

ans =

   36.1359

>>
```

Note:

Q for all the three scenarios given in the problem were calculated and maximum Q was considered.

References

Francis S. Manning, Richard E.Thompson, Oilfield Processing of Petroleum: Vol 2, Crude oil, Pennwell Books.

6

Heat Equation – Stability Analysis using Explicit and Implicit Euler method

Objective: Emphasize the following principle

Perform stability analysis for Heat equation – understand the significance on stability by *explicit, implicit methods*.

Problem Statement

Study and find the temperature profile in heat rod for one dimensional heat transfer

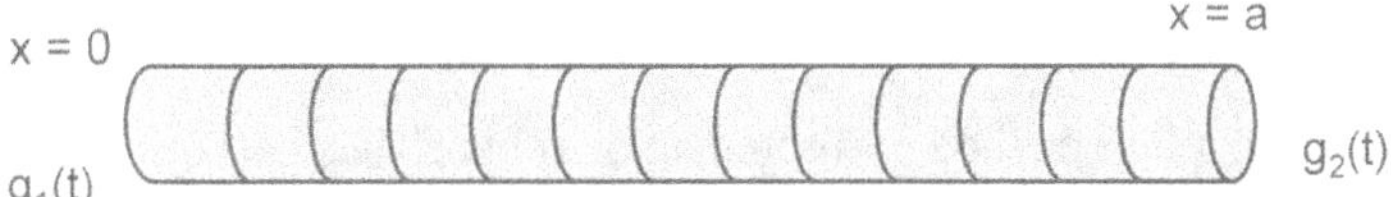

Fig. 6.1 Descritized view of heating rod

Theory

Governing equations:

$$\frac{\partial u}{\partial t} = c\frac{\partial^2 u}{\partial x^2}; 0 < x < a, 0 \le t \le T$$

$$u(x,0) = f(x), \quad 0 < x < a$$

$$\begin{cases} u(0,t) = g_1(t) \\ u(a,t) = g_2(t) \end{cases}, 0 \le t \le T$$

Discretize the solution domain in space and time with $h = \Delta x$ and $k = \Delta t$

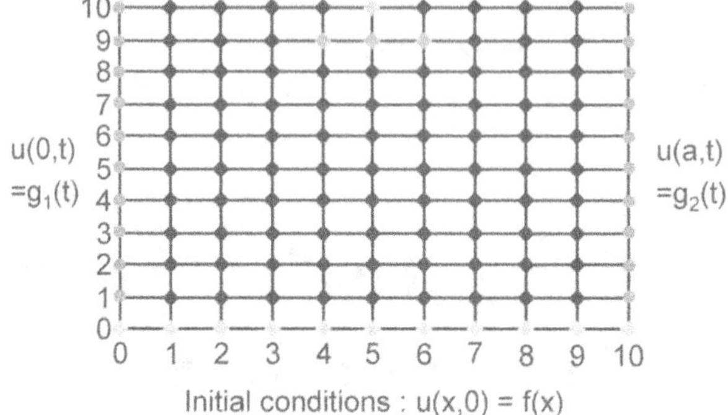

Fig. 6.2 Descretized grid

Forward difference is used to discretize the time derivative and central difference is used to discretize the space derivative:

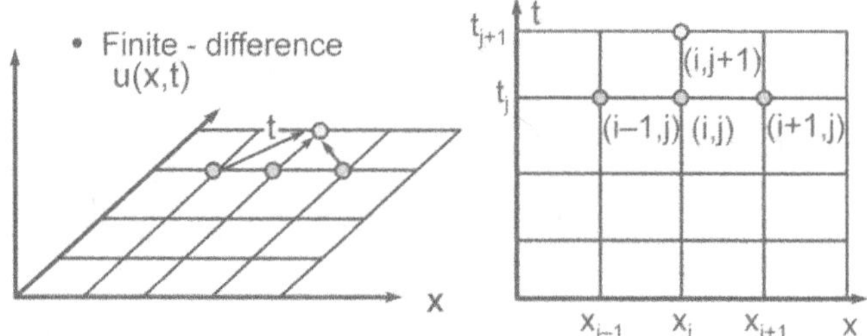

Fig. 6.3 Explicit method of discretization (obtain Ti_{j+1} based on T_j data)

Forward-difference $u_t = \dfrac{1}{k}(u_{i,j+1} - u_{i,j})$

Central-difference at time level j $cu_{xx} = \dfrac{c}{h_2}(u_{i-1,j} - 2u_{i,j} + u_{i+1,j})$

For Explicit Euler method,

$$\text{let}\begin{cases} h = \Delta x = a/n, & x_i - ih \\ k = \Delta t = T/m, & t_j - jk \end{cases}$$

Taking the heat equation and substituting the discretized form of derivatives, the equation becomes:

$$u_t = cu_{xx} \Rightarrow \frac{1}{k}(u_{i,j+1} - u_{i,j}) = \frac{c}{h^2}(u_{i-1,j} - 2u_{i,j} + u_{i+1,j})$$

Upon re-arranging, the above equation can be written as

$$u_{i,j+1} = u_{i,j} + \frac{ck}{h^2}(u_{i-1,j} - 2u_{i,j} + = u_{i+1,j})$$

$$= ru_{i-1,j} + (1-2r)u_{i,j} + ru_{i+1,j}$$

where

$$r = \frac{ck}{h^2} = \frac{c\Delta t}{\Delta x^2} \text{ and the equation is stable when } 0 < r \leq 0.5$$

Governing equation

$$u_{i,j+1} = ru_{i-1,j} + (1-2r)u_{i,j} + ru_{i+1,j}$$

Stable region:

$$r = 0.01 \Rightarrow u_{i,j+1} = 0.01u_{i-1,j} + 0.98u_{i,j} + 0.01u_{i+1,j}$$

$$r = 0.1 \Rightarrow u_{i,j+1} = 0.1u_{i-1,j} + 0.8u_{i,j} + 0.1u_{i+1,j}$$

$$r = 0.4 \Rightarrow u_{i,j+1} = 0.4u_{i-1,j} + 0.2u_{i,j} + 0.4u_{i+1,j}$$

$$r = 0.5 \Rightarrow u_{i,j+1} = 0.5u_{i-1,j} + 0.5u_{i+1,j}$$

Unstable region: You can observe negative coefficients in the equations when r > 0.5

$$\begin{cases} r = 1 \Rightarrow u_{i,j+1} = u_{i-1,j} - u_{i,j} + u_{i+1,j} \\ r = 10 \Rightarrow u_{i,j+1} = 10u_{i-1,j} - 19u_{i,j} + 10u_{i+1,j} \\ r = 100 \Rightarrow u_{i,j+1} = 100u_{i-1,j} - 199u_{i,j} + 100u_{i+1,j} \end{cases}$$

Case Study:

$$u_t = cu_{xx}; 0 \leq x \leq 1$$

$$\begin{cases} u(x,0) = 20 + 40x \\ u(0,t) = 20e^{-t}, u(1,t) = 60e^{-2t} \end{cases}$$

Consider c = 0.5, h = 0.25, k = 0.05

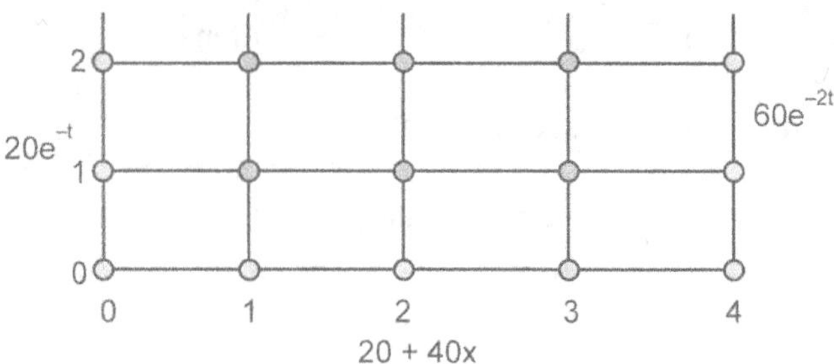

Fig. 6.4 Grid view with initial and boundary conditions

$$r = \frac{ck}{h^2} = \frac{(0.5)(0.05)}{(0.25)^2} = 0.4$$

$$u_{i,j+1} = ru_{i-1,j} + (1-2r)u_{i,j} + ru_{i+1,j}$$

$$= 0.4u_{i-1,j} + 0.2u_{i,j} + 0.4u_{i+1,j}$$

First step: t = 0.05

$$\begin{cases} u_{0.d} = 20e^{-0.05} = 19.02458849 \\ u_{1.d} = 0.4u_{0.0} + 0.2u_{1.0} + 0.4u_{2.0} = 0.4(20) + 0.2(30) + 0.4(40) = 30 \\ u_{2.d} = 0.4_{1.0} + 0.2u_{2.0} + 0.4u_{3.0} = 0.4(30) + 0.2(40) + 0.4(50) = 40 \\ u_{3.d} = 0.4u_{2.0} + 0.2u_{3.0} + 04.u_{4.0} = 0.4(40) + 0.2(50) + 0.4(60) = 50 \\ u_{4.d} = 60e^{-0.10} = 54.29024508 \end{cases}$$

Second step: t = 0.10

$$\begin{cases} u_{0.2} = 20e^{-0.10} = 18.09674836 \\ u_{1.2} = 0.4u_{0.1} + 0.2u_{1.1} + 0.4u_{2.d} \\ \quad\quad = 0.4(19.02458849) + 0.2(30) + 0.4(40) = 29.6098354 \\ u_{2.2} = 0.4_{1.d} + 0.2u_{2.d} + 0.4u_{3.d} = 0.4(30) + 0.2(40) + 0.4(50) = 40 \\ u_{3.2} = 0.4u_{2.d} + 0.2u_{3.d} + 0.4u_{4.d} \\ \quad\quad = 0.4(40) + 0.2(50) + 0.4(54.2924508) = 47.71609803 \\ u_{4.2} = 60e^{-0.20} = 49.12384518 \end{cases}$$

Explicit method is conditionally stable. i.e., $r \le \dfrac{1}{2}$ or $\Delta t \le \dfrac{1}{2}\dfrac{\Delta x^2}{c}$.

Implicit method, on the other side is unconditionally stable and is explained below:

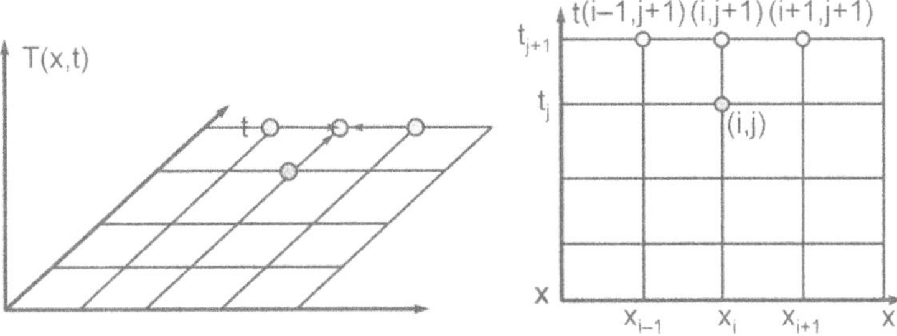

Fig. 6.5 Implicit method of discretization

Forward-difference $\qquad u_t = \dfrac{1}{k}(u_{i,j+1} - u_{i,j})$

Central-difference at time level j+1 $\qquad cu_{xx} = \dfrac{c}{h^2}(u_{i-1,j+1} - 2u_{i,j+1} + u_{i+1,j+1}$

Substituting the descretised forms into the governing equation,

$$\frac{1}{k}(u_{i,j+1} - u_{i,j}) = \frac{c}{h^2}(u_{i-1,j+1} - 2u_{i,j+1} + u_{i+1,j+1})$$

$$ru_{i-1,j+1} + (1+2r)u_{i,j+1} - ru_{i+1,j+1} = u_{i,j}$$

which can NOT be solved explicit way mentioned earlier.

Tridiagonal matrix is obtained by substituting the values of i and j:

$$\begin{bmatrix} 1+2r & -r & & & \\ -r & 1+2r & -r & & \\ & -r & 1+2r & -r & \\ & & \ddots & \ddots & -r \\ & & & -r & 1+2r \end{bmatrix} \begin{Bmatrix} u_{1,j+1} \\ u_{2,j+1} \\ u_{3,j+1} \\ \vdots \\ u_{n-1,j+1} \end{Bmatrix} = \begin{Bmatrix} u_{1,j} + ru_{0,j+1} \\ u_{2,j} \\ u_{3,j} \\ u_{n-1,j} + ru_{n,j+1} \end{Bmatrix}$$

Matrix inversion method can be used to find the solution for the above and the temperature profile is obtained for any value of r and is unconditionally stable.

Considering the same case study where explicit method is unstable,

$$r = 0 \qquad \frac{ck}{h^2} = \frac{(0.5)(0.1)}{(0.25)^2} = 0.8$$

Governing equation:

$$(-r)u_{i-1,j+1} + (1+2r)u_{i,j+1} + (-r)u_{i+1,j+1} = u_{i,j}$$

$$(-0.8)u_{i-1,j+1} + (2.6)u_{i,j+1} + (-0.8)u_{i+1,j+1} = u_{i,j}$$

Writing the equations for each i and j values and represent them as matix, the above equation becomes

$$\begin{bmatrix} 1+2r & -r & 0 \\ -r & 1+2r & -r \\ 0 & -r & 1+2r \end{bmatrix} \begin{Bmatrix} u_{1,1} \\ u_{2,1} \\ u_{3,1} \end{Bmatrix} = \begin{Bmatrix} u_{1,0} + ru_{0,1} \\ u_{2,0} \\ u_{3,0} + ru_{4,1} \end{Bmatrix}$$

Substituting the initial values and the boundary conditions, this becomes

$$\begin{bmatrix} 2.6 & -0.8 & 0 \\ -0.8 & 2.6 & -0.8 \\ 0 & -0.8 & 2.6 \end{bmatrix} \begin{Bmatrix} u_{1,1} \\ u_{2,1} \\ u_{3,1} \end{Bmatrix} = \begin{Bmatrix} 30 + 0.8(20e^{-0.1}) \\ 40 \\ 50 + 0.8(60e^{-0.2}) \end{Bmatrix}$$

leading to the solution

$$\Rightarrow \begin{Bmatrix} u_{1,1} \\ u_{2,1} \\ u_{3,1} \end{Bmatrix} = \begin{Bmatrix} 28.95515793 \\ 38.50751457 \\ 46.19426454 \end{Bmatrix}$$

This proves that the implicit method is unconditionally stable.

Fluid Flow

"[To] mechanical progress there is apparently no end: for as in the past so in the future, each step in any direction will remove limits and bring in past barriers which have till then blocked the way in other directions; and so what for the time may appear to be a visible or practical limit will turn out to be but a bend in the road."

— Osborne Reynolds

Opening address to the Mechanical Science Section, Meeting of the British Association, Manchester. In Nature (15 Sep 1887), 36, 475.

7

Multistage Compressor

Objective: Perform 2 stage compressor simulation and find the inter-stage pressure for optimum power consumption.

Problem Statement

Estimate the power required to compress 1000 m3/h air from ambient conditions to 700 kN/m^2 gauge, using a two-stage reciprocating compressor with an intercooler. Consider cp/cv = 1.4 and efficiency as 84%.

Adiabatic Compression

Compressibility is expressed as the multiplier for the perfect gas law to account for deviation from the ideal. At a given set of conditions of temperature and pressure:

$PV = ZNRT$ where Z = compressibility factor, usually less than 1.0

If the compression or expansion is isothermal (but not realized under actual conditions) (at constant temperature) then for unit mass of an ideal gas: $P_1 V_1 = P_2 V_2 = $ constant $= C$;

For the adiabatic condition of no heat lost or gained: $P_1 V_1^k = P_2 V_2^k$

Work done
$$-w = P_1 v_1 \ln \frac{P_2}{P_1} = \frac{RT_1}{M} \ln \frac{P_2}{P_1}$$

In actual practice compressors generate friction heat, give off heat, have valve leakage and have piston ring leakage. These deviations from the true adiabatic condition results in the process known as "polytropic." It is defined as an internally reversible change of state where

For a polytropic process the change of state does not take place at constant entropy, but for an adiabatic process, it does. Heat may be added to or rejected from a gas in a polytropic process. For a polytropic process, the correlating exponent for $P_1 v_1^n$ component is the exponent "n", which becomes important part of the compressor design. "n" values are determined from performance testing.

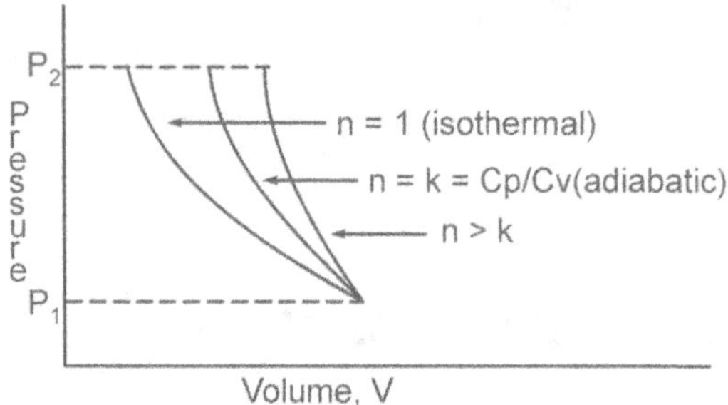

Fig. 7.1 Typical P Vs. V curve for gases

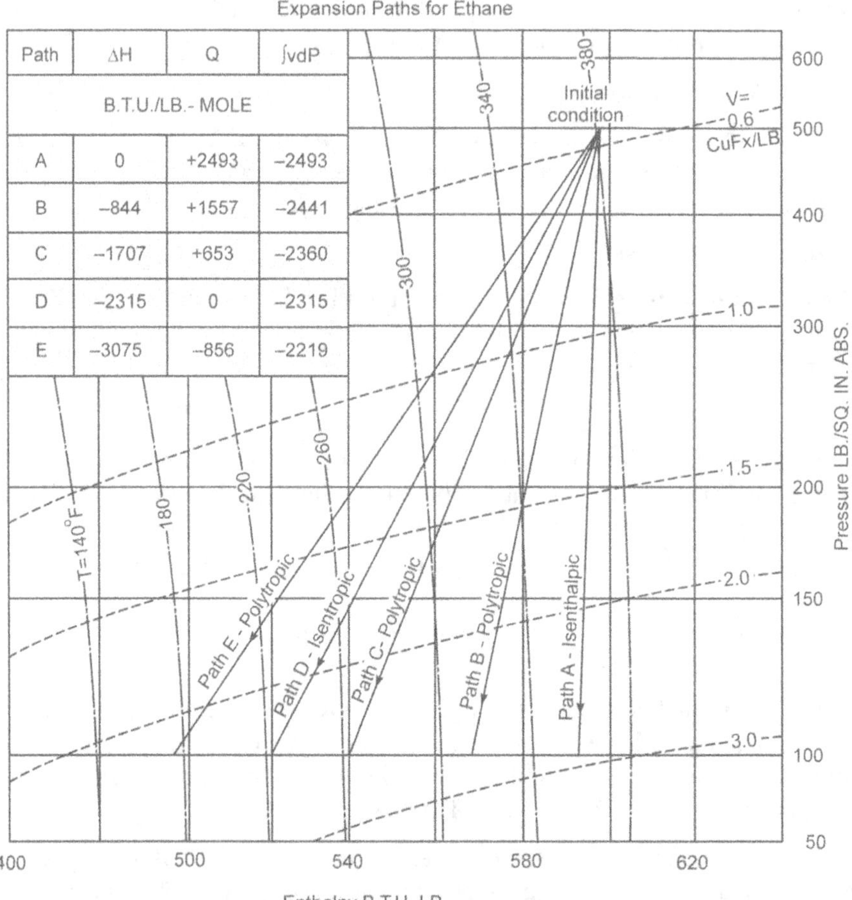

Path	ΔH	Q	∫vdP
	B.T.U./LB.- MOLE		
A	0	+2493	−2493
B	−844	+1557	−2441
C	−1707	+653	−2360
D	−2315	0	−2315
E	−3075	−856	−2219

Fig. 7.2 Expansion paths for ethane gas

$$\text{Work done: } -W = P_1 v_1 \frac{n}{n-1}\left[\left(\frac{P_2}{P_1}\right)^{(n-1)/n} - 1\right] = Z\frac{RT_1}{M}\frac{n}{n-1}\left[\left(\frac{P_2}{P_1}\right)^{(n-1)/n} - 1\right]$$

For a two-stage compressor the interstage pressure is given by:

$$P_i = \sqrt{(P_1 \times P_2)}$$

Adiabatic Calculations

$$\text{Adiabatic head, } H_{ad} = \frac{(k)}{(k-1)}RT_1\left[\left(\frac{P_2}{P_1}\right)^{\frac{k-1}{k}} - 1\right]$$

where

H_{ad} = adiabatic head, ft

R = gas constant, ft–lb$_{(force)}$/lb*R = 1,545/(mol wt)

T_1 = inlet gas temperature, *R

P_1 = absolute inlet pressure, $lb_{(f)}$/in.^{2}

P_2 = absolute discharge pressure, $lb_{(f)}$/in.^{2}

W = Weight of gas, lb/sec

Work on the gas during compression = H_{ad} (W)

k = ratio of specific heats, c_p/c_v

$$HP_{ad} = \frac{WH_{ad}}{550}$$

HP_{ad} = horsepower, hp

Procedure

Create compress.m file with the formula mentioned above to compute the power.

function horsepower = compress (k, pin, Tin, pout, MW, mflow)

 R = 1545/MW;

 Head = k*R*Tin*(((pout/pin)^(k–1)/k)–1)/(k–1);

Horsepower = Head*mflow/550;

From command prompt, calculate the power required to compress air from 14.7 psi to 147 psi in single stage.

Also calculate in two stage compression by assuming any inter stage pressure and calculate the total power required from 2 stages. Note that the inlet temperature is always room temperature as the first stage discharge is cooled to room temperature using interstage cooler using cooling water.

Repeat the same procedure for various interstage pressures assumed.

Finally, find the optimum interstage pressure for which the power requirement is lowest.

Solution for Air Compressor

Take the inlet pressure, P_1, as 1 atmosphere = 101.33 kN/m^2, absolute outlet pressure, P_2 = 700 + 101.33 = 801.33 kN/m^2, absolute. For equal work in each stage the intermediate pressure, P_i.

$$= \sqrt{\left(1.0133 \times 8.0133 \times 10^5\right)} = 2.8495 \times 10^5 \, N/m^2$$

For air, take ratio of the specific heats, y, to be 1.4. For equal work in each stage the total work will be twice that in the first stage. Take the inlet temperature to be 20 °C. At that temperature the specific volume is given by

$$v_1 = \frac{29}{22.4} \times \frac{293}{273} = 1.39 m^3/kg$$

$$\text{Work done, } -W = 2 \times 1.0133 \times 10^5 \times 1.39 \times \frac{1.4}{1.4-1}\left[\left(\frac{2.8495}{1.0133}\right)^{(1.4-1)/1.4} - 1\right]$$

$$= 338.844 \text{ J/kg} = 339 \text{ kJ/kg}$$

So work required considering the efficiency of 84%

$$= 339/0.84 = 404 \text{ kJ/kg}$$

$$\text{Mass flow-rate} \quad = \frac{1000}{1.39 \times 3600} = 0.2 \text{kg/s}$$

$$\text{Power required} \quad = 404 \times 0.2 = 80 \text{ kw}$$

Approximate Ratio of Specific Heats ("k" values) for various gases

Gas	Symbol	Mol wt	k @ 14.7 psia 60°F	150°F	Density @ 14.7 psi & 60°F lb/ft³
Monatomic	He, Kr, Ne, Hg		1.67		
Most diatomic	O_2, N_2, H_2, etc.		1.4		
Acetylene	C_2H_2	26.03	1.3	1.22	0.0688
Air		28.97	1.406	1.40	0.0765
Ammonia	NH_3	17.03	1.317	1.29	0.0451
Argon	A		1.667		0.1056
Benzene	C_6H_6	78.0	1.08	1.09	0.2064
Butane	C_4H_{10}	58.1	1.11	1.08	0.1535
Isobutane	C_4H_{10}	58.1	1.11	1.08	0.1578
Butylene	C_4H_8	56.1	1.1	1.09	0.1483
Iso-butene	C_4H_8	56.1	1.1	1.09	0.1483
Carbon dioxide	CO_2	44.0	1.3	1.27	0.1164
Carbon monoxide	CO	28.0	1.4	1.4	0.0741
Carbon tetrachloride	$C Cl_4$	153.8	1.18		0.406
Chlorine	Cl_2	70.9	1.33		0.1875
Dichlorodifluoromethane	$C Cl_2F_2$	120.9	1.13		
Dichloromethane	CH_2Cl_2	84.9	1.18		0.2245
Ethane	C_2H_6	30.0	1.22	1.17	0.0794
Ethylene	C_2H_4	28.1	1.25	1.21	0.0741
Ethyl chloride	C_2H_5Cl	64.5	1.13		0.1705
Flue gas			1.4		
Helium	He	4.0	1.667		0.01058
Hexane	C_6H_{14}	86.1	1.08	1.05	0.2276
Heptane	C_7H_{16}	100.2		1.04	0.264
Hydrogen	H_2	2.01	1.41	1.40	0.0053
Hydrogen chloride	HCl	36.5	1.48		0.09650
Hydrogen sulfide	H_2S	34.1	1.30	1.31	0.0901
Methane	CH_4	16.03	1.316	1.28	0.0423
Methyl chloride	CH_3Cl	50.5	1.20		0.1336
Natural gas (approx.)		19.5	1.27		0.0514
Nitric oxide	NO	30.0	1.40		0.0793
Nitrogen	N_2	28.0	1.41	1.40	0.0743
Nitrous oxide	N_2O	44.0	1.311		0.1163
Oxygen	O_2	32.0	1.4	1.39	0.0846
Pentane	C_5H_{12}	72.1	1.06	1.06	0.1905
Propane	C_3H_8	44.1	1.15	1.11	0.1164
Propylene	C_3H_6	42.0	1.16		0.1112
Sulfur dioxide	SO_2	64.1	1.256		0.1694
Water vapor (steam)	H_2O	18.0	1.33*	1.32	0.04761

*At 212°F

Matlab Programme for the given Problem:

```
function power=calcpower(p1,p2,t1,vflow,k, eff)
% estimates compression power required in kW
% p1, p2 : Inlet and outlet pressures in N/m2
% vflow: volumetric flow rate in m3/h
```

```
molwt=29;
spvol=(molwt/22.4)*(t1/273);
interstagepr=sqrt(p1*p2);
head=2*(k/(k-1))*spvol*p1*((interstagepr/p1)^((k-1)/k)-1);
mflow=vflow/(spvol*3600); % in kg/s
power=mflow*head/eff/1000;
```

Execution Procedure:

```
>> power=calcpower(101300,801330,293,1000,1.4,0.84)

power =

   80.6034
```

Note: Repeat the execution by considering the interstage pressures as 2*14.7, 3*14.7, 4*14.7, 5*14.7, 6*14.7, 7*14.7, 8*14.7, 9.14.7 and exectute the program two times and sum up the power to get the toal power consumption.

Also, note the power consumption by considering the sqrt(14.7*147) as interstage pressure.

Write your obervations and find the best interstage pressure.

Excel Solution for the Problem

	C29		f_x				
	A	B	C	D	E	F	G	H
1							
2							
3	inlet pressure, psi	14.7					
4	outlet pressure, psi	147					
5	Inlet Temperature, F	300					
6	mass flow rate, lb/h	1000					
7	k	1.4	=$B$9*A13^$B$10/$B$3-1				
8							
9	(k/(k-1))*R*T=	1623.3		=$B$9*$B$4^$B$10/A13-1			
10	(k-1)/k=	0.285714286					
11					=B13+C13		
12	Interstage pressure, psi	1st stage head	2nd stage head	Total head	power, hp		
13	29.4	289.1	228.8	517.9	941.6	=$B$6*D13/550	
14	44.1	324.8	152.2	477.0	867.2		
15	58.8	352.7	113.9	466.6	848.3		
16	73.5	376.0	90.9	466.9	848.9		
17	88.2	396.1	75.6	471.7	857.7		
18	102.9	414.0	64.6	478.7	870.3		
19	102.9	414.0	64.6	478.7	870.3		
20	132.3	444.9	50.1	495.0	899.9		
21							
22							
23	NOTE: After completing the row 13 formulae, cells B3 to E3 can be copied & then paste them in B14 to E20 to copy						

Note: The changes in units as compared to the MATLAB Program.

Additional problem for CHEMCAD

Design the compression system for syngas as shown in the diagram

Stream No	301		Stream No	307
Stream Name	SynGas		Stream Name	MEOH-Feed
Temp C	149.0601		Temp C	184.5853
Pres bar	5.000		Pres bar	48.0000
Enth kcal/h	− 1.4033E+007		Enth kcal/h	−4.0963E + 006
Vapor mole fraction	1.0000		Vapour mole fraction	1.0000
Total kmol/h	596.0195		Total kmol/h	423.0918
Total kg/h	7008.7631		Total kg/h	3893.4256
Total std L m^3/h	16.4535	Compressor Section	Total std L m^3/h	13.3381
Total std v m^3/h	13358.97		Total std v m^3/h	9483.03
Flowrates in kmol/h			Flowrates in kmol/h	
Methane	1.9903		Methane	1.9903
Hydrogen	322.0384		Hydrogen	322.0380
Carbon Monoxide	70.0005		Carbon Monoxide	70.0005
Carbon Dioxide	28.0092		Carbon Dioxide	28.0073
Water	173.9811		Water	1.0557
Oxygen	0.0000		Oxygen	0.0000
Nitrogen	0.0000		Nitrogen	0.0000
Methanol	0.0000		Methanol	0.0000

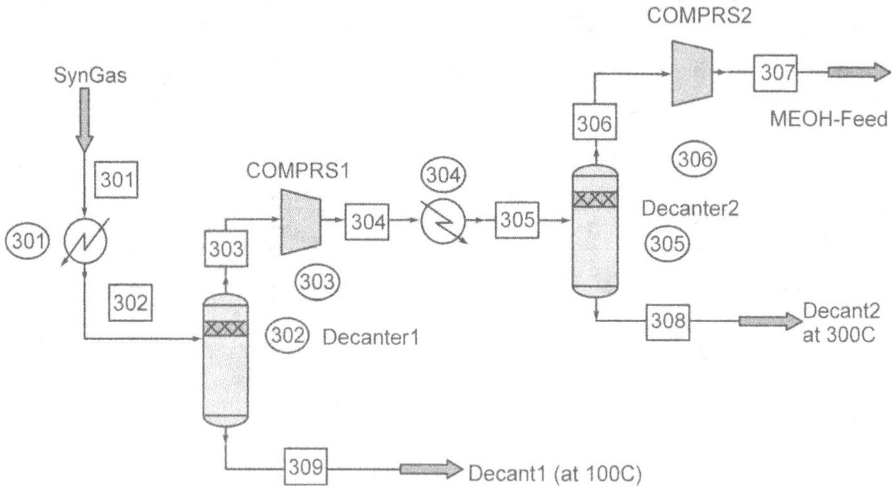

Fig. 7.3 Two stage compression model of syngas

References

Coulson & Richardson, Chemical Engineering, Vol.1, Butterworth Heinmann Publishers.

Ludwig, Applied Process Design, Vol.3, Gulf Professional Publishing.

8

Flare Sizing

Objective

Emphasize the following principle:

Estimation of flare header size, Relief valve and the flare stack height.

Problem Description

The example given is the minimum-hp case for the multi-stage separation system. The conditions for relief calculations are those indicated by Hafnor and Huaun (1977). These conditions apply for the first three separators and are reviewed now.

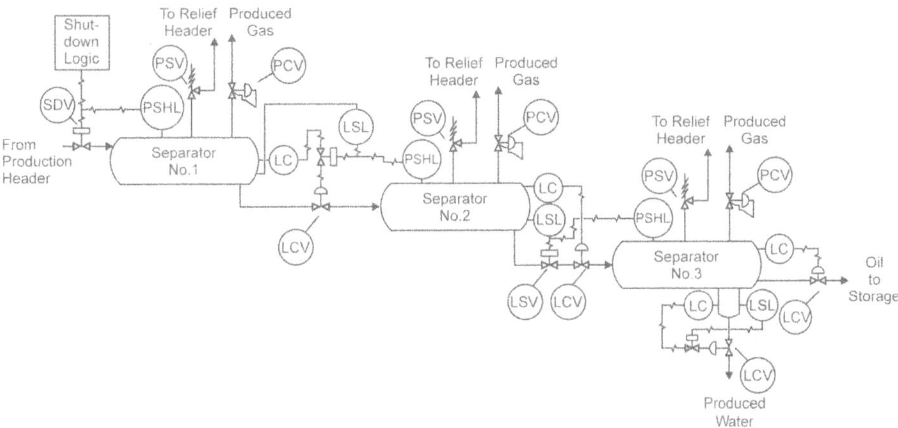

Fig. 8.1 Safety system for crude oil separator train

First, the safety system must never allow the main oil stream to go to flare! This can be avoided by making sure that the inlet line to the high-stage separator never fails open, no matter the cause of the malfunction (power failure, instrument air failure, cooling water failure, or malfunction of a controller). Given this fact, relief capacities are based on the following:

1ˢᵗ separator   The wellhead choke may open above normal setting. Use a 30% increase in total flow and assume that the oil flows out through Level Control Valve (LCV) and that gas flows at the normal rate to the sales line (flow controlled). Then the relief system must handle the additional gas at relieving pressure. Assume a Maximum Allowable Working Pressure (MAWP) of 1,055 psig, which is the normal operating pressure plus 50 psi.

2ⁿᵈ separator   Suppose the LCV controller malfunctions, so that not only oil goes out, but also gas. The second stage separator must handle the total gas load at its relieving pressure. Assume normal flow from second separator; oil to the third-stage and gas to the compressor section. The remainder of the gas must pass through the relief system. Assume a MAWP of 440 psig, which is the normal operating pressure 400 psig plus 40 psi.

3ʳᵈ separator   Suppose the vapour outlet is closed with normal inlet flow (no flow to the compressor); oil continues to flow to the low-pressure separator. The relief system must handle all gas outflow at relieving pressure. Assume a MAWP of 100 psig which is the normal operating pressure of 90 psig plus 10 psi.

Three simulations were performed for the conditions noted previously. In each case the relieving pressure is the MAWP (relief-valve set point) plus 10%, which gives 1,160.5 psig, and 110 psig, respectively for the separators. The data for the three cases are summarized.

Table 8.1 Relief conditions from 3 stages of separators

Case	First Stage	Second Stage	Third Stage
Vapor to Relief, lbmole/hr	201.5	2.327.2	858.6
Qs, scfh	86,500	883,200	325,800
W, /h/hr	3,829	45,788	21,168
P, psia*	1,175.2	498.7	124.7
T, *F	120.0	116.9	115.2
MW, lb/lbmole	19.00	19.68	24.65
ρ, lb/fth³	4.226	1.728	0.5169
Z	0.8493	0.9179	0.964
μ, cP	0.0144	0.0127	0.0116
C_p, Bru/(lbmole.*F)	12.21	10.99	12.15
K	1.194	1.221	1.195
LHV, Bru/scf	1,011.9	1,056.0	1,329.9

* Barometric pressure is assumed to be 14.7 psia.

The second stage case appears to control the design; calculations will be made for that case.

Relief Valve

This example uses various equations and charts from the reference book.

Assume critical flow for relief valve, since built-up back-pressure will be limited to 30% of the set pressure.

$$C = 520\sqrt{1.221(2/2.221)^{2.221/0.221}} = 339.3$$

Assume $K_d = 0.975$ (preliminary estimate)

$Kb = 1.0$ (for 30% built-up backpressure)

$$A = 45788\sqrt{576.6\times0.9179/19.68/(339.3\times0.975\times498.7)}$$

Select 3K4 since A = 1.838 > 1.44 required

A manufacturer's catalog lists a 3K4 valve with inlet flange rating of 720 psig and outlet flange rating of 150 psig, both satisfactory.

If a rupture disk is used in series with the relief valve, it must be derated to 80% of flow or area.

$$0.8 \times 1.838 = 1.47 \text{ in}^2 > 1.44$$

Flare Stack

Assume gas temperature unchanged to end of relief line.

$$V_s = 223 \ \sqrt{1.221\times576.6/19.68}$$

$$= 1334 \text{ ft/s}$$

$$V_a = 0.5 \ V_s = 0.5 \times 1334$$

$$= 667 \text{ ft/s}$$

$$Q = (883200) \ (576.6/14.7)/127300$$

$$= 272 \text{ ft}^3/\text{s}$$

$$D = \left(\sqrt{4(272)/(3.14159\times667)}\right)$$

$$= 0.721 \text{ ft} = 8.65 \text{ in}$$

Use $\qquad$ D = 10.01 in = 0.834 (10 in standard pipe)

$$A = 3.14159\,(0.834)^2/4 = 0.547\ \text{ft}^2$$

$$V_a = 272/0.547 = 498\ \text{ft/s}$$

$$M_a = 498/1334 = 0.37$$

$$K = 1500\ \text{Bru/(hr-ft}^2)\ \text{(for closest equipment)}$$

Deduct 250 Btu/(hr-ft^2) for solar radiation.

$$K = 1500 - 250 = 1250\ \text{Btu/(hr-ft}^2)$$

Heat Release $\qquad$ $Q = Q_5\ \text{LHV} = 883200 \times 1056$

$$= 9.33 \times 10^8\ \text{Btu/hr}$$

$$F = 0.25\ \text{(Conservative value)}$$

$$D = \sqrt{0.25 \times 9.33 \times 10^8 / (4 \times 3.14159 \times 1250))}$$

$$= 122\ \text{ft}$$

Set flare height at 50 ft

$$L = 130\ \text{ft}$$

Assume a design wind velocity of 60 miles = 88 ft/s

Ratio of wind velocity to flare tip velocity = 88/498 = 0.18

From charts $\qquad$ X/L = 0.87; $\qquad\qquad$ X = 0.87(130) = 113 ft

$\qquad\qquad\qquad$ Y/L = 0.33; $\qquad\qquad$ Y = 0.33(130) = 43 ft

$$X_c = X/2 = 57\ \text{ft}$$

$$Y_c = Y/2 = 22\ \text{ft}$$

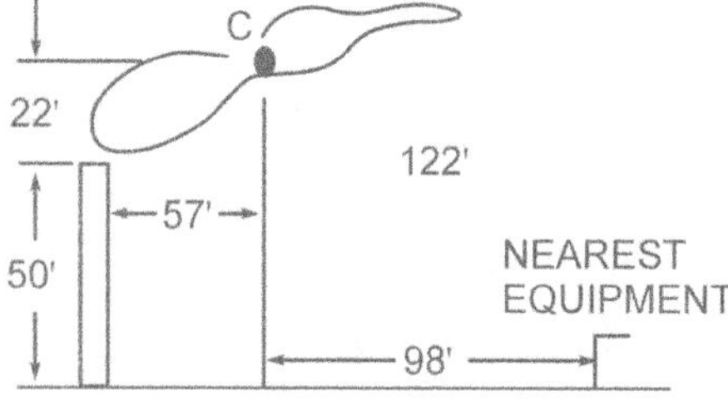

Fig. 8.2 Physical layout of the system

The horizontal distance from the center of the flame to any equipment and the total distance from the flare bottom is seen to be:

$$\sqrt{122^2 - 72^2} = 98\ \text{ft}$$

Total distance $= 98 + 57 = 155$ ft

The frictional pressure drop in the flare is calculated next. Calculate the density at 16.3 psia, assuming ideal behaviour.

$$\rho = PM/(RT) = 16.3 \times 19.68/(10.73 \times 576.6)$$
$$= 0.0518 \text{ lb/ft}^3$$

$$C_1 = 0.001 \, (45788/1000)^2 = 2.10$$
$$C_2 = 4990/8^{4.984} = 0.157$$
$$\Delta P_{100} = (2.1) \, (0.157)/0.0518 = 6.36$$
$$\Delta P = (50/100) \, (6.36) = 3.2 \text{ psi}$$

Average pressure $= 14.7 + 3.2/2 = 16.3$ psi (check)

Assume that a layout provides a total length of the relief line of 250 ft.

$$\Delta P = 0.3 \times \text{set pressure} - 3.2$$
$$= 0.3 \, (100) - 3.2$$
$$= 26.8 \text{ psi}$$
$$\Delta P_{100} = 26/8/(100/250) = 10.7 \text{ psi}$$

Average pressure $= 14.7 + 3.2 + 10.7/2 \times 23.3$ psis

$$\rho = 23.3 \times 19.68/(10.73 \times 576.6)$$
$$= 0.0741 \text{ lb/ft}^3$$
$$C_2 = 0.0741 \times 10.7/2.1 = 0.378$$
$$0.378 = 4990/D_n^{4.984}$$
$$D_n = 6.71, \text{ use an 8-in, pipe}$$

Useful Equations

Sonic velocity, V_s, is given by (Oenbring and Sifferman, 1980);
$$V_s = 223 \, \sqrt{kT/MW} \text{ ft/s}$$

where $T =$ gas temperature, $^\circ$R

The allowed velocity, V_a, is then

$$V_a = 0.5 \, Vs$$

Next, calculate the volumetric flow rate of the gas at the flare tip (where the gas can be treated as ideal).

$$Q = Q_s \, (14.7/P) \, (T/520)/3600$$

$$= Q_s \, (T/P)/1.273 \times 10^5$$

Where Q = gas flow, actual $ft^3/5$

Q_s = gas flow, scfh

The diameter, D_a, of the stack is:

$$D_a \, (ft) = \sqrt{4Q / \pi V_a}$$

Round up to the nearest available standard diameter, D_a. Calculate the pipe area, A, for that size and then the actual velocity, V_a, and Mach number, Ma.

$$A(ft^2) = \pi D_a^2 / 4$$

$$V_a = Q/A$$

$$Ma = V_a/V_s$$

Pressure drop in low-pressure gas lines can be estimated by the following equation (GPSA, 1987);

$$\Delta P_{100} = C_1 \, C_2/\rho \ (16\text{-}21)$$

where DP_{100} = pressure drop per 100 ft of pipe, psi

C_1 = constant depending on gas flow rate

C_2 = constant depending on nominal pipe diameter

ρ = gas density at average pressure, lb/ft^3

The data for C_1 and C_2 presented in graphical form in the GPSA Engineering Data Bank (GPSA, 1987) can be represented by the following two equations.

$$C_1 = 0.001 \, (W/1000)^2$$

where W = gas flow rate, lb/hr

$$C_2 = 4990/D_n^{4.984}$$

where D_n = nominal pipe diameter for Sch 40 pipe, in.

This method is adequate for relief valve sizing for gas flow only

Solution Provided in CD/Publisher Website

References

Francis S. Manning, Richard E. Thompson, Oilfield Processing of Petroleum: Crude oil, Vol. 2, Penn well Books.

9

Pressure Drop in Pipelines

Objective:

- Pressure drop calculations for liquids
- Pressure drop calculations for gases
- Pressure drop calculations for flashing liquids (steam condensate)
- Pressure drop calculations for 2 phase flows

Calculation of pressure drop for liquid flowing in a pipe

For the flow of a fluid in a pipe of length l and diameter d, the total frictional force at the walls is the product of the shear stress R and the surface area of the pipe $(R\pi dl)$. The head lost due to friction is:

$$h_f = \frac{-\Delta P_f}{\rho g} = 4\frac{R}{\rho u^2}\frac{l}{d}\frac{u^2}{g} = 4\phi\frac{l}{d}\frac{u^2}{g} = 8\phi\frac{l}{d}\frac{u^2}{2g} \qquad .....(9.1)$$

$$= f'\frac{l}{d}\frac{u^2}{2g} = 4f\frac{l}{d}\frac{u^2}{2g}$$

Moody friction factor $f' = 8R/\rho u^2$ the Fanning or Darcy friction factor $f = 2R/\rho u^2$ is often used.

Eq. (9.1)
$$-\Delta P_1 = 4\phi\frac{l}{d}\rho u^2$$

$$h_r = 8\phi\frac{l}{d}\frac{u^2}{2g}$$

Reynolds number $(Re = \frac{ud\rho}{\mu})$

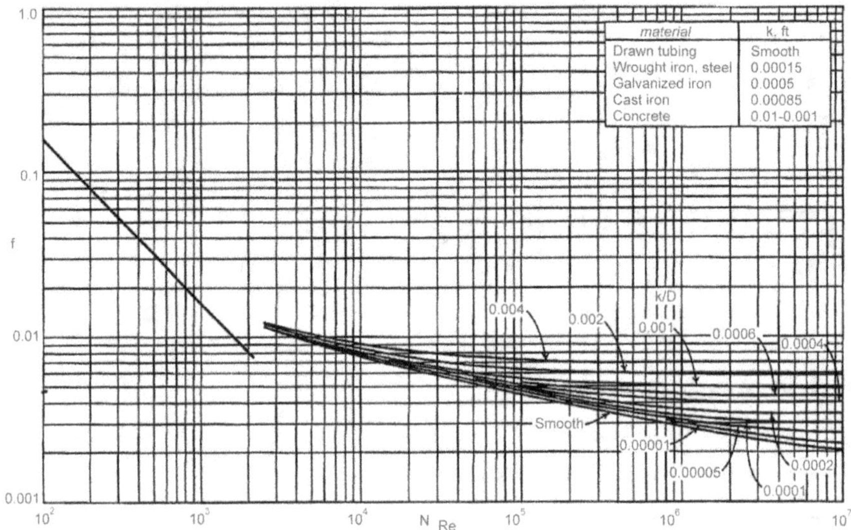

Fig. 9.1 Pipe friction chart ϕ versus Re

Scenario 1

0.015 m³/s of acetic acid is pumped through a 75 mm diameter horizontal pipe 70 m long. What is the pressure drop in the pipe?

Viscosity of acid = 2.5 mNs/m², density of acid = 1060 kg/m³, and roughness of pipe surface = 6×10^{-5} m.

Complex Pipe Systems Handling Natural (or similar) Gas

The method suggested in the Bureau of Mines Monograph No. 6 has found wide usage, and is outlined here using the Weymouth Formula as a base.

1. Equivalent lengths of pipe for different diameters $L_1 = L_2 (d_1 / d_2)^{16/3}$ where L_1 = the equivalent length of any pipe of length L_2, and diameter, d_2, in terms of diameter, d_1.

   $d_1 = d_2 (L_1 / L_2)^{3/16}$ where d_1 = the equivalent diameter of any pipe of a given diameter, d_2, and length, L_2, in terms of any other length, L_1

2. Equivalent diameters of pipe for parallel lines

   $d_0 = (d_1^{8/3} + d_2^{8/3} ... + d_n^{8/3})^{3/8}$ where d_0 is the diameter of a single line with the same delivery capacity as that of the individual parallel lines of diameters d_1, d_2 .. and d_n,. Lines of same length.

Senario 2

Below Figure shows a complex-series pipeline made up of four lengths of different size pipe. Determine the equivalent length of this pipe if each size of pipe has the same friction factor.

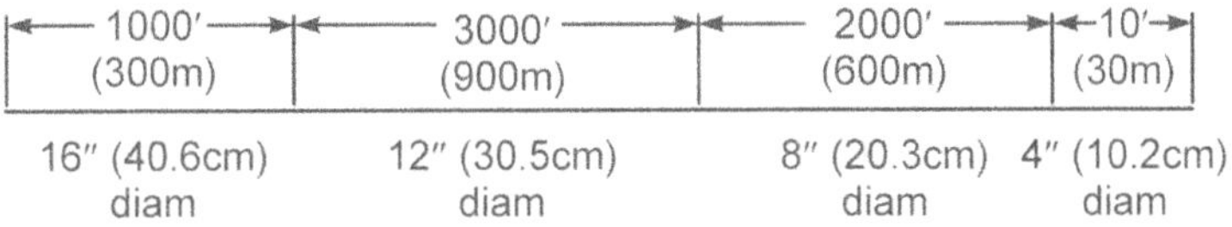

Fig. 9.2 Complex-series pipeline

Scenario 3: Looped System

Determine the equivalent length of 25 miles of 10-in. (10.136-h. I.D.) which has a parallel loop of 6 miles of 8- in. (7.981-in. I.D.) pipe tied in near the midsection of the 10-in. line.

Pipe Sizing using the suggested velocities:

Table 9.1 Suggested Fluid Velocities in Pipe and Tubing: Liquids, Gases, and Vapors at low/moderate pressure to 50 psig and 50° to 100°F.

The velocities are suggestive only and are to be used to Approximate line size as a starting point for pressure drop calculations.			The final line size should be such as to give as economical balance between pressure drop and reasonable velocity		
Fluid	**Suggested Trial Velocity**		**Pipe Material**	**Fluid**	**Suggested Trial Velocity**	**Pipe Material**
Acetyience (Observe Pressure limitations)	4000	fpm	Steel	Sodium Hydroxide 0–30 Percent	6 fps	Steel
Air, 0 to 30 psig	4000	fpm	Steel	30–50 Percent 50–73 Percent	5 fps 4	and Nickel
Ammonia Liquid	6	fps	Steel	Sodium Chloride Sol'n.		
Gas	6000	fpm	Steel	No Solids	5 fps	Steel
Benzene	6	fps	Steel	With Solids	(6 Min, – 15 Max)	Monel or nickel
Bromine Liquid	4	fps	Glass		7.5 fps	
Gas	2000	fpm	Glass	Perchloret hylene	6 fps	Steel
Calcium Chloride	4	fps	Steel	Steem		

Table 9.1 *Contd...*

Fluid	Suggested Trial Velocity		Pipe Material	Fluid	Suggested Trial Velocity	Pipe Material
Carbon Tetrachloride	6	fps	Steel	0–30 psi Satu-	4000–6000 fpm	Steel
Chlorine (Dry) Liquid	5	fps	Steel, Sch, 80	rated or super-		
Gas	2000–5000	fpm	Steel, Sch, 80	heated 150 psi up	6000–10000 fpm	
Chloroform Liquid	6	fps	Copper & Steeel	superheated	6500–15000 fpm	
Gas	2000	fpm	Copper & Steeel	Short lines	15,000 fps (max)	
Ethylene Gas	6000	fpm	Steel			
Ethylene Dibromide	4	fps	Glass	Sulfuric Acid		
Ethylene Dichloride	6	fps	Steel	88–93 Percent	4 tps	S.S –316. Lead
Ethylene Glycol	6	fps	Steel	93–100 Percent	4 fps	Cast Iron & Steel,
Hydrogen	4000	fpm	Steel			Sch, 80
Hydrochloric Acid Liquid	5	fps	Rubber Lined	Sulfuric Dioxide Styrene	4000 fpm 6 fps	Steel Steeel
Gas	4000	fpm	R. L, Saran, Haveg	Trichloroethyle Vinyl Chloride Vinylidene Chloride ne	6 fps 6 fps 6 fps	Steel Steel Steel
Methyl Chloride Liquid	6	fps	Steel	Water		
Gas	4000	fpm	Steel	Average service	3 – 8 (avg, 6) fps	Steel
Natural Gas	6000	fpm	Steel	Boiler feed	4–12 fps	Steel
Oils, lubrication	6	fps	Steel	Pump suction lines	4–5 fps	Steel
Oxygen (ambient temp.) (Low temp.)	1800 4000	fpm Max	Steel (300 psig Max.) Type 304 SS	Maximum economical (usual) Sea and brackish	7–10 fps	Steel R.L concrete,
Propylene Glycol	5	fps	Steel	water, lined pipe Concrete	5–8 fps 3 5–12 fps (Min.)	asphalt-line, saran-lined, transite

Note: R.L Rubber lined steel

Table 9.2 Typical Design Vapor Velocities* (ft/sec)

Fluid	≤ 6"	Line Sized 8"-12"	≥ 14"
Saturated Vapor 0 to 50 psig	30-115	50-125	60-145
Gas or Superheated Vapor 0 to 10 psig 11 to 100 psig 101 to 900 psig	50-140 40-115 30- 85	90-190 75-165 60-150	110-250 95-225 85-165

* Values listed are guides, and final line sizes and flow velocities must be determined by appropriate calculations to suit circumstances. Vacuum lines are not included in the table, but usually tolerate higher velocities. High vacuum conditions require careful pressure drop evaluation.

Table 9.3 Usual Allowable Velocities for Duct and Piping Systems*

Service/Application	Velocity, ft./min.
Forced draft ducts	2,500 – 3,500
Induced-draft flues and breeching	2,000 – 3,000
Chimneys and stacks	2,000
Water lines (max)	600
High pressure steam lines	10,000
Low pressure steam lines	12,000 – 15,000
Vacuum steam lines	25,000
Compressed vapor lines	2,000
Refrigerant vapor lines High pressure Low pressure	 1,000 – 3,000 2,000 – 5,000
Refrigerant liquid	200
Brine lines	400
Ventilating ducts	1,200 – 3,000
Register grilles	500

* Chemical Engineer's Handbook, 3rd Ed., McGraw-Hill Book Co., New York, N.Y., P. 1642.

Table 9.4 Typical Design* Velocities for Process System Applications

Service	Velocity, ft./sec.
Average liquid process	4 - 65
Pump suction (except boiling)	1 – 5
Pump suction, boiling	0.5 – 3
Boiler feed water (disch., pressure)	4 – 8
Drain lines	15 – 4
Liquid to reboiler (no pump)	2 – 7
Vapor-liquid mixture out reboiler	15 – 30
Vapor to condenser	15 – 80
Gravity separator flows	0.5 – 1.5

* To be used as guide, pressure drop and system environment govern final selection of pipe size.

For heavy and viscous fluids, velocities should be reduced to about ½ values shown.

Fluids not to contain suspended solid particles.

Table 9.5 Suggested Steam Pipe Velocities in Pipe Connection to Steam Turbines

Service–Steam	Typical range, ft./sec.
Inlet to turbine	100 – 150
Exhaust, non-condensing	175 – 200
Exhaust, condensing	500 – 400

Two-phase Liquid and Gas Flow

Flow Patterns

Six or seven types of flow patterns are usually considered in evaluating two-phase flow. Only one type can exist in a line at a time, but as conditions change (velocity, roughness, elevation, etc.) the type may also change. The unit pressure drop varies significantly between the types. Fig.9.3 illustrates the typical flow regimes recognized in two-phase flow.

Fig.9.2 typically represents a graphical illustration of the various flow patterns of as the two phase mixture flows through the piping. Long gas transport lines may have hydrocarbon or other liquids form (condense) as the fluid flows, and this becomes a real problem for offshore or long buried onshore raw gas transmission.

Bubble or Froth: Bubbles dispersed in liquid

Stratified: Liquid and gas flow in stratified layers

Wave: Gas flows in top of pipe section, liquid in waves in lower section

Slug: Slugs of gas bubbles flowing through the liquid

Annular: Liquid flows in continuous annular ring on pipe wall, gas flows through center of pipe

Plug: Plugs of liquid flow followed by plugs of gas

Dispersed: Gas and liquid dispersed

Beggs-Brill-Moody (BBM) Method

For the pressure drop elevation term, the friction factor, f, is computed from the relationship:

$$f / f_n = f_{tp} / f_n = \exp(s)$$

The exponent s is given by:

$$s = y / (-0.0523 + 3.182y - 0.8725y^2 + 0.01853y^4)$$
$$s = \ln(22e^y - 1.2), \ 1 < e^y < 1.2$$
$$y = \ln(\lambda_L / H_L^2)$$

where

f_n = friction factor obtained from the Moody diagram for a smooth pipe.

λ_L = No-slip liquid holdup = vsl/ (vsl + vsg)

v_{sl} = Superficial liquid velocity

v_{gl} = Superficial gas velocity

The liquid holdup term, H_L, is computed using the following correlations:

$$H_{L_0} = (a\lambda_L^b / NF_y^c)$$
$$H_L = H_{L0}, \text{ when } \phi = 0$$
$$H_L = H_{L0}\, \varphi, \text{ when } \phi \neq 0$$

where $\quad \varphi = 1 + (1 - \lambda L)Ln[d\lambda_L^e\ N_{LV}\ f N_{fr}^g]\left[\sin(1.8\phi) - 0.333\ \sin^3(1.8\phi)\right]$

where: N_{F_r} = Froude number; N_{Lv} = liquid velocity number; a, b, c, d, e, f, g = constants

Baker Graph Method

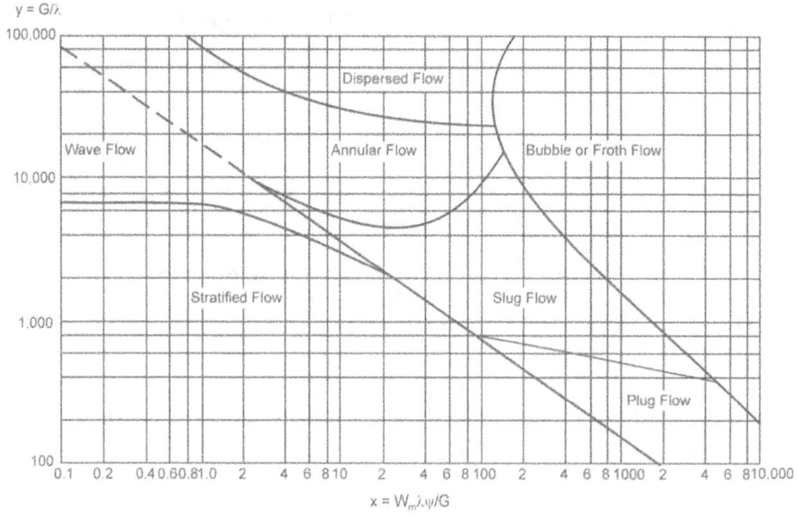

Fig. 9.3 Flow patterns for horizontal two-phase flow. By permission, O. Baker, Oil and Gas journal, Nov. 10. 1958, P.156

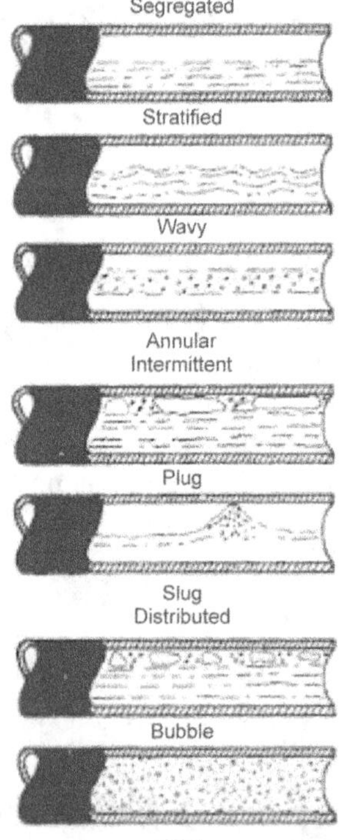

Fig. 9.4 Representative forms of horizontal two-phase flow patterns; By permission, Heim, H., Oil and Gas Journal, Aug-2, 1982, p.132

G = Mass flow rate of gas phase, pounds per hour per square foot of total pipe cross-section area

λ = Two-phase flow term to determine probable type of flow = $\left[(\rho_g / 0.075)(\rho_L / 62.3)\right]^{1/2}$. where both liquid and gas phases are in turbulent flow (two-phase flow)

W_m = Mass flow rate of liquid phase, pounds per hour per square foot of total pipe cross-section area

Ψ = Two-phase term = $(73 / \gamma)\ [\mu_L (62.3 / \rho_L)^2]^{1/3}$

γ = Surface tension of liquid, dynes/centimeter

Total System Pressure Drop

The pressure drop for a system of horizontal and vertical (or inclined) pipe is the sum of the horizontal pressure drop plus the additional drop attributed to each vertical rise, regardless of initial and final elevations of the line.

$$\Delta P_{TPh} = \Delta P_{PT} \text{ (horizontal pipe)} + nhF_e\ \rho_L / 144$$

A. To determine most probable type of two-phase flow using Fig.(9.3).

1. Calculate Wm $\lambda\Psi$ /G

2. Calculate G/λ

3. Read intersection of ordinate and abscissa to identify probable type of flow. Since this is not an exact, clear-cut position, it is recommended that the adjacent flow types be recorded also.

B. Calculate the separate liquid and gas flow pressure drops.

1. For general process application both ΔP_L and ΔP_g may be calculated by the general flow equation:

ΔP_L or ΔP_g (using proper values respectively)

$$= \frac{3.36LW^5(10^{-6})}{d^5\rho} \text{ where f is obtained from Reynolds-Friction Factor}$$

chart for an assumed line size, d.

2. For gas transmission, in general form $\Delta P_g = \dfrac{(q_{d14.65})LS_g TZf}{20,000\,d^5 P_{avg}}$ where

$q_{d14.65}$ is the thousands of standard cubic feet of gas per day, measured at 60°F and 14.65 psia, and Pavg, is the average absolute pressure in the pipe system between inlet and outlet. This is an estimated value and may require correction and recalculation of the final pressure drop if it is very far off.

For oil flow in natural gas transmission lines $\Delta P_L = \dfrac{fLQ^2 \, b\rho}{181,916 d^5}$

3. Calculate $\mathbf{X} = (\Delta P_L / \Delta P_g)^{1/2}$
4. Calculate θ for types of flow selected from Table 9.6

Type Flow	Equation for ϕ GTT
Froth or Bubble	$\Phi = 14.2 \ X^{0.75}/W_m^{\,0.1}$
Plug	$\Phi = 27.315 \ X^{0.855}/W_m^{\,0.17}$
Stratified	$\Phi = 15,400 \ X//W_m^{\,0.8}$
Slug	$1190 \ X^{0.185}/W_m^{\,0.5}$
Annular*	$\Phi = (4.8 - 0.3125d) \ X^{0.348 - 0.021d}$

*Set d = 10 for any pipe larger than 10-in.

Table 9.6 ϕ estimation for various flow patterns

5. Calculate two-phase pressure drop, horizontal portions of lines.

For all types of flow, except wave and fog or spray: $\Delta P_{TP} = \Delta P_G \phi_{GTT}^2$, psi per foot

For wave $\Delta P_{TP} = f_{TP} \left(G_g'\right)^2 / 193.2 d\rho_g$, psi / foot

where $f_{TP} = 0.0043 \left(W_{in}\mu L / G\mu_g\right)^{0.214}$

6. Total two-phase pressure drop, including horizontal and vertical sections of line. Use calculated value times 1.1 to 2.0, depending upon critical nature of application.

$\Delta P_{TPh} = \Delta P_{TP} 1. + n \, h \, F_e \rho_L / 144$ where ρ_L is the density, lb/cu ft, of the liquid flowing in the line, and Fe, elevation factor using gas velocity, v. $F_e = 1.7156 \ Vg^{-0.702}$ for v > 10. Use Fig.9.6 for v less than 10. Most gas transmission lines flow at from 1-15 ft/sec. For fog or spray type flow, Baker suggests using Martinelli's correlation and multiplying results by two.

Scenario 4

A liquid-vapor mixture is to Row in a line having *358* feet of level pipe and three vertical rises of 10 feet each plus one vertical rise of 50 feet. Evaluate the type of flow and expected pressure drop.

Vapor = 3,000 lbs/hr;   Liquid = 1,000 lbs/hr   Density: Ibs/cu ft; Vapor = 0.077 Liquid = 63.0

Viscosity, centipoise; Vapor = 0.00127 Surface tension liquid = 15 dynes/cm

Pipe to be schedule 40, steel Use maximum allowable vapor velocity = 15,000 ft/min.

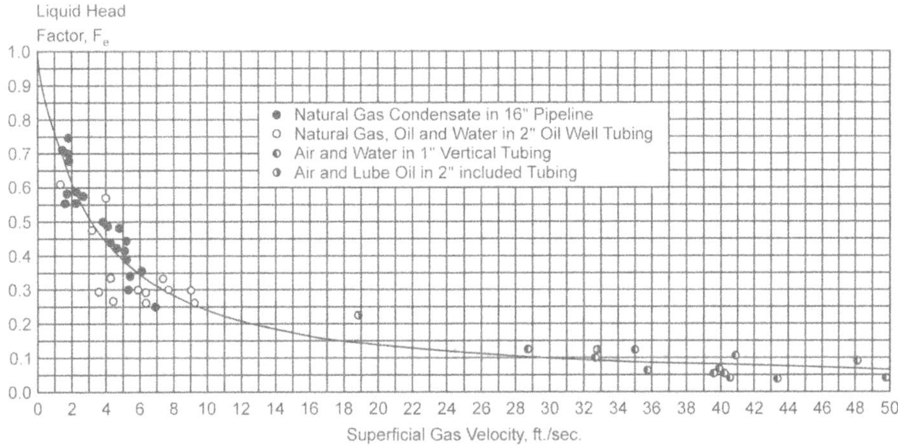

Fig. 9.5 Estimating pressure drops in uphill sections of pipeline for two-phase flow
(O.Flanigan,Oil and Gas Journal, Mar. 10, 1958, p. 132)

Solution provided in CD/Publisher Website

Problems for CHEMCAD

- Calculate the pressure drop in 1 inch 5 km long pipe for 100 kg/hr water flow at 2bar, 120.2C. Write your observations.

1. Chemical Engineering, Vol 1, Fluid Flow, Heat Transfer and Mass Transfer J. M. Coulson and J. F. Richardson with J. R. Backhurst and J. H. Harker.

2. Applied Process Design – Vol. 1, Ernest E. Ludwig, Gulf Professional Publishing.

3. Johnson, T.W. and Mi. B. Berwald, *Flow* of *Natural Gas Through High Pressure Transmission Lines,* Monograph No. 6, U.S. Department of the Interior, Bureau of Mines, Washington, D.C.

4. Baker, *O.,* "Multiphase Flow in Pipe Lines," *Oil and Gas Jour,* Nov. 10, 1958, p. 1516.

Solutions

Scenario 1

Cross-sectional area of pipe = $(\pi/4)\,(0.075)^2 = 0.0044$ m^2.

Velocity of acid in the pipe, u = (0.015/0.0044) = 3.4 m/s.

Reynolds number $= \rho u d / \mu = (1060 \times 3.4 \times 0.07)/ (2.5 \times 10^{-3}) = 1.08 \times 10^5$

Pipe roughness $e = 6 \times 10^{-5}$ m and $e/d = (6 \times 10^{-5}) / 0.075 = 0.0008$

The pressure drop is calculated from equation 3.18 as: $- \Delta P f = 4(R/\rho u^2)(l/d)(\rho u^2)$

From Fig.3.7, when $Re = 1.08 \times 10^5$ and $e/d = 0.0008$, $R/\rho u^2 = 0.0025$.

Substituting: $- \Delta Pf = (4 \times 0.0025)(70/0.075)(1060 \times 3.4^2)$

$$= 114,367 \text{ N/m}^2 \text{ or: } 114, 4 \text{ k N/m}^2$$

Scenario 2

1. *Select the pipe size for expressing the equivalent length.* The usual procedure when analyzing complex pipelines is to express the equivalent length in terms of the smallest, or next-to-smallest, diameter pipe. Choose the 8-in size as being suitable for expressing the equivalent length.

2. *Find the equivalent length of each pipe.* For any complex-series pipeline having equal friction factors in all the pipes, Le = equivalent length, ft, of a section of constant diameter = (actual length of section, ft) (inside diameter, in, of pipe used to express the equivalent length/inside diameter, in, of section under consideration)$^{5.33}$.

For the 16-in pipe, $Le = (1000) (7.981/15.000)^{5.33} = 34.6$ ft. The 12-in pipe is next; for it, $Le = (3000)(7.981/12.00)^{5.33} = 341.2$ ft. For the 8-in pipe, the equivalent length = actual length = 2000 ft.

For the 4-in pipe, $Le = (10) (7.981/4.026)^{5.33} = 383.7$ ft. Then, the total equivalent length of 8-in pipe = sum of the equivalent lengths = $34.6 + 341.2 + 2000 + 383.7 = 2760$ ft, or rounding off, 2760 ft of 8-in pipe (835 m of 0.2-m pipe) will have a frictional resistance equal to the complex-series pipeline shown in Figure.

Scenario 3

Figure the looped section as parallel lines with 6 miles of 8-in. and 6 miles of 10-in. The equivalent diameter for one line with the same carrying capacity is:

$$d_o = \left[(7.981)^{8/3} + (10.136)^{8/3} \right]^{3/8} = 11.9 - \text{in.}$$

This simplifies the system to one section 6 miles long of 11.9-in. I.D. (equivalent) pipe, plus one section of 25 minus 6, or 19 miles of 10-in.

(10.136-in. I.D.) pipe. Now convert the 11.9-in. pipe to a length equivalent to the 10-in. diameter.

$$L_1 = 6(10.136/11.9)^{5.33} = 2.58 \text{ miles}$$

Total length of 10-in. pipe to use in calculating capacity is 19 + 2.58 = 21.58 miles.

Scenario 4

Step1. Determine probable types of flow:

$$\lambda = \left[(\rho_g / 0.075)(\rho_L / 62.3)\right]^{0.5} = \left[\left(\frac{0.077}{0.075}\right)\left(\frac{63.0}{62.3}\right)\right]^{0.5} = 1.017$$

$$\psi = (73/\gamma)[\mu_L (62.3/\rho_L)^2]^{1/3} = \left(\frac{73}{15}\right)\left[1.0\left(\frac{62.3}{63.0}\right)^2\right]^{1/3} = 4.86$$

Try 3-in. pipe, 3.068411. I.D., cross-section area = 0.0513 square feet.

W_m = 1,000/0.0513 = 19,494 Ibs/hr (sq ft)

G = 3,000/0.0513 = 58,482 Ibs/hr (sq ft)

Wm $\lambda\Psi$ /G = 19,494 (1.017) (4.86)/58,482 = 1.641

G/λ = 58,482/1.017 = 57,500

Reading Fig.9.2 type flow pattern is probably annular, but could be wave or dispersed, depending on many undefined and unknown conditions.

Liquid Pressure drop ΔP_L = 3.36 fLW^2 (10^{-6})/$d^5\rho$

Determine Re for 3-in. pipe:

From Moody's chart; ε/d = 0.0006 for steel pipe

$$v = \frac{1000}{63(3600)(0.0513)} = 0.086 \, \text{ft} / \text{sec}$$

$\mu_e = 1 \text{cp} / 1488 = 0.000672 \, \text{lbs} / \text{ft sec}$

D = 3.068/12 = 0.2557 ft ρ = 63.0

$R_e = D v\rho / \mu_e = 0.2557(0.086)(63.0) / 0.000672$

Re = 2060 (this is borderline, and in critical region)

Reading Moody's Chart, approximate f = 0.0576

Substituting

$$\Delta P_L = 3.36\ (10\text{-}6)\ (0.0576)\ (1000)^2\ (1\ \text{foot}) / (3.068)^5\ (63)$$

$$= 1.1\ (10^{-5})\ \text{psi/foot}$$

Gas pressure drop

$$v = \frac{3000}{0.077\ (3600)\ (0.0513)} = 211\,\text{ft / sec}$$

$$\mu_e = 0.00127/1488 = 0.000000854\ \text{lbs/ft sec}$$

$$R_e = Dv\rho / \mu_e = 0.2557\ (211)\ (0.077)/0.000000854$$

$$= 4,900,000$$

Reading Moody's chart f = 0.0175

$$\Delta P_G = 3.36\ (10^{-6})\ (0.0175)\ (1\ \text{foot})\ (3000)^2/(3.068)^5\ (0.077)$$

$$= 0.0254\ \text{psi/foot}$$

Steps 3, 4, 5 & 6:

3. $X = (\Delta P_L / \Delta P_g)^{1/2} = (1.1\ (10^{-5})\ /2.54 \times 10^{-2})^{1/2}$

$$= 2.10\ (10^{-2})$$

4. For annular flow:

$$\Phi_{GTT} = (4.8 - 0.3125d)\ X^{0.343 - 0.021d}$$

$$= [4.8 - 0.3125\ (3.068)\ (2.10 \times 10^{-2})^{0.343 - 0.021\ (3.068)}$$

$$= 1.31$$

5. Two –phase flow for horizontal flows:

$$\Delta P_{TP} = \Delta P_G \Phi^2{}_{GTT} = (0.0254)\ (1.31)^2 = 0.0438\ \text{psi/ft}$$

6. $Fe = 0.00967\ (W_m)^{0.5}/v^{0.7}$

$$= 0.00967\ (19,494)^{0.5}/\ (211)^{0.7}$$

$$= 0.032$$

Vertical elevation pressure drop component:

$$= n\,h\,F_e\rho_L\ /144 = [(3)\ (10) + (1)\ (50)]\ (0.032)\ (63)\ /144$$

$$= 1.125\ \text{psi total}$$

Total: $\Delta P_{TPh} = (0.0438)\,(358) + 1.125$

$$= 16.7 \text{ psi, total for pipe line}$$

Because these calculations are somewhat uncertain due to lack of exact correlations, it is best to calculate pressure drop for other flow patterns, and apply a generous safety factor to the results.

Excel Solution for the Problem

	A	B	C	D	E
1					
2	**Two Phase Pressure drop calculations**				
3					
4	Input Data				
5	Pipe line length, ft	358			
6	Vertical height, ft	80			
7					
8					
9					
10	Vapor flow rate, lb/hr	3000			
11	Liquid flow rate, lb/hr	1000			
12	vapor density, lb/ft3	0.077			
13	liquid density, lb/ft3	63			
14	vapor visosity, cP	0.00127			
15	liquid visosity, cP	1			
16	liquid surface tension, Y, dynes/cm	15			
17	Max vapor velocity, ft/min	15000			
18		250	ft/s		
19	Selected Pipe size in inch (Assume)	3			
20	Calculations:				
21	Step 1. Determine probable flow type:				
22	Two phase flow term, λ	1.019			
23					
24	$\lambda = \left[(\rho_g / 0.075)(\rho_L / 62.3) \right]^{0.5}$				
25					
26	Two phase flow term, Ψ	4.831			
27					
28	$\psi = (73 / \gamma)\left[\mu_L (62.3 / \rho_L)^2 \right]^{1/3}$				
29					
30	Pipe cross sectional area, ft2	0.049			
31	Wm	20371.8			
32	G	61115.4981			
33	x axis value: WmλΨ/G	1.64077366			
34	y axis value: G/ λ	59980.5403		Flow Types:	
35				Bubble or Froth	
36	Refer Baker Graph and select the flow type	Annular		Plug	
37	NOTE: Adjust the pipe size such that Flow is Annualr (Best option)			Stratified	
38				Slug	
39				Annular	

40				
41				
42	**Step 2. Calculate the separate liquid and gas flow pressure drops**			
43	friction factor-Liquid-pipe surface (Assumed)	0.0576		
44			$\Delta P_L = 3.36\, f\, L\, W^2\, (10^{-6})/d^5\rho$	
45	Liquid phase pressure drop, ΔPL, psi/ft	1.2642E-05		
46				
47	friction factor-Liquid-gas surface (Assumed)	0.0175		
48	gas phase pressure drop,	0.02828283		
49				
50				
51	**Step 3. Calculate X**	0.021142	$X = (\Delta P_l / \Delta P_g)^{1/2}$	
52				
53				
54	**Step 4. Calculate θ for types of flow**	1.2869	**Flow Types:**	
55			Bubble or Froth	2.6974
56			Plug	0.1870
57			Stratified	0.1163
58			Slug	4.0849
59			Annular	1.2869
60				
61	**Step 5. Calculate Twp phase pressuredrop - horizontal flow**			
62				
63	ΔP TP	0.04683743 psi/ft		
64				
65		$\Delta P_{TP} = \Delta P_G \Phi^2_{GTT}$		
66				
67	**Step 6. Total two-phase pressure drop**			
68				
69	Gas velocity, v	220.47 ft/s		
70	NOTE: This is below the specified limit, else revise the pipe size			
71				
72	Fe	0.0316		
73				
74	Vertical elevation pressure drop	1.1058 psi		
75				
76	Total pressure drop (Horizontal+Vertical)	17.874 psi		
77				
78				

Two Phase Pressure Drop Calculations formulae used:

Input Data

Pipe line length, ft	358
Vertical height, ft	80
Vapor flow rate, lb/hr	3000
Liquid flow rate, lb/hr	1000
Vapor density, lb/ft^3	0.077
Liquid density, lb/ft^3	63
Vapour viscosity, cP	0.00127
Liquid viscosity, cP	1
Liquid surface tension, Y, dynes/cm	15

Max vapour velocity, ft/min   15000

$$= B17/60 \text{ ft/s}$$

Selected Pipe size in inch (Assume)   3

Calculations:

Step 1. Determine probable flow type:

Two phase flow term, λ $= ((B12/0.075) \times (B13/62.3)) \wedge 0.5$

$$\lambda = [(\rho_g/0.075)(\rho_L/62.3)]^{0.5}$$

Two phase flow term, ψ $= (73/B16) \times (B15 \times (62.3/B13) \wedge 2) \wedge 0.33$

$$\psi = (73/\gamma)[\mu_L(62.3/\gamma_L)^2]^{1/3}$$

Pipe cross sectional area, ft^2 $= P1(\ ) \times B19 \wedge 2/(4 \times 12 \wedge 2)$

Wm $= B11/B30$

G $= B10/B30$

x axis value: $W_m \lambda\, \psi\, /G$ $= B31 \times B22 \times B26/B32$

y axis value: G/λ $= B32/B22$

Flow Types:

Bubble or Froth

Plug

Stratified

Slug

Annular

Step 2. Calculate the separate liquid and gas pressure drops

Friction factor-Liquid-pipe surface (Assumed)   0.0576

Luquid phase pressure drop, ΔPL, psi/ft

$$= 3.36 \times B43 \times 1 \times B11 \wedge 2 \times 10 \wedge .6/(B19{\wedge}5 \times B13)$$

Friction factor-liquid gas surface (Assumed)   0.0175

Gas phase pressure drop

$$= 3.36 \times B47 \times 1 \times B10 \wedge 2 \times 10 \wedge .6/(B19{\wedge}5 \times B12)$$

Step 3. Calculate X $= (B45/B48) \wedge 0.5$

Step 4. Calculate θ for types of flow = D59

Flow Types:

Bubble or Froth

$$= 14.2 \times B51 \wedge 0.75/B10 + B31 \wedge 0.1$$

Plug　　　　　　　　　　　　$= 27.315 \times B51^{\wedge}0.855/B31^{\wedge}0.17$

Stratified　　　　　　　　　$= 15400 \times B51/B31^{\wedge}0.8$

Slug　　　　　　　　　　　　$= 1190 \times 851^{\wedge}0.185/B31^{\wedge}0.5$

Annular　　　　　　　　　　$= (4.8 - 0.3125 \times B19) \times B51 \wedge$

$$(0.348 – 0.021 \times 819)$$

Step 5. Calculate Two phase pressure drop – horizontal flow

ΔP TP (psi/ft)　　　　　　　$= B48 \times B54^{\wedge}2$

$$\Delta P_{TP} = \Delta P_G \, \phi^2 GTT$$

Step 6. Total two-phase pressure drop

Gas velocity, v in ft/s　　　　$= 810/(B12 \times 3600 \times B30)$

Note: This is below the specified limit, else revise the pipe size

Fe　　　　　　　　　　　　　$= 0.00967 \times B31^{\wedge}0.5/B69^{\wedge}0.7$

Vertical elevation pressure drop, psi　　$= B6 \times B72 \times B13/144$

Total pressure drop (Horizontal + Vertical), psi　$= B63 \times B5 + B74$

References

Ernest E. Ludwig, Applied Process Design for Chemical and Petrochemical Plants, Volume 1, 3rd Edition.

10

Flow in Tube – PDE Solver in MATLAB

Objective: This session emphasis the following principle:

- Solution for Partial differential equation using MATLAB pdepe function;

Problem Statement

A fluid of viscosity μ and density ρ fills the inside of a very long vertical cylindrical pipe of radius a. We suddenly remove the plug initially placed at the bottom of the pipe at time t' = 0. Find the velocity profile. The only driving force is gravity. The radial and angular components of the velocity field are 0, and only velocity component is in z-direction. vz' is a function of both time t' and radius r'.

$$V_r = V_\theta = 0$$

$$V_{z'} = V_{z'} = (t'r')$$

Theory

Step 0: Develop a mathematical model. The following partial differential equation (PDE) describes the development of velocity in the z-direction. Note that the differential equation describes the laws of nature / physics -- the same laws of physics apply to all problems and never change; they remain the same beyond the day we die. Whereas, the boundary conditions are problem-specific (man-made) and depends on the particular physical set up or initial condition.

$$\rho \frac{d}{dt'} v = \rho \cdot g \cdot h + \frac{\mu}{r'} \cdot \frac{d}{dr'} \left(r' \cdot \frac{d}{dr'} v_{z'} \right) \qquad \begin{array}{l} \text{B.C at } t' = 0 \; v_z = 0 \\ \text{at } r' = a \; v_z = 0 \end{array}$$

Step 1: Non-dimensionalization. We usually scale everything to O (1).

$$r = \frac{r'}{a} \quad t = \frac{t'}{\dfrac{a^2 \cdot \rho}{\mu}} \quad v_z = \frac{v_{z'}}{\dfrac{\rho \cdot g \cdot h \cdot a^2}{4 \cdot \mu}}$$

$$\frac{d}{dt}v_z = 4 + \frac{1}{r}\cdot\frac{d}{dr}\left(r\cdot\frac{d}{dr}v_z\right) \qquad \text{B.C at } t=0 \;\; v_{z0}(r):=0$$
$$\text{at } r=1 \;\; v_z=0$$

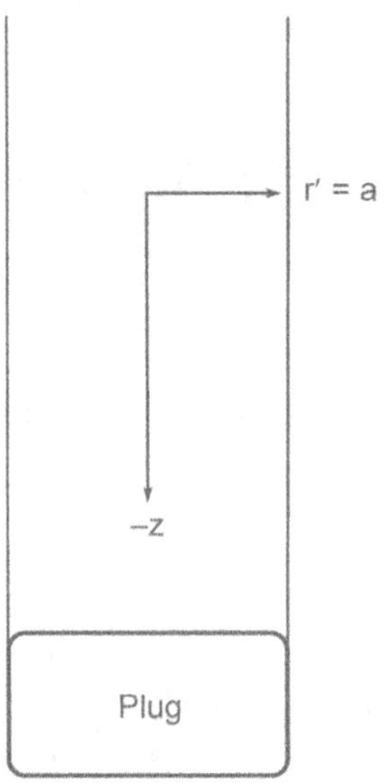

Fig. 10.1 Vertical pipe flow with plug to study unsteady behaviour

Step 2: Find steady-state solution.

$$V_{ss}(r):=r^2-1$$

Step 3: Find solution. Almost all PDE problems have no "neat" analytical solution, and we have to settle for an approximation or a numeric one.

pdepe Solve initial-boundary value problems for systems of parabolic and elliptic partial differential equations (PDEs) in one space variable and time
Syntax

    sol = pdepe(m,pdefun,icfun,bcfun,xmesh,tspan)

Arguments

m	A parameter corresponding to the symmetry of the problem. m can be slab = 0, cylindrical = 1, or spherical = 2.
pdefun	A function that defines the components of the PDE.
icfun	A function that defines the initial conditions.
bcfun	A function that defines the boundary conditions.
xmesh	A vector [x0, x1, ..., xn] specifying the points at which a numerical solution is requested for every value in tspan. The

elements of xmesh must satisfy x0 < x1 < ... < xn. The length of xmesh must be > = 3.

tspan   A vector [t0, t1, ..., tf] specifying the points at which a solution is requested for every value in xmesh. The elements of tspan must satisfy t0 < t1 < ... < tf. The length of tspan must be > = 3.

pdepe   solves PDEs of the form:

$$c\left(x,t,u,\frac{\partial u}{dx}\right)\frac{\partial u}{\partial t} = x^{-m}\frac{\partial}{\partial x}\left(x^m f\left(x,t,u,\frac{\partial u}{\partial x}\right)\right) + s\left(x,t,u,\frac{\partial u}{\partial x}\right)$$

where $f(x,t,u,\partial u/\partial x)$ is a flux term and $s(x,t,u,\partial u/\partial x)$ is a source term.

The solution components satisfy boundary condition of the following form:

$$p(x,t,u) + q(x,t)f\left(x,t,u,\frac{\partial u}{\partial x}\right) = 0$$

Example:

Start-up flow in a pipe

$$\frac{dv}{dt} = -4 + \frac{1}{r}\frac{d}{dr}\left(r*\frac{dv}{dr}\right)$$

$v = v\,(t, \gamma)$

IC   at $t = 0$, $v = 0$ (no flow before $t = 0$)

BC   at $r = 0$, $dv/dr = 0$;   (no sink at center)

BC   at $r = 1$, $v = 0$ (no slip at pipe wall)

Procedure

Step 1: Define the coordinate system and the space and time discretisation from MATLAB command:prompt:

```
>> m=1;
>> r=0:0.01:1;
>> t=0:0.1:10;
>>
```

Step 2: Define three different functions for PDE expression, initial condition and boundary condition:

```
function [c,f,s]=pdefun(r,t,v,dvdr)
c=1;
f=dvdr;
s=-4;
end
```

```
function vt0=icfun(r)
vt0=0;
end
function [p0,q0,p1,q1] = bcfun(r0,v0,r1,v1,t)
p0=0; q0=1;
p1=v1; q1=0;
end
```

Step 3

Execute *pdepe* from MATLAB command prompt to get solution:

> v = pdepe(m, @pdefun, @icfun, @bcfun, r, t);

Step 4:

Execute *plot* commands from MATLAB command prompt to get various graphs:

```
>> v = pdepe(m, @pdefun, @icfun, @bcfun, r, t);
>> plot(r,v(1,:,1), r,v(2,:,1), r,v(3,:,1), r,v(4,:,1), r,v(5,:,1), r,v(6,:,1), r,v(max
(size(t)),:,1))
     xlabel('Radius r (dimensionless)')
     ylabel('Velocity in z-direction, vz (dimensionless)')
     legend('t=0', 't=0.1', 't=0.2', 't=0.3', 't=0.4', 't=0.5', 't=10
(infinity)','Location','NorthEastOutside')
     title('Suddenly Start Flow in a Cylinder')
>>
>>
```

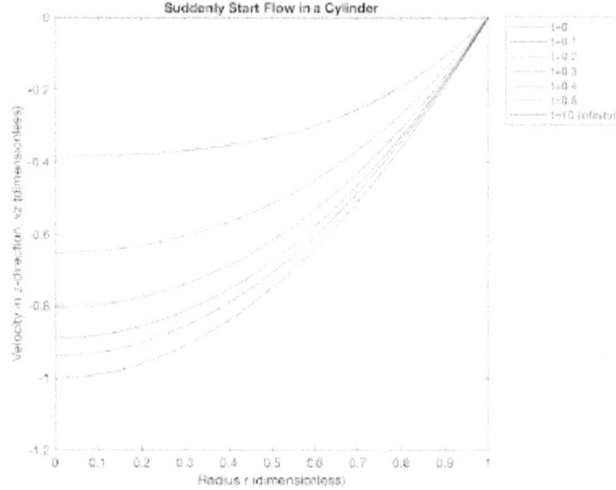

Analytical Solution

Steedy state: $\qquad\qquad 0 =$

Convective flow: $\qquad\quad \rho v_z v_z (2\pi r \Delta r)\big|z=0 - \rho v_z v_z (2\pi r \Delta r)\big|z=L$

Molecular transport: $\qquad +\tau_{rz}(2\pi r L)\big|_r - \tau_{rz}(2\pi r L)\big|_{r+\Delta r}$

Gravity force: $\qquad\qquad +Pg_z(2\pi\tau L\Delta\tau)$

Pressure force: $\qquad\qquad +P_0(2\pi\tau\Delta\tau)PL(2\pi\tau\Delta\tau)$ $\qquad$ .....(10.1)

Dividing by Δr and taking the limit as $\Delta r \to 0$ gives

$$0 = -2\pi L\frac{d\tau_{\tau z}\tau}{d\tau} + \rho g_z 2\pi\tau L + 2\pi\tau(P_0 - P_L) \qquad \text{.....(10.2)}$$

or

$$\frac{d}{d\tau}(\tau\tau_{\tau z}) = \left(\frac{P_0 - P_L}{L} + \rho g\right)\tau \qquad \text{.....(10.3)}$$

Integrating (10.3) gives

$$\tau\tau_{\tau z} = \left(\frac{P_0 - P_L}{L} + \rho g\right)\frac{\tau^2}{2} + c_1 \qquad \text{.....(10.4)}$$

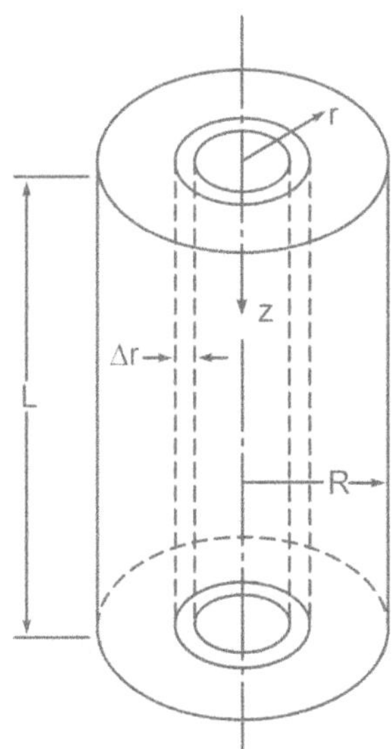

Fig. 10.2 Pipe Flow of a Newtonian Fluid

where c_1 is a constant of integration. Using the definition of a Newtonian fluid,

$$\tau_{\tau z} = -\mu\frac{dv_z}{d\tau} \qquad \text{.....(10.5)}$$

gives

$$\frac{dv_z}{d\tau} = -\left(\frac{P_0 - P_L}{L} + \rho g\right)\frac{\tau}{2\mu} + \frac{c_1}{\mu\tau} \qquad \text{....(10.6)}$$

The boundary conditions for this problem are

    B.C.1 at $t = 0$ the velocity is finite

    B.C.2 at $t = R$ the velocity is zero $(v_z|R = 0)$

(This is the no-slip-at-the-wall condition).

Using B.C.1 we find that the constant c_1 in eq.(10.6) must be zero. This gives

$$\frac{dv_z}{d\tau} = -\left(\frac{P_0 - P_L}{L} + \rho g\right)\frac{\tau}{2\mu} \qquad \text{.....(10.7)}$$

which upon integration becomes

$$v_z = \left(\frac{P_0 - P_L}{L}\rho g\right)\frac{\tau^2}{4\mu} + c_2 \qquad \text{.....(10.8)}$$

Using B.C.2 we can evaluate c_2 and the final expression for the velocity

$$v_z = \left(\frac{P_0 - P_L}{L}\rho g\right)\frac{R^2}{4\mu} + \left[1 - \left(\frac{\tau}{R}\right)^2\right] \qquad \text{.....(10.9)}$$

This is the well-known parabolic velocity for laminar pipe flow.

References

Start-up flow in a pipe by Nam Sun Wang

    http://eng.umd.edu/~nsw/chbe250/chbe250.htm

    http://www.mathworks.in/help/matlab/ref/pdepe.html

W. F. Ramirez, Computational Methods for Process Simulation, Chapter 6.

11

Gravity Flow Tank

Objective:

Emphasizes the following principle:

Model the gravity flowtank and simulate the height variations

Problem Statement

Simulate the gravity tank with various inflow conditions.

Physical dimensions, parameter values, and steady-state flow rate and liquid height are given in Table 11.1.

Table 11.1 Gravity flow tank data

Gravity-flow Tank Data
Pipe: ID = 3ft   Area = 7.06 ft^2 Length = 3000 ft   Tank: ID = 12 ft   Area = 133 ft^2 Height = 7 ft
Steady-state values: $\bar{F}$ = 35.1 ft^3/s (15.700 gpm) $\bar{h}$ = 4.72 ft $\bar{v}$ = 4.97 ft/s
Parameters: Reynolds number = 1,380,000   Friction factor = 0.0123 $K_F = 2.81 \times 10^{-2}$ lb$_f$/(ft/s)^2 ft

Gravity Flow Mode

Fig.11.1 shows a tank into which an incompressible (constant density) liquid is pumped at a variable rate $F_0 (\text{ft}^3 / \text{s})$. This inflow rate can vary with time because of changes in operations upstream. The height of liquid in the vertical cylindrical tank is h(ft). The flow rate out of the tank is $F(\text{ft}^3 / \text{s})$.

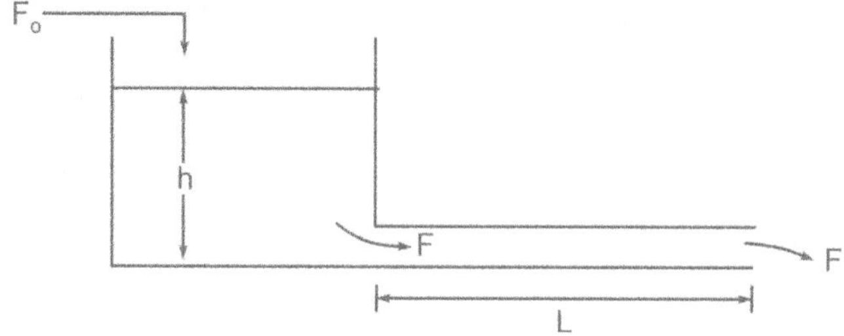

Fig. 11.1 Gravity-flow tank

Now F_0, h, and f will all vary with time and are therefore functions of time t.

Consequently we use the notation $F_{0(t)}, h_{(t)}$ and $F_{(t)}$. Liquid leaves the base of the tank via a long horizontal pipe and discharges into the top of another tank. Both tanks are open to the atmosphere.

Let us look first at the steady-state conditions. By steady-state we mean, in most systems, the conditions when nothing is changing with time. Mathematically this corresponds to having all time derivatives equal to zero, or to allowing time to become very large, i.e., go to infinity. At steady-state the flow rate out of the tank must equal the flow rate into the tank. In this book we will denote steady-state values of variables by an overscore or bar above the variables. Therefore at steady-state in our tank system $\overline{F}_0 = \overline{F}$.

For a given $\overline{F}$, the height of liquid in the tank at steady-state would also be some constant $\overline{h}$. The value of $\overline{h}$ would be that height that provides enough hydraulic pressure head at the inlet of the pipe to overcome the frictional losses of liquid flowing down the pipe. The higher the flow rate $\overline{F}_r$, the higher $\overline{h}$ will be.

In the steadystate design of the tank, we would naturally size the diameter of the exit line and the height of the tank so that at the so that at the maximum flow rate expected the tank would not overflow. And as any good, conservative design engineer knows, we would include in the design a 20 to 30 percent safety factor on the tank height.

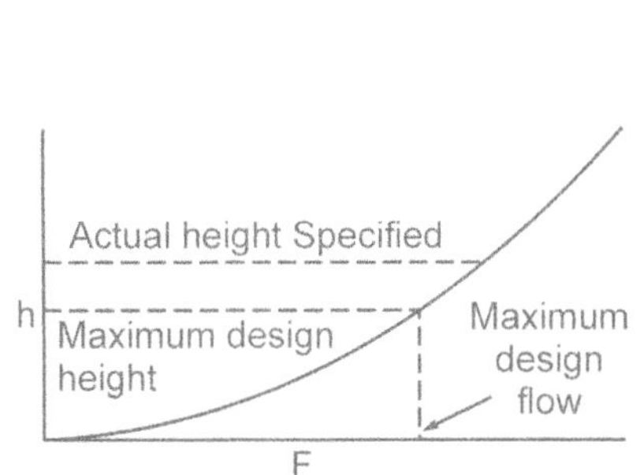

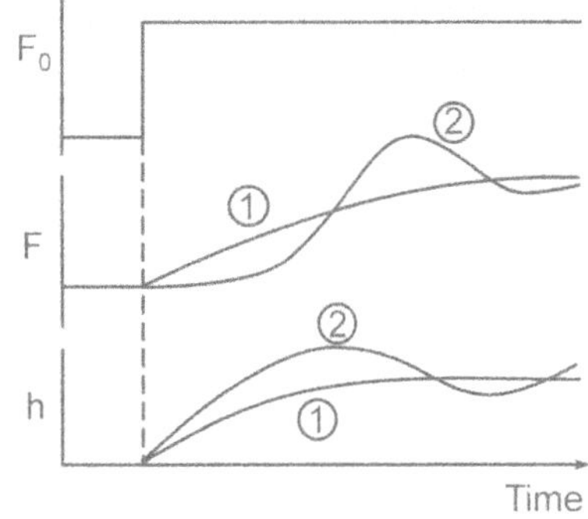

Fig. 11.2 Steady height vs. flow   **Fig. 11.3** Actual paths of F and h

Steady-state Height versus Flow

The design of the system would involve an economic balance between the cost of a taller tank and the cost of a bigger pipe, since the bigger the pipe diameter the lower is the liquid height. Fig.11.2 shows the curve of $\bar{h}$ versus $\bar{F}$ for a specific numerical case.

So far we have considered just the traditional steady-state design aspects of this fluid flow system. Now let us think about what would happen dynamically if we changed F_0. How will $h_{(t)}$ and $F_{(t)}$ vary with time? Obviously F eventually has to end up at the new value of F1. We can easily determine from the steady-state design curve of Fig. 1.2 where h will go at the new steady-state. But what paths will $h_{(t)}$ and $F_{(t)}$ take to get to their new steady-state?

Fig.11.3 sketches the problem. The question is which curves (1 or 2) represent the actual paths that F and h will follow. Curves 1 show gradual increases in h and F to their new steady-state values However, the paths could follow curves 2 "overshoot." Clearly, if the peak of the overshoot in h is above the top of the tank, we would be in trouble.

Our steady-state design calculations tell us nothing about what the dynamic response to the system will be.

Referring to Fig.11.1 let the length of the exit line be L (ft) and its cross-sectional area be A. (ft^2). The vertical, cylindrical tank has a cross-sectional area of A, (ft^2).

The part of this process that is described by a force balance is the liquid flowing through the pipe. It will have a mass equal to the volume of the pipe (A_pL) times the density of the liquid r. This mass of liquid will have a velocity v (ft/s) equal to the volumetric flow divided by the cross-sectional area of the pipe. Remember we have assumed plug-flow conditions and incompressible liquid, and therefore all the liquid is moving at the same velocity, more or less like a solid rod. If the flow is turbulent, this is not a bad assumption.

$$M = A_p L \rho$$

$$v = \frac{F}{A_p} \qquad \qquad .....(11.1)$$

The amount of liquid in the pipe will not change with time, but if we want to change the rate of outflow, the velocity of the liquid must be changed. And to change the velocity or the momentum of the liquid we must exert a force on the liquid.

The direction of interest in this problem is the horizontal, since the pipe is assumed to be horizontal. The force pushing on the liquid at the left end of the pipe is the hydraulic pressure force of the liquid in the tank.

$$\text{Hydraulic force} = A_p \rho h \frac{g}{g_c} \qquad \qquad .....(11.2)$$

The units of this force are (in English engineering units):

$$ft_2 \, \frac{lb_m}{ft^3} \, ft \, \frac{32.\,ft.\,/\,s^2}{32.2\,lb...ft\,/\,lb_f s^2} \, lb,$$

Where g is the acceleration due to gravity and is 32.2 ft/s2 if the tank is at sea level. The static pressures in the tank and at the end of the pipe are the same, so we do not have to include them.

The only force pushing in the opposite direction from right to left and opposing the flow is the frictional force due to the viscosity of the liquid. If the flow is turbulent, the frictional force will be proportional to the square of the velocity and the length of the pipe.

$$\text{Frictional force} = K_F \, Lv^2 \qquad \qquad .....(11.3)$$

The time rate of change of momentum (mass times velocity) is equal to net sum of forces pushing in that direction.

As per this, the equation becomes,

$$\frac{1}{gc} \frac{d(A_p L \rho v)}{dt} = A_p \rho h \frac{g}{g_c} - K_F \, Lv^2$$

The sign of the frictional force is negative because it acts in the direction opposite the flow. We have defined left to right as positive direction.

The force balance on the outlet line gave us the nonlinear ODE

$$\frac{dv}{dt} = \frac{g}{L}h - \frac{K_F g_c}{\rho A_p}v^2 \qquad \qquad .....(11.4)$$

To describe the system completely a total continuity equation on the liquid in the tank is also needed.

$$A_T \frac{dh}{dt} F_0 - F \qquad \qquad .....(11.5)$$

We have to pick a specific numerical case to solve these two coupled ordinary differential equations. Eq.(11.1) is nonlinear because of the v^2 term.

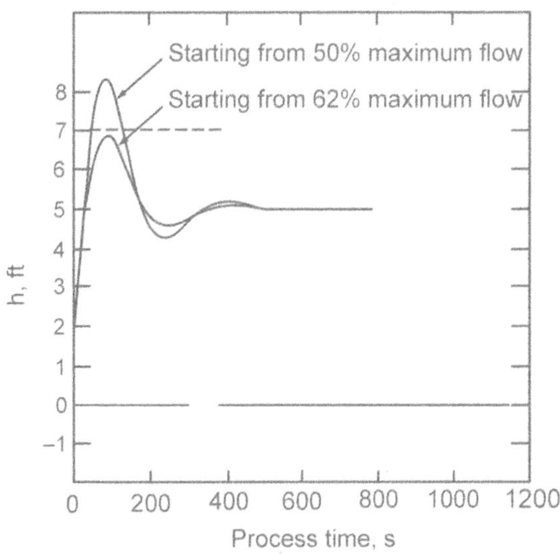

Fig. 11.4 Gravity-flow tank, level variation with time

Using the relationship $F = vA_p$ and substituting the numerical values of parameters into Eqs. (11.1) and (11.2) give

$$\frac{dv}{dt} = 0.0107h - 0.000205v^2 \qquad \qquad .....(11.6)$$

$$\frac{dh}{dt} = 0.311 - 0.0624v \qquad \qquad .....(11.7)$$

Fig.11.4 gives a MATLAB, program that numerically integrates the two ODEs describing this system for two different initial conditions of flow and liquid level in the tank: (1) when the initial flow rate is 50 percent of the design rate, and (2) when the initial flow rate is 67 percent of the maximum design flow rate. At time equal zero, the flow rate into the tank is increased to the maximum design flow rate of $35.1 \text{ft}^3/\text{s}$.

$$\frac{dv}{dt} = 0.0107h - 0.00205v^2$$

$$\frac{dh}{dt} = 0.311 - 0.00624v$$

MATLAB Program for the gravity flow tank problem:

Step 1 Create .m file as shown below and save it as gravity.m

```
function uprime=gravity(t,u)
uprime=zeros(2,1);
uprime(1)=0.311-0.0624*u(2);
uprime(2)=0.0107*u(1)-0.00205*u(2)^2;
```

Step 2 Solve the ordinary differential equations using initial conditions

```
>> u0 = [2.05  3.4];
[t,u] = ode45(@gravity,[0, 120], u0)

u =

    2.0500    3.4000
    2.1530    3.3987
    2.2561    3.3987
    2.3592    3.3997
```

Step 3 Plot the necessary Graphs

```
>> plot(t,u(:,1),'+',t,u(:,2));
>> legend('Tank Height','Velocity');
>> xlabel('Time');
>> ylabel('Height and velocity');
>> title('Gravity Tank Simulation');
>>
```

```
Note: gravity is the file name in script file,
      U(1) = height of the tank
      U(2) = velocity of the fluid
      [0,120] = time period
```

Ode45 is the inbuilt program of Runge Kutta 4th order method.

ode45 is based on an explicit Runge-Kutta formula, the Dormand-Prince pair. It is a *one-step* solver – in computing $y(t_n)$, it needs only the solution at the immediately preceding time point, $y(t_{n-1})$. In

general, ode45 is the best function to apply as a
first try for most problems.

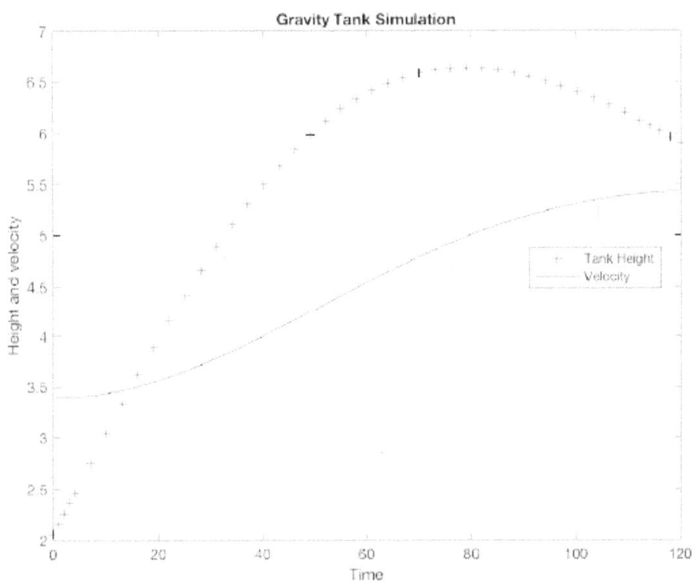

Also try the following options:

```
options = odeset('RelTol',1e-4,'AbsTol',[1e-4 1e-4]);

[t,u] = ode45(@gravity,[0, 120], u0, options);
```

Also, try the following options:

```
dsolve  -  Symbolic  solution  of  ordinary  differential
equations
```

Syntax

```
dsolve('eq1','eq2',...,'cond1','cond2',...,'v')
dsolve(...,'IgnoreAnalyticConstraints',value)

[h  v]  =  dsolve  ('Dh=0.311-0.0624*v','Dv=0.0107*h-
0.00205*v*v',-

'h(0)=2.05','v(0)=3 .4')

Hf=@(t,a,b)eval(vectorize(h));

vf=@(t,a,b)eval(vectorize(v));
```

the functions hf, vf can be evaluated for a particular choice of initial
conditions of h(0)=a, v(0)=b at a particular value of t.

Description

```
dsolve('eq1','eq2',...,'cond1','cond2',...,'v')
```
symbolically solves the ordinary differential equations eq1, eq2,... using v as the independent variable. Here cond1,cond2,... specify boundary or initial conditions or both. You also can use the following syntax: `dsolve('eq1, eq2',...,'cond1,cond2',...,'v')`.

The default independent variable is t.

The letter D denotes differentiation with respect to the independent variable. The primary default is d/dx. The letter D followed by a digit denotes repeated differentiation. For example, D2 is d^2/dx^2. Any character immediately following a differentiation operator is a dependent variable. For example, D3y denotes the third derivative of y(x) or y(t).

You can specify initial and boundary conditions by equations like y(a) = b or Dy(a) = b, where y is a dependent variable and a and b are constants. If the number of the specified initial conditions is less than the number of dependent variables, the resulting solutions contain the arbitrary constants C1, C2,....

For more details on ode45 & examples, refer to the following website:

http://www.mathworks.com/help/techdoc/ref/ode23.html

Excel solution for the Problem

	A	B	C	D	E	F	G	H	I
1	GravityFlow Tank data							Goal Seek Variable	
2								input variable	
3	Pipe:							calculated variable	
4	ID	0.062499998 ft		inch:	0.75				
5	Area	0.00307 ft^2							
6	Length	16.4 ft		m	4.99872				
7									
8	Tank:								
9	ID	1 ft		m	0.3048				
10	Area	1.00 ft^2		m2	0.0929 =		929.03 CM2		
11	Height	7 ft		m	2.1336 =		198.218 lit volume		
12									
13	Steady state values:								
14	Flow	0.00475 ft^3/s			0.1345 lit/s				
15	h	5.77 ft							
16	v	1.55 ft/s							
17									
18	Parameters:								
19	N$_{Re}$	8964.407842			Hydraulic frictional force	force	difference		
20	f	0.0123			11.3242	11.3242	0		
21	Kf	0.0281 lbf/(ft/s)^2 ft							

```
22
23
24
25  Model equation:
27  dv/dt =1.9634*h-4.72410*v2
29  dh/dt=0.0048-0.003067961330033436*v
30
```

Note: After entering the details, 'Goal Seek' operation should be performed to get solution. Select cell 'G20' i.e., error and use Tools/Goal seek option from tool bar.

A window will pop-up asking the value to be changed i.e., height (cell B15) inorder to set the error i.e., cell G20 to zero

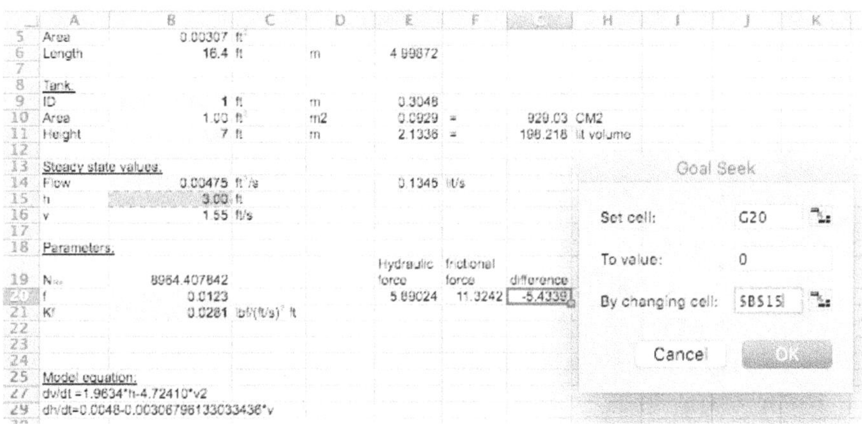

Another Case study

ft	lit/s	lit/min	cm (for min holdup)
0.03	0.00283	0.1698	0.18
0.73	0.01416	0.8496	0.91
1.86	0.02265	1.359	1.46
2.91	0.02832	1.6992	1.83
3.85	0.03256	1.9536	2.10
4.19	0.03398	2.0388	2.19
5.31	0.03823	2.2938	2.47
6.55	0.04248	2.5488	2.74
.5 inch pipe 5 m length			
1x1 ft tank, 7ft hight = 200lit volume			

ft	lit/s	lit/min	cm (for min holdup)
0.34	0.03256	1.95	2.103
1.02	0.05663	3.40	3.657
2.3	0.08495	5.10	5.486
3.13	0.09911	5.95	6.401
4.09	0.11327	6.80	7.315
5.18	0.12743	7.65	8.230
5.77	0.1345	8.07	8.686
6.39	0.14158	8.49	9.144
.75 inch pipe 5 m length			
1x1 ft tank, 7ft hight = 200lit volume			
1.14	0.14	8.49	9.14
1.64	0.17	10.19	10.97
2.23	0.20	11.89	12.80
2.91	0.23	13.59	14.63
3.69	0.25	15.29	16.46
4.55	0.28	16.99	18.29
5.51	0.31	18.69	20.12
6.55	0.34	20.39	21.95
1 inch pipe 5 m length			
1 × 1 ft tank, 7 ft height = 200 lit volume			

References

Dormand, J. R. and P. J. Prince, "A family of embedded Runge-Kutta formulae," *J. Comp. Appl. Math.*, Vol. 6, 1980, pp 19-26.

Process Modeling, Simulation and Control for Chemical Engineers by W.L.Luyben, McGraw Hill Publishers.

Thermodynamics

It will be perfectly clear that in all my studies I was quite convinced of the real existence of molecules, that I never regarded them as a figment of my imagination, nor even as mere centres of force effects. I considered them to be the actual bodies, thus what we term "body" in daily speech ought better to be called "pseudo body". It is an aggregate of bodies and empty space. We do not know the nature of a molecule consisting of a single chemical atom. It would be premature to seek to answer this question but to admit this ignorance in no way impairs the belief in its real existence. When I began my studies I had the feeling that I was almost alone in holding that view. And when, as occurred already in my 1873 treatise, I determined their number in one gram-mol, their size and the nature of their action, I was strengthened in my opinion, yet still there often arose within me the question whether in the final analysis a molecule is a figment of the imagination and the entire molecular theory too. And now I do not think it any exaggeration to state that the real existence of molecules is universally assumed by physicists. Many of those who opposed it most have ultimately been won over, and my theory may have been a contributory factor. And precisely this, I feel, is a step forward. Anyone acquainted with the writings of Boltzman and Willard Gibbs will admit that physicists carrying great authority believe that the complex phenomena of the heat theory can only be interpreted in this way. It is a great pleasure for me that an increasing number of younger physicists find the inspiration for their work in studies and contemplations of the molecular theory.

*— **Johannes D. van der Waals's** notes in Nobel Lecture, The equation of state for gases and liquids (12 December 1910).*

12

Molar Volume Estimation using Equation of State Methods

Objective: Estimate molar volume of mixture from first principles.

Problem Scenario: Molar volume estimation

Glanville measured specific volumes of vapor and liquid mixtures of propane and benzene over wide ranges of temperature and pressure. Use the R-K equation to estimate specific volume of a vapor mixture containing 26.92 wt% propane at 400°F (477.6 K) and a saturation pressure of 410.3 psia (2,829 kPa). Compare the estimated and experimental values.

The critical constants for propane and benzene are given by Poling

	Propane	Benzene
T_c, K	369.8	562.2
P_c, KPa	4,250	4,890

Glanville reported experimental values of Z = 0.7128 and V/M = 0.2478 ft3/lb.

Theory: Equations of state for phase-equilibrium calculations are applicable to wide ranges of temperature and pressure conditions. They can also be used to calculate all the related thermodynamic properties such as enthalpy and entropy. The reference state for both the vapor and liquid phase is the ideal gas, and deviations from the ideal-gas state are determined by calculating fugacity coefficients for both phases. For cubic equations of state in particular, critical and super-critical conditions can be predicted quite accurately.

A general two-parameter cubic equation of state can be expressed by the equation:

$$P = \frac{RT}{(v-b)} - \frac{a(T)}{v^2 + ubv + wb^2}$$

where

P = the pressure

T = the absolute temperature

v = the molar volume

u, w = constants, typically integers

Constants for four cubic equation of state:

Equation	u	w	b	a
van der Waals	0	0	$\dfrac{RT_c}{8P_c}$	$\dfrac{27}{64}\dfrac{R^2T_c^2}{P_c}$
Redlich-K wong	1	0	$\dfrac{0.08664RT_c}{P_c}$	$\dfrac{0.42748R^2T_c^{25}}{P_cT^{1/2}}$
Sove	1	0	$\dfrac{0.08664RT_c}{P_c}$	$\dfrac{0.42748R^2T_c^2}{P_c}[1+f\omega(1-T_r^{1/2})]^2$ where $f\omega=0.48+1.574\omega-0.176\omega^2$
Peng-Robinson	2	-1	$\dfrac{0.07780RT_c}{P_c}$	$\dfrac{0.45724R2T_c^2}{P_c}[1+f\omega(1-T_r^{1/2})]^2$ where $f\omega=0.37464+1.54226\omega-0.26992\omega^2$

Name	Equation	Equation Constants and Functions
Ideal gas law	$P=\dfrac{RT}{v}$	None
Generalized	$P=\dfrac{ZRT}{v}$	$Z = Z[P_c,T_c,Z_c$ or $\omega]$ as derived from data
Redlich-K wong (R-K)	$P=\dfrac{RT}{v-b}-\dfrac{a}{v^2+bv}$	$b=0.08664RT_c/P_c$ $a=0.42748R^2T_c^{25}/p_CT^{0.5}$
Soave-Redlich-K wong (S-R-K or R-K-S)	$P=\dfrac{RT}{v-b}-\dfrac{a}{v^2+bv}$	$b=0.08664RT_c/P_c$ a $=0.42748R^2T_c^2[1+f\omega(1-T_r^{0.5})]^2/P_c$ $f_\omega=0.48+1.574\omega-0.176\omega^2$
Peng-Robinson (P-R)	$P=\dfrac{RT}{v-b}-\dfrac{a}{v^2+2bv-b^2}$	$b=0.07780RT_c/P_c$ $a=$ $0.45724R2T_c^2[1+f_\omega(1-T_r^{0.5})]^2/P_c$ $f_\omega=0.37464+1.54226\omega-0.2699\omega^2$

Compressibility factor Z, which is a function of P_r, T_r, and the critical compressibility factor, Z_c, or the acentric factor ω, which is determined from experimental P-V-T data. The acentric factor, introduced by Pitzer, accounts for differences in molecular shape and is determined from the vapor pressure curve.

$$\omega = \left[-\log\left(\frac{P^s}{P_c}\right)_{Tr=0.7} \right] - 1.000$$

Another equivalent form is

$$Z^3 - (1+B^* - uB^*)\, Z^2 + (A^* + wB^{*2} - uB^* - uB^{*2})Z$$

$$A^*B^* - wB^{*2} - wB^{*3} = 0$$

where $$A^* = \frac{aP}{R^2T^2}$$

and $$B^* = \frac{bP}{RT}$$

At any given pressure and temperature, Cubic equation for Z can be obtained. This equation may be solved analytically for the largest and smallest values of Z to give Z_L and Z_V. From this, molar volumes, and other properties can be estimated.

Mixing rule for EOS

The mixing rules recommended for all two-constant cubic equations of state are

$$a_m = \sum_i \sum_j y_i y_j (a_i a_j)^{1/2} (1 - \bar{k}_{ij})$$

$$b_m = \sum_i y_i b_i$$

kij are empirical and are provided for various component mixtures by Knapp et.al [1]

Solution for Molar volume estimation

Let propane be denoted by P and benzene by B. The mole fractions are

$$y_P = \frac{0.2692 / 44.097}{(0.2692 / 44.097) + (0.7308 / 78.114)} = 0.3949$$

$$y_B = 1 - 0.3949 = 0.6051$$

From the equations for the constants b and a in Table for the R–K equation, using SI units,

$$b_p = \frac{0.08664(8.3144)2(369.8)2.5}{4,250} = 0.06268\, m^3 / kmol$$

$$a_p = \frac{0.42748(8.3144)^2(369.8)^{2.5}}{(4,250)(477.59)^{0.5}}$$

$$= 836.7\, kPa - m^6 / kmol^2$$

Similarly, $b_B = 0.08263\, m^3 / kmol$

$$a_B = 2,072\, kPa - m^6 / kmol^2$$

Using Mixing Rule,

$$b = (0.3949)(0.06268) + (0.6051)(0.08263) = 0.07475\ m^3/kmol$$

$$a = y_p^2 a_p + 2 y_p y_B (a_p a_B)^{0.5} + y_B^2 a_B$$

$$= (0.3949)^2(836.7) + 2(0.3949)(0.6051)[(836.7)(2,072)]^{0.5}$$

$$+ (0.6051)^2(2,072) = 1,518\ kPa\text{-}m^6/kmol^2$$

From (2-47) and (2-48) using SI units,

$$A = \frac{(1,5,18)(2,829)}{(8.314)^2(477.59)^2} = 0.2724$$

$$B = \frac{(0.07475)(2,829)}{(8.314)(477.59)} = 0.05326$$

we obtain the cubic Z form of the R–K equation:

$$Z^3 - Z^2 + 0.2163Z - 0.01451 = 0$$

Solving this equation gives one real root and a conjugate pair of complex roots:

$$Z = 0.7314, \qquad 0.1314+0.04233i, \qquad 0.1314 - 0.04243i$$

The one real root is assumed to be that for the vapor phase. the molar volume is

$$v = \frac{ZRT}{P} = \frac{(0.7314)(8.314)(477.59)}{2,829} = 1.027\, m^3 / kmol$$

The average molecular weight of the mixture is computed to 64.68 kg/kmol. The specific volume is

$$\frac{v}{M} = \frac{1.027}{64.68} = 0.01588 \, m^3 \,/\, kg = 0.2543 \, ft^3 \,/\, lb$$

Glanville et al. report experimental values of $Z = 0.7128$ and $v/M = 0.2478 \, ft^3/lb$, which are within 3% of the above estimated values.

MATLAB Program for the Problem:

Write a MATLAB function to get the EOS constants namely u, w, a, b by calculating from the critical properties and save the file as get EOS constants.m. contents of the file are shown below:

```
function [u w a b]=getEOSconstants(Tc,Pc, T, methodtype)
% This function calculates the Equation of State constants namely
% u,w,a,b for the generic EOS:
%          RT              a
%  P= ----------- -- -----------------
%        (v-b)        v^2+ubv+w b^2
%
% Tc: Critical Temperature, K
% Pc: Critical Pressure, kPa
% T: Actual Temperature, K
% Methodtype: 0 for van-derWaals; 1 for Redlich-Kwong
%
R=8.314;
switch methodtype
    case 0
        u=0;w=0;
        b=R*Tc/(8*Pc);
        a=(27/64)*R^2*Tc^2/Pc;
    case 1
        u=1;w=0;
        b=0.08664*R*Tc/Pc;
        a=0.42748*R^2*Tc^2.5/(Pc*T^0.5);
    % Users are encourages to add other EOS methods
    otherwise
        u=0;w=0;a=0;b=0; % Ideal Gas assumed
end
```

Pure Component Molar Volume Estimation

Verify the code by estimating the constants for Propane and Benzene from MATLAB command prompt.

```
>> [u w aP bP]=getEOSconstants(369.8,4250, 477.6, 1)

u =

     1

w =

     0

aP =

   836.6273

bP =

    0.0627
>> [u w aB bB]=getEOSconstants(562.2,4890, 477.6, 1)

u =

     1

w =

     0

aB =

    2.0722e+03

bB =

    0.0828

>>
```

Component Mixture Molar Volume estimation:

Write a MATLAB function to estimate a, b for given mixture based on their individual constants and the composition: (Save the file as CalcMixture Constants.m) contents of the file are shown below:

```
function [amix bmix]=calcMixtureConstants(a,b,y)
% This function estimates the EOS constans a, b for mixtures using the
% mixing Rule
% a: Array of a values for all components
% b: Array of b values for all components
% y: Composition of the components
NOC=length(y);
amix=0;
bmix=0;
for i=1:NOC
    for j=1:NOC
        amix=amix+y(i)*y(j)*sqrt(a(i)*a(j));
    end
    bmix=bmix+y(i)*b(i);
end
```

Execution Procedure from MATLAB command prompt:

```
>> y=[0.3949 0.6051];
>> a=[aP aB];
>> b=[bP bB];
>> [amix bmix]=calcMixtureConstants(a,b,y)

amix =

   1.5184e+03


bmix =

   0.0749

>>
```

Define the composition and a,b values of the mixing components at command prompt and execute the function to estimate mixture a, b values:

```
>> y=[0.3949 0.6051];
>> a=[aP aB];
>> b=[bP bB];
>> [amix bmix]=calcMixtureConstants(a,b,y)
amix =
1.5184e+03
bmix =
7.4865e-02
```

Once the a, b values are found, the generic cubic equation of R-K method can be used to find the mixture molar volume:

The Equation of state method for R-K when it is written as cubic equation. By re-writing the generic equation with substituting the values of u, w constants and converting to polynomial of v, it is represented as

```
>> P=2829;
>> R=8.314;
>> T=477.6;
>> roots([P -R*T (amix-bmix*R*T-P*bmix^2) -amix*bmix])

ans =

     1.0268 + 0.0000i
     0.1884 + 0.0603i
     0.1884 - 0.0603i

>>
```

As molar volume can NOT be an imaginery number or negative number, Molar volume of the given Propane, benzene mixture is 1.0268 m^3/kmol.

Verification of Solution in Excel sheet:

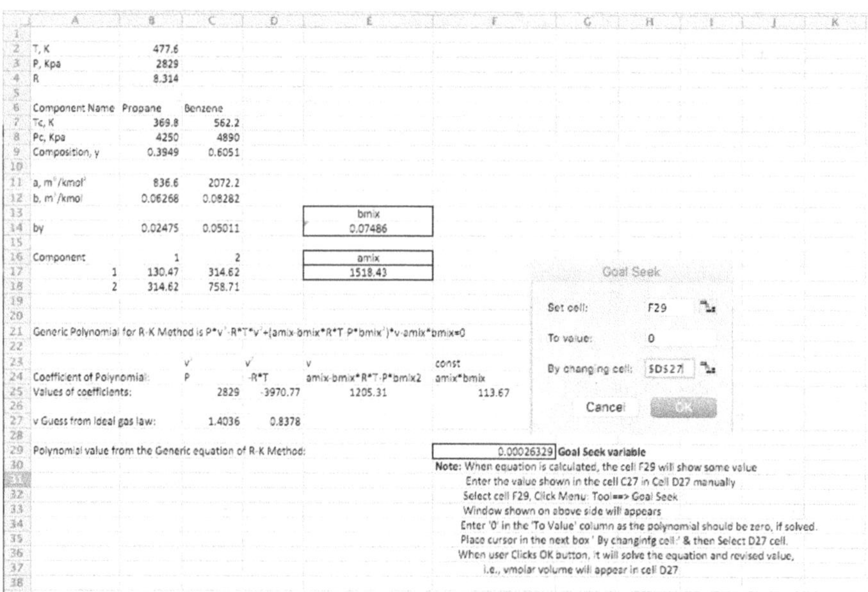

	A	B	C	D	E	F
1						
2	T, K	477.6				
3	P, Kpa	2829				
4	R	8.314				
5						
6	Component Name	Propane	Benzene			
7	Tc, K	369.8	562.2			
8	Pc, Kpa	4250	4890			
9	Composition, y	0.3949	0.6051			
10						
11	a, m^6/kmol^2	836.6	2072.2			
12	b, m^3/kmol	0.06268	0.08282			
13					bmix	
14	by	0.02475	0.05011		0.07486	
15						
16	Component	1	2		amix	
17	1	130.47	314.62		1518.43	
18	2	314.62	758.71			

Generic Polynomial for R-K Method is P*v³-R*T*v²+(amix-bmix*R*T-P*bmix²)*v-amix*bmix=0

	v³	v²	v	const
Coefficient of Polynomial:	P	-R*T	amix-bmix*R*T-P*bmix2	amix*bmix
Values of coefficients:	2829	-3970.77	1205.31	113.67

v Guess from ideal gas law: 1.4036   0.8378

Polynomial value from the Generic equation of R-K Method: 0.00026329   Goal Seek variable

Note: When equation is calculated, the cell F29 will show some value
Enter the value shown in the cell C27 in Cell D27 manually
Select cell F29, Click Menu: Tool==> Goal Seek
Window shown on above side will appears
Enter '0' in the 'To Value' column as the polynomial should be zero, if solved.
Place cursor in the next box ' By changing cell ' & then Select D27 cell.
When user Clicks OK button, it will solve the equation and revised value,
i.e., vmolar volume will appear in cell D27

Formulae used in the spread sheet

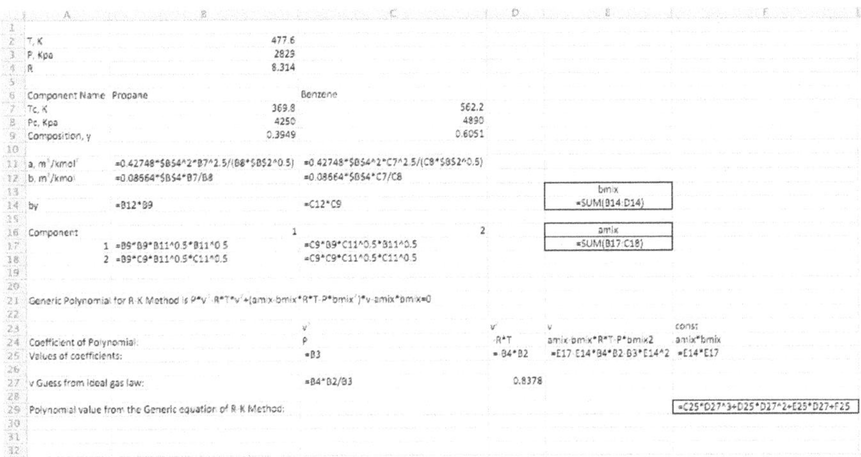

	A	B	C	D	E	F
1						
2	T, K	477.6				
3	P, Kpa	2829				
4	R	8.314				
5						
6	Component Name	Propane	Benzene			
7	Tc, K	369.8	562.2			
8	Pc, Kpa	4250	4890			
9	Composition, y	0.3949	0.6051			
10						
11	a, m³/kmol	=0.42748*$B$4^2*B7^2.5/(B8*$B$2^0.5)	=0.42748*$B$4^2*C7^2.5/(C8*$B$2^0.5)			
12	b, m³/kmol	=0.08664*$B$4*B7/B8	=0.08664*$B$4*C7/C8			
13					bmix	
14	by	=B12*B9	=C12*C9		=SUM(B14:D14)	
15						
16	Component		1	2	amix	
17	1	=B9*B9*B11^0.5*B11^0.5	=C9*B9*C11^0.5*B11^0.5		=SUM(B17:C18)	
18	2	=B9*C9*B11^0.5*C11^0.5	=C9*C9*C11^0.5*C11^0.5			
19						
20						
21	Generic Polynomial for R-K Method is P*v³-R*T*v²+(amix-bmix*R*T-P*bmix²)*v-amix*bmix=0					
22						
23			v³	v²	v	const
24	Coefficient of Polynomial:		P	-R*T	amix-bmix*R*T-P*bmix2	amix*bmix
25	Values of coefficients:		=B3	=-B4*B2	=E17-E14*B4*B2-B3*E14^2	=-E14*E17
26						
27	v Guess from ideal gas law:		=B4*B2/B3	0.8378		
28						
29	Polynomial value from the Generic equation of R-K Method:				=C25*D27^3+D25*D27^2+E25*D27+F25	
30						
31						
32						

References

Knapp, H., R. Doring, L. Oellrich, U. Plocker, and J. M. Prausnitz: "Vapor-Liquid. Equilibria for Mixtures of Low Boiling Substances," Chem. Data. Ser., vol. VI A982), DEHEMA.

Separation process principles, J. D. Seader, Ernest J. Henley, John Wiley & Sons.

13

Bubble and Dew Point Estimation for Multi Component System

Objective: Emphasizes the following principle:

Bubble points and Dew points for Propane-propylene system.

Theory

Phase Equilibrium: Equilibrium between two phases occurs when the chemical potential of each component is the same in the two phase:

$$\mu_j^I = \mu_j^{II}$$

where μ_j^I = chemical potential of the jth component in phase I

μ_j^{II} = chemical potential of the jth component in phase II

Basically we need a relationship that permits us to calculate the vapor composition if we know the liquid composition, or vice versa. The most common problem is a *bubble point* calculation: calculate the temperature T and vapour composition y_j, given the Pressure P and the liquid composition x_j. This usually involves a trial-and-error, iterative solution because the equations can be solved explicitly only in the simplest cases.

We will assume ideal vapor-phase behaviour in out examples, i.e., the partial pressure of the jth component in the vapor is equal to the total pressure P times the mole fraction of the jth component in the vapour y_j (Dalton's law): $p_j = Py_j$

Correction may required at high pressures.

Liquids that obey Raoult's are called *ideal*.

We are given the pressure P and the liquid composition x. We want to find the bubble point temperature and the vapor composition.

For simplicity let us assume a binary system of components 1 and 2.

Component 1 is the more volatile, and the mole fraction of component 1 in the liquid is x and in the vapor is y. Let us assume also that the system is ideal: Raoult's and Dalton's laws apply.

The partial pressures of the two components $(P_1 \text{ and } P_2)$ in the liquid and vapor phases are:

In liquid: $\quad p_1 = xP_1^S \qquad p_2 = (1-x)P_2^S$ $\hspace{3cm}$ ....(13.1)

In vapor $\quad p_1 = yP \qquad p_2 = (1-y)P$ $\hspace{3cm}$ .....(13.2)

where P_j^S = vapor pressure of pure component j which is a function of only temperature

$$\text{In } P_1^S = \frac{A_1}{T} + B_1 \qquad \text{In } P_1^S = \frac{A_2}{T} + B_2 \qquad .....(13.3)$$

Equating partial pressures in liquid and vapor phases gives

$$P = xP_1^S + (-x)P_2^S \hspace{3cm} .....(13.4)$$

$$y = \frac{xP_1^S}{P} \hspace{3cm} ....(13.5)$$

Our convergence problem is to find value of temperature T that will satisfy Eq. (13.4). The procedure is as follows:

1. Guess a temperature T.
2. Calculate the vapor pressures of components 1 and 2 from Eq.(13.3).
3. Calculate a total pressure P^{calc} using Eq. (13.4)

$$P^{calc} = xP_{1(T)}^S + (1-x)P_{2(T)}^S \hspace{2cm} .....(13.6)$$

4. Compare P^{calc} with the actual total pressure given, P. If it is sufficiently close to P (perhaps using a relative convergence criterion of 10^{-6}) the guess T is correct. The vapor composition can then be calculated from Eq. (13.5).

5. If P^{calc} is greater then be P, the guessed temperature was too high and we must make another guess of T that is lower. If P^{calc} is too low, we must guess a higher **T**.

CHEMCAD Procedure for the given Problem:

Bubble and Dew point calculations for Propylene-Propane system

1. Launch CHEMCAD

2. Drag & drop a stream from the PFD palette
3. Change the mode to "Simulation", Add components and supply the stream conditions:

   Flow rate: 100 lb moles/h

   Mole fraction of Propylene as 0.5

   Temperature: 107F

   Pressure: 206.9 psi

4 Select the stream; click Plot/TPXY menu option on the toolbar.
5 Select Propylene as first component and propane as second component
6 TPXY Diagram when Temperature, x supplied: Select the constant temperature option, provide temperature as 107F and get the TPXY data
7 TPXY Diagram when Pressure, x supplied: Select the constant pressure option, provide pressure as 233psia and get the TPXY data

Results from CHEMCAD

XY data for Propylene / Propane when T and x were provided.

K value model: SRK

Mole Fractions			
T Deg F	**P psia**	**XI**	**YI**
107.00000	209.17870	0.00000	0.00000
107.00000	214.60156	0.10000	0.11585
107.00000	219.78278	0.20000	0.22628
107.00000	224.72491	0.30000	0.33214
107.00000	229.42937	0.40000	0.43415
107.00000	233.89687	0.50000	0.53300
107.00000	238.12656	0.60000	0.62929
107.00000	242.11670	0.70000	0.72360
107.00000	245.86391	0.80000	0.81646
107.00000	249.36250	0.90000	0.90842
107.00000	252.60535	1.00000	1.00000

XY data for Propylene / Propane when P and x were provided.

K value model: SRK

Mole Fraction			
T Deg F	**P psia**	**XI**	**YI**
115.57414	233.00000	0.00000	0.00000
113.5446	233.00000	0.10000	0.11518
111.64842	233.00000	0.20000	0.22551
109.87840	233.00000	0.30000	0.33156
108.22893	233.00000	0.40000	0.43390
106.69450	233.00000	0.50000	0.53306
105.27025	233.00000	0.60000	0.62959
103.95232	233.0000	0.70000	0.71401
102.73686	233.0000	0.80000	0.81686
101.62151	233.0000	0.90000	0.90867
100.60435	233.0000	1.00000	1.00000

Also, generate PTxy data using ideal, one activity coefficient method (NRTL or Wilson), another Equation of state method (PR or BWRS) & analyse the results.

Bubble and Dew point calculations for ternary mixture using MATLAB:

Create sat_temp_antoine.m file to calculate Saturation Temperature for a given Pressure.

```
function vt = sat_temp_antoine(A,B,C,p_sat)
% Saturation Temperature for a given Pressure
% Input: Antoine Constants (A,B,C), p_sat
% Output: Saturation Temperature
% ln P = A - B/(T+C)
% T-temperature, P-saturation pressure
% Saturation Temperature from Antoine Equation
% A, B, C constants of Antoine equation.
% T temperature
% vt = Saturation Temperature.
% Vectorized (the dot next to B)

vt = B./(A - log(p_sat)) - C;

% end
```

Create sat_pr_antoine.m file to calculate Vapor Pressure.

```
function vp = sat_pr_antoine(A, B, C, T)
% Vap_Pressure = sat_pr_antoine(A, B, C, T)
% Vapor Pressure for a given Temperature
% Input: Antoine Constants (A,B,C), T
% Output: Vapor Pressure
% ln P = A - B/(T+C)
% A, B, C constants of Antoine equation.
% T temperature

% Vapor pressure from Antoine Equation
% vp = Vapor pressure
% Vectorized (the dot next to B)
vp = exp(A - B./(T+C));
% end
```

Create BUBL_T.m file to calculate Bubble Temperature and Composition of Vapor.

```
function bt_y = BUBL_T(A, B, C, P,x)
% [T y] = BUBL_T(A, B, C, P,x)
% Bubble Temperature and Composition of Vapor
% Input: Antoine Constants (A,B,C), Total Pressure, x
% Output: Bubble Temperature & Composition of vapor
n = size(A,2); % No of components
T_init = sum(x.*sat_temp_antoine(A,B,C,P));
p_sat = sat_pr_antoine(A,B,C,T_init);
alpha = p_sat./p_sat(n);
p_sat(n) = P/sum(x.*alpha);
T = sat_temp_antoine(A(n),B(n),C(n),p_sat(n));
%
while abs(T-T_init) > 0.001
    T_init = T;
    p_sat = sat_pr_antoine(A,B,C,T_init);
    alpha = p_sat./p_sat(n);
    p_sat(n) = P/sum(x.*alpha);
    T = sat_temp_antoine(A(n),B(n),C(n),p_sat(n));
end
y = x.*p_sat/P;
bt_y = [T y];
% end
```

Calculate the bubble point temperature for ternary system
(from MATLAB command window)

```
function bp_y = BUBL_P(A,B,C,T,x)
% [P y] = BUBL_P(A,B,C,T,x)
% Bubble Pressure and Composition of Vapor
% Input: Antoine Constants (A,B,C), Temperature of liquid, x
% Output: Bubble Pressure & Composition of vapor

p_sat = sat_pr_antoine(A,B,C,T);
P =sum(x.*p_sat);
y = x.*p_sat/P;
bp_y = [P y];
%end

>> A = [14.5463 14.2724 14.2043];
>> B = [2940.46 2945.47 2972.64];
>> C = [237.22 224 209];
>> x = [0.3 0.45 0.25];
>> P = 80;
>> BUBL_T(A, B, C, P, x)

ans =

    68.6007    0.5196    0.3773    0.1031

>>
```

Excel Solution for the given Problem

	A	B	C	D	E	F	G	H	I	J	K
1						Temp. in K	Vapor pressurein mmHg	Vapor pressure in psi			
2	Propylene	15.893	1875.3	-22.91		314.816667	12947.86634	250.369279			
3	Propane	15.725	1872.8	-25.101		314.816667	10516.81843	203.3607838			
4											
5	Pressure	233 psia									
6											
7	T in K	x	P1sat	P2sat	x1*P1sat	x2*P2sat	P Calculated		T in F	y1	y2
8	315	0.9	251.3809	204.1938	226.2427903	20.419383		246.66	107.33		
9	310	0.9	239.4869	197.0072	215.5381819	19.7007169		235.24	98.33		
10	309	0.9	234.4126	192.8344	210.9713511	19.2834421		230.25	96.53		
11	309.5	0.9	236.9397	194.9124	213.2457165	19.4912386		232.74	97.43		
12	309.55	0.9	237.1935	195.1211	213.4741474	19.5121106		232.99	97.52	0.916252	0.083748
13											
14	315	0.8	251.3809	204.1938	201.1047025	40.8387661		241.94	107.33		
15	310	0.8	239.4869	197.0072	191.589495	39.4014338		230.99	98.33		
16	311	0.8	244.6419	201.2474	195.7134915	40.2494886		235.96	100.13		
17	310.5	0.8	242.0542	199.1188	193.64339	39.8237667		233.47	99.23		
18	310.4	0.8	241.5391	198.6951	193.2313163	39.7390294		232.97	99.05	0.829425	0.170575
19											
20	315	0.2	251.3809	204.1938	50.27617562	163.355064		213.63	107.33		
21	320	0.2	294.7651	242.5305	58.95301764	194.024373		252.98	116.33		
22	317	0.2	277.2998	228.1351	55.45996258	182.508082		237.97	110.93		
23	316	0.2	271.6483	223.4792	54.32965121	178.783377		233.11	109.13		
24	315.9	0.2	271.0877	223.0175	54.21754788	178.414014		232.63	108.95	0.233062	0.766938
25											

References

William L. Luyben, Process Modelling, Simulation and Control for Chemical Engineers, 2nd Edition, McGraw Hill Publishers.

M. Subrahmanyam, www.svce.ac.in.

14

VLE Calculation for Multi-component System

Objective

Emphasizes the following principle:

Perform vapour liquid equilibrium for multi component mixture.

Problem Statement

Perform single stage flash for 65% n-Pentane, 20% n-Hexane and 15% n-Heptane mixture at the given 760mm Hg & 330K & estimate the liquid and vapour flows and compositions.

Vapor-Liquid Equilibrium

Following Figure describes the isothermal flash drum with the multi-component feed.

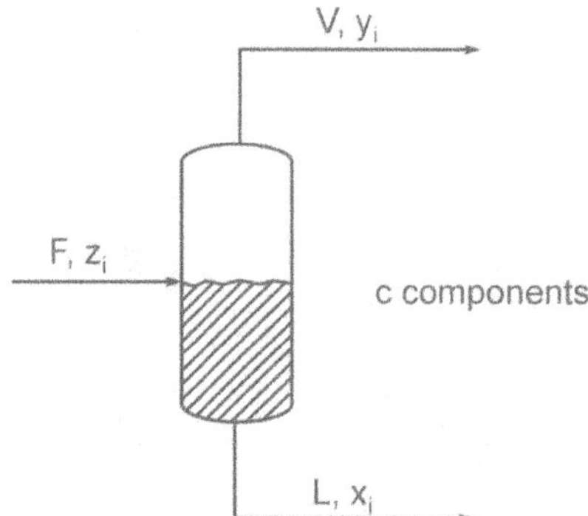

Fig. 14.1 Simple flash calculation

Given data for flash problem:

$$F = 100 \text{ mol} \qquad T = 330 \text{ K}$$

$Z_1 = 0.65$ $P = 760$ mmHg

$Z_2 = 0.2$

Comp.	Name	A_i	B_i	C_I
1	n-Pentane	6.85221	1064.63	232
2	n-Hexane	6.87776	1171.53	224.366
3	n-Heptane	6.90242	1268.115	216.9

A_i, B_i, C_i are Antoine vapor pressure constants

Model Equations

Material Balance

$$V + L = F \qquad .....(14.1)$$

$$y_i V + x_i L = z_i F \qquad .....(14.2)$$

Equilibrium

$$y_i = K_i x_i \qquad .....(14.3)$$

Physical Properties

$$K_i = P_i^o/P \qquad .....(14.4)$$

$$\log_{10}(P_i^o) = A_i - B_i/(C_i + T - 273.15) \qquad ....(14.5)$$

Other

$$\Sigma x_i - \Sigma y_i = 0 \qquad .....(14.6)$$

$$\Sigma z_i = 1 \qquad .....(14.7)$$

where the notation is as follows:

x_i, y_i, z_i	mole fractions of component i
V, L, F	molar flow rates
K_i	vapor-liquid equilibrium K-Value for component i
P_i^o	vapor pressure of component 1
P	total system pressure
T	absolute temperature

Eq.(14.4) makes it evident that we are assuming that equilibrium is defined by Raoult's law.

By combining eq.(14.2) and (14.3) to eliminate all y_i variables, we get

$$(K_i V + L) x_i = z_i F \; i = 1, 2, 3, ... c \qquad .....(14.8)$$

which then allows us to see our way to a solution procedure.

1. Given that values are available for $z_i, (i = 1, 2, 3.....c\text{-}1)$, P, T, F

2. (a) Using eq. (14.7), $z_c = 1-(z_1 + z_2 + z_3 + ... + z_{c\text{-}1})$

   (b) Using eq. (14.5), $P_i^° = 10^{\wedge}[A_i - B_i/(C_i + T - 273.15)]$

   (c) Using eq. (14.4), $K_i = P_i^° / P$

   Above 2 steps are performed once to calculate $P_i^°$ and K_i values. Step 3-5 are iterative.

3. Guess vapor fraction V/F, V then equals to (V/F).F

4. (a) Using eq. (14.1), $L = F\text{-}V$

   (b) Using eq. (14.8), $x_i = z_i F / (K_i V + L)$ $i = 1, 2, 3,..... c$

   (c) Using eq. (14.3), $y_i = K_i x_i$ $i = 1, 2, 3,..... c$

5. Evaluate eq. (14.6), $\Sigma\, x_i - \Sigma y_i$

   (a) If not essentially zero, reguess V/F and repeat from step 4.

   (b) Otherwise, Exit

Investigating steps 3 to 5, we see that we have, infact, been finding the root of a single nonlinear function in a single unknown. The unknown is V/F and the function we wish to drive to zero is $\Sigma\, x_i - \Sigma y_i$. The first five steps of the bounded secant method are illustrated. Initial guesses of $(V/F)^{(1)} = 0.5$ and $(V/F)^{(2)} = 0.5\ V/F)^{(1)} + 0.05$ were used. Below table gives the results of secant search.

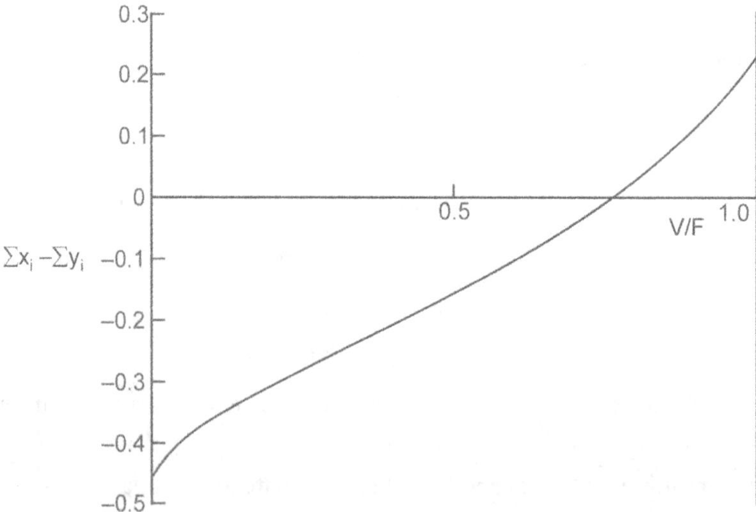

Fig. 14.2 Plot of error function for simple flash calculation specified in table 14.1

Table 14.1 Points evaluated during secant search for root for flash unit example

i	$VIF^{(i)}$	Error
1	0.50000	−0.15397
2	0.55000	−0.12802
3	0.79666	0.02358
4	0.75829	−0.00436
5	0.76428	−0.00016
6	0.76450	0.00000

MATLAB Program for the given Problem

Step 1 Create flash.m file as shown below

```
% Perform Single stage flash calculations for multi component mixture
% Input Arguments
% F: Feed flow Moles              z: Feed Composition
% A, B, C: Antoine constants of components
% Results from the function
% Vf: Vapor flow in moles         Lf: Liquid flow in moles
% x: Liquid composition           y: Vapor Composition
function [V,L,y,x] = flash(F, z, A,B,C,P,T)
%define the function variables
noc=length(z);
P0=zeros(noc,1);
K=zeros(noc,1);
x=zeros(noc,1);
y=zeros(noc,1);
for i=1:noc
    P0(i)=10^(A(i)-B(i)/(C(i)+T-273.15));
    K(i)= P0(i)/P;
end
z=z./sum(z);
vf= 0.5;
error = 0.1;
while (error >0.001)
    V=(vf)*F;
    L=F-V;
    for j=1:noc
        x(j)=(z(j)*F)/(K(j)*V+L);
        y(j)=K(j)*x(j);
    end
    vf=vf+0.005;
    error = abs(sum(x)-sum(y));
end
```

Step 2 From MATLAB command prompt, use the following instructions to provide the mixture composition and solve the flash algorithm.

```
>> F=100;
>> z=[0.65 0.2 0.15];
>> A=[6.85221 6.87776 6.90242];
>> B=[1064.63 1171.530 1268.115];
>> C=[232 224.366 216.9];
>> P=760;
>> T=330;
>> [Vf,Lf,yc,xc]=flash(F,z,A,B,C,P,T)

Vf =

    76.5000


Lf =

    23.5000


yc =

     0.7330
     0.1799
     0.0870


xc =

     0.3797
     0.2655
     0.3551

>>
```

Excel Calculations for the Given Problem:

Physical Properties

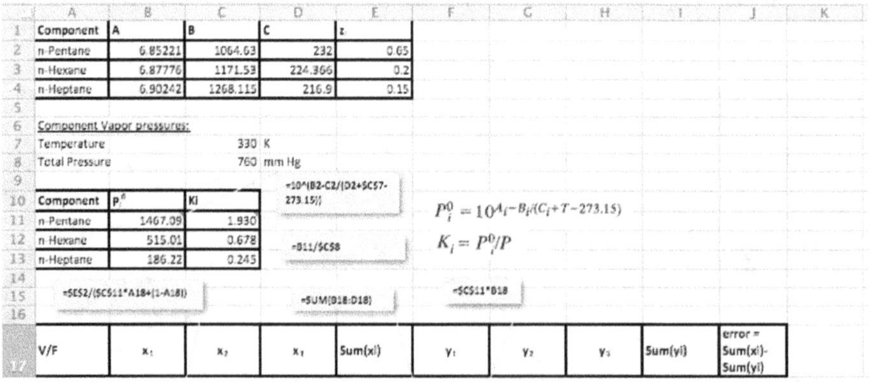

	A	B	C	D	E	F	G	H	I	J	K
1	Component	A	B	C	z						
2	n-Pentane	6.85221	1064.63	232	0.65						
3	n-Hexane	6.87776	1171.53	224.366	0.2						
4	n-Heptane	6.90242	1268.115	216.9	0.15						
5											
6	Component Vapor pressures:										
7	Temperature			330 K							
8	Total Pressure			760 mm Hg							
9				=10^(B2-C2/(D2+$C$7-273.15))							
10	Component	P_i^0	Ki								
11	n-Pentane	1467.09	1.930								
12	n-Hexane	515.01	0.678	=B11/$C$8							
13	n-Heptane	186.22	0.245								
14											
15	=$E$2/($C$11*A18+(1-A18))		=SUM(D18:D18)		=$C$11*B18						
16											
17	V/F	x_1	x_2	x_3	Sum(xi)	y_1	y_2	y_3	Sum(yi)	error = Sum(xi)-Sum(yi)	

$$P_i^0 = 10^{A_i - B_i/(C_i + T - 273.15)}$$

$$K_i = P_i^0/P$$

Contd...

18	0.5	0.444	0.2384	0.2410	0.923	0.8564	0.1616	0.059	1.0770	-0.1540
19	0.55	0.430	0.2431	0.2565	0.930	0.8300	0.1647	0.0629	1.0576	-0.1280
20	0.6	0.417	0.2480	0.2742	0.939	0.8052	0.168	0.0672	1.0404	-0.1011
21	0.65	0.405	0.2530	0.2945	0.953	0.7819	0.1715	0.0722	1.0256	-0.0730
22	0.7	0.394	0.2583	0.3181	0.970	0.7599	0.175	0.0779	1.0128	-0.0427
23	0.75	0.383	0.2638	0.3438	0.992	0.7390	0.1787	0.0847	1.0024	-0.0100
24	0.8	0.373	0.2695	0.3788	1.021	0.7193	0.1826	0.0928	0.9947	0.0262
25	0.775	0.378	0.2666	0.3615	1.006	0.7291	0.1807	0.0886	0.9984	0.0075
26	0.7625	0.380	0.2652	0.3535	0.999	0.7340	0.1797	0.0866	1.0003	-0.0014
27	0.76875	0.379	0.2659	0.3575	1.002	0.7315	0.1802	0.0876	0.9993	0.0030
28	0.765625	0.380	0.2655	0.3555	1.001	0.7328	0.1799	0.0871	0.9998	0.0008
29										
30										
31										
32		$x_i = z_i F/(K_i V + L)$				$y_i = K_i x_i$				
33										
34										

Additional Practice Data from various Books:

Table 14.2 Vapor-Liquid Equilibrium Data for the Methonal-Water System at Temperatures of 50, 150, and 250°C

a. Methanol (A)-Water (B) System

$T = 50°C$

Data of McGlashan and Williamson. J. Chem. Eng. Data, 21, 196 (1976)

Pressure, psia	y_A	x_A	$\alpha_A.B$
1.789	0.0000	0.0000	
2.373	0.2661	0.0453	7.64
2.838	0.4057	0.0863	7.23
3.369	0.5227	0.1387	6.80
3.764	0.5898	0.1854	6.32
4.641	0.7087	0.3137	5.32
5.163	0.7684	0.4177	4.63
5.771	0.8212	0.5411	3.90
6.122	0.8520	0.6166	3.58
6.811	0.9090	0.7598	3.16
7.280	0.9455	0.8525	3.00
7.800	0.9817	0.9514	2.74
8.032	1.0000	0.0000	

Additional Practice Data from various Books:

Data Set 1: Methanol-Water system data at 50, 150, 200C

b. Methanol (A)-Water (B) System

$T = 150°C$

Data of Griswold and Wong. Chem. Eng. Prog. Symp. Ser., 48 (3), 18 (1952)

Pressure, psia	y_A	x_A	$\alpha_A.B$
73.3	0.060	0.009	7.03
79.0	0.135	0.022	6.94
85.7	0.213	0.044	5.88
93.9	0.286	0.079	4.67
114.9	0.459	0.186	3.71
139.7	0.610	0.374	2.62
148.6	0.662	0.459	2.31
160.4	0.731	0.578	1.98
177.4	0.832	0.748	1.67
193.5	0.929	0.893	1.57
194.5	0.943	0.913	1.58
196.5	0.960	0.936	1.64
197.7	0.972	0.953	1.71
199.2	0.982	0.969	1.75

c.Methanol (A)-Water (B) System

T = 250°C

Data of Griswold and Wong, Chem. Eng. Prog. Symp. Ser., 48 (3), 18 (1952)

Pressure, psia	y_A	x_A	$\alpha_A.B$
681	0.163	0.066	2.76
764	0.280	0.132	2.56
818	0.344	0.180	2.39
889	0.423	0.254	2.15
949	0.487	0.331	1.92
994	0.542	0.404	1.75
1049	0.596	0.483	1.58
1099	0.643	0.553	1.46
1159	0.698	0.631	1.35
1204	0.756	0.732	1.13
1219	0.772	0.772	1.00
1234	0.797	0.797	1.00

Data Set 2:

A 100-knol/h feed consisting of 10, 20, 30, and 40 mol% of propane (3), n-butane (4), n-pentane (5), and n-hexane (6), respectively, enters a distillation column at 100 psia (689.5 kPa) and 200°F   (366.5°K). Assuming equilibrium, what mole fraction of the feed enters as liquid, and what are the liquid and vapor compositions?

	x	y
Propane	0.0719	0.3021
n-Butane	0.1833	0.3207
n-Pentane	0.3098	0.2293
n-Hexane	0.4350	0.1479
	1.0000	1.000

Data Set 3:

Propylene (P) is to be separated from T-butene (B) by distillation into a vapor distillate containing 90 mol% propylene. Calculate the column operating pressure assuming the exit temperature from the partial condenser is 100°F (37.8°C), the minimum attainable temperature with cooling water. Determine the composition of the liquid reflux.

An operating pressure of 186 psia (1,282 kPa) at the partial condenser outlet is indicated. The composition of the liquid reflex is obtained from $xi = zi/Ki$ with the result.

Equilibrium Mole Fraction		
Component	Vapor Distillate	Liquid Reflux
Propylene	0.90	0.76
I-Butene	0.10	0.24
	1.00	1.00

Data set 4:

kmol/h			
Component	Feed 120oF 485 psia	Vapor 112oF 165 psia	Liquid 112oF 165 psia
Hydrogen	1.0	0.7	0.3
Methane	27.9	15.2	12.7
Benzene	345.1	0.4	344.7
Toluene	113.4	0.04	113.36
Total	487.4	16.34	471.06
Enthalpy	−1,089,000	362,000	−1,451,000

Data set 5: A liquid containing 50 mol % benzene (A), 25 mol % toluene (B), and 25 mol % o-xylene (C) is flashl-vaporized at 1 std atm pressue and 100°C. Compute the amounts of liquid and vapor products and their composition.

Substantace	P = vapor pressure, mmHg	$m = \dfrac{P}{760}$	$2F$	$\dfrac{W}{D} = 3.0$ $\dfrac{2F(W/D+1)}{1+W/D_m} = y_D^*$	$\dfrac{W}{D} = 2.08$ y_D^*	$x_W = \dfrac{F_{2_F} - Dy_D^*}{W}$ $= \dfrac{y_D^*}{m}$
A	1370	1,803	0.50	$\dfrac{0.5(3+1)}{1+3/1.803} = 0.750$	0.715	0.397
B	550	0.724	0.25	0.1940	0.1983	0.274
C	200	0.263	0.25	$\Sigma = \dfrac{0.0805}{1.0245}$	$\dfrac{0.0865}{0.9998}$	$\dfrac{0.329}{1.000}$

The properties of Gases and Liquids by Reid, McGrawHill

Data set 6:

Table 14.3 Vapor-Liquid Euilibrium for 2,3-dimethylbutane (1)-chloroform (2) at 760 mmHg

Calculated			Experimental		
x_1	y_1	$t(°C)$	x_1	y_1	$t(°C)$
0.0	0.000	61.3	0.087	0.130	59.2
0.1	0.163	58.7	0.176	0.230	28.1
0.2	0.268	57.5	0.275	0.326	57.0
0.3	0.352	56.7	0.367	0.406	56.5
0.4	0.430	56.3	0.509	0.525	56.0
0.5	0.508	56.1	0.588	0.588	56.0
0.6	0.590	56.1	0.688	0.671	56.1
0.7	0.678	56.3	0.785	0.760	56.5
0.8	0.774	56.7	0.894	0.872	57.0
0.9	0.881	57.3			
1.0	1.000	58.0			

Table 14.4 Vapor-Liquid Equilibrium Data for Acetone (1)-Methanol (2)-Cyclohexane (3) at 45 °C

		Calculated			Experimental		
x_1	x_2	y_1	y_2	P(mm Hg)	y_1	y_2	P(mmHg)
0.117	0.127	0.254	0.395	558	0.276	0.367	560
0.118	0.379	0.169	0.470	559	0.191	0.452	568
0.123	0.631	0.157	0.474	554	0.176	0.471	561
0.249	0.120	0.400	0.272	567	0.145	0.252	594
0.255	0.369	0.296	0.375	574	0.132	0.367	585
0.250	0.626	0.299	0.435	551	0.325	0.412	566
0.382	0.239	0.418	0.281	580	0.433	0.267	593
0.379	0.497	0.405	0.367	563	0.414	0.343	579
0.537	0.214	0.521	0.222	581	0.526	0.202	597
0.669	0.076	0.656	0.098	569	0.654	0.088	584
0.822	0.054	0.772	0.064	557	0.743	0.076	574

Table 14.5 Vapor-Liquid Equilibria for Acetone (1)-2 Butone (2)- Ethyl acetate (3) at 760

		Calculated			Experimental		
x_1	x_2	y_1	y_2	$t(^oC)$	y_1	y_2	$t(^oC)$
0.200	0.640	0.341	0.508	72.5	0.290	0.556	72.6
0.400	0.480	0.583	0.322	67.5	0.525	0.370	67.6
0.600	0.320	0.761	0.185	63.2	0.720	0.215	63.5
0.800	0.160	0.896	0.081	59.4	0.873	0.095	59.8
0.200	0.480	0.336	0.374	71.8	0.295	0.420	71.8
0.400	0.360	0.576	0.238	67.1	0.535	0.285	67.3
0.600	0.240	0.755	0.137	63.0	0.725	0.170	63.3
0.800	0.120	0.893	0.060	59.3	0.880	0.075	59.6
0.200	0.320	0.334	0.248	71.2	0.302	0.276	71.3
0.400	0.240	0.571	0.157	66.6	0.540	0.180	67.0
0.600	0.160	0.750	0.091	62.7	0.729	0.105	62.9
0.200	0.160	0.339	0.125	70.7	0.295	0.145	71.1
0.400	0.120	0.570	0.078	66.2	0.530	0.095	62.7
0.600	0.080	0.746	0.045	62.5	0.720	0.095	62.7
0.800	0.040	0.887	0.020	59.1	0.873	0.024	59.6

References

Separation process principles by Seader & Henley, 2nd Ed., John Weiley & Sons.

Process Flowsheeting by Westerberg, Hutchison, Motard & Winter, Cambridge university press.

15

CO_2 Vapor Pressure Regression

Objective: Fit experimental data to polynomial or nonlinear equation.

Problem Statement Table 15.1 gives a set of pressure versus temperature data of equilibrium CO_2 hydrate formation conditions that were obtained from a series of experiments in a 20 wt% aqueous glycerol solution. The objective is to fit a function of the form

$$ln\,P = A + BT + CT^{-2} \qquad\qquad .....(15.1)$$

or $\qquad ln\,P = A + BT \qquad\qquad\qquad .....(15.2)$

Table 15.1 Incipient Equilibrium Data on CO_2 Hydrate Formation in 20 (wt%) Aqueous Glycerol Solution

Temperature (K)	Experimental Pressure (MPa)
270.4	1.502
270.6	1.556
272.3	1.776
273.6	2.096
274.1	2.281
275.5	2.721
276.2	3.001
277.1	3.556

Procedure for using Microsoft Excel ™ for Windows

Step 1: First the data are entered in columns. The single dependent variable is designated by y in this case vapour pressure whereas the independent ones by x_1, x_2, x_3 etc in this problem temperature.

Step 2: Select cells below the entered data to form a rectangle [5 × p] (cells A_{24} to C_{28} in this problem scenario) where p is the number of parameters sought after, e.g. in the equation $y = k_1x_1 + k_2x_2 + k_3$ you are looking for k_1, k_2 and k_3. Therefore p would be equal to 3.

Note: Excel casts the p-parameter model in the following form:

$$y = m_1x_1 + m_2x_2...+ m_{p-1}x_{p-1} + b$$

Step 3:   Now that you have selected an area [5 × p] on the spreadsheet, go to the f_x (paste Function button) and click.

Step 4:   Click on Statistical on the left scroll menu and click on LINEST on the right scroll menu; then hit OK.

A box will now appear asking for the following
```
known Y's
known X's
Const
Stats
```

Step 5:   Click in the next box for known values for the single response variable y; then go to the Excel sheet and highlight the y values. (Cells A15 to A22 in this problem scenario).

Step 6:   Repeat Step 5 for the known values for the independent variables x_1, x_2 etc. by clicking on the box containing these values. This the program lets you highlight the area that encloses all the x values (x_1, x_2, x_3 etc....). (Cells B15 to C22 in this problem scenario).

Step 7:   Set the logical value *Const=true* if you wish to calculate a y intercept value.

Step 8:   Set the logical value *Stats–true* if you wish the program to return additional regression statistics.

At this stage the function syntax should be {= LINEST (A15 : A22, B15 : C22, 1, 1)}

Step 9:   Now that you have entered all the data you do not hit the OK button but instead press *Control-Shift-Enter*.

This command allows all the elements in the array to be displayed. If you bit OK you will only see one element in the array.

Once the above steps have been followed, the program returns the following information on the worksheet. The information is displayed in a [5 × p] table where p is the number of parameters

I^{s}t row: parameter values

$$m_{p-1} \quad m_{p-2} \quad \ldots \quad m_2 \quad m_1 \quad b$$

or $\quad k_{p-1} \quad k_{p-2} \quad \ldots \quad k_2 \quad k_1 \quad k_p$

2nd row: Standard errors for the estimated parameter values

$$se\ (k_{p-1}) \qquad se(k_{p-2}) \ \ldots \ se(k_2)\ se(k_1)\ se\ (k_p)$$

3rd row: Coefficient of determination and standard error of the y value

$$R^2 \quad sev$$

Note: The coefficient of determination ranges in value from 0 to 1. If it is equal to one then there is a perfect correlation in the sample. If on the other head it has a value of zero then the model is not useful in calculating a y-value.

4th row: F statistic and the number of degrees of freedom (d.f)

5th row: Information about the regression

$$ssreg\ (regression\ sum\ of\ squares)\ ssresid$$
$$(residual\ sum\ of\ squares)$$

The above procedure will be followed in this problem with two and three parameter linear single response models.

Excel Solution for this Problem:

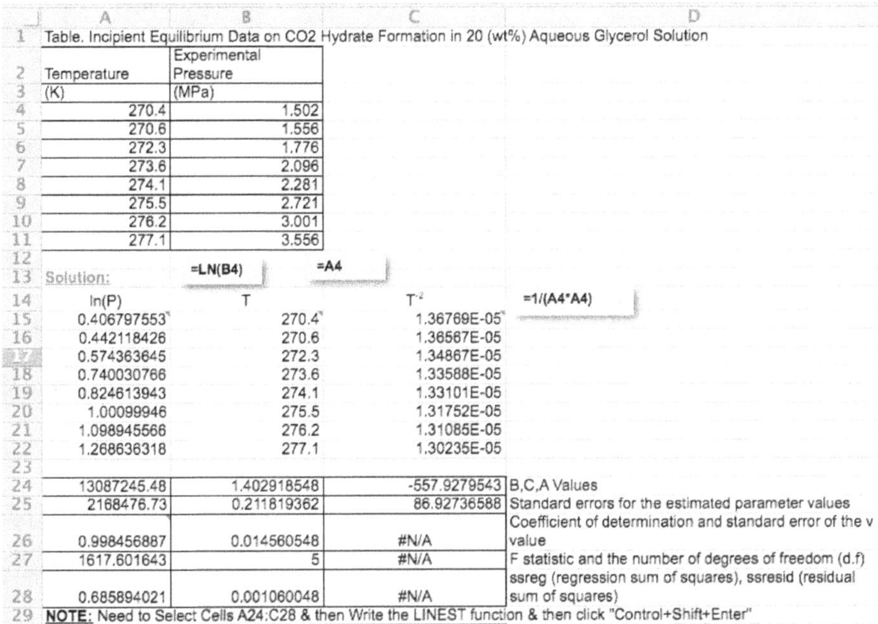

	A	B	C	D
1	Table. Incipient Equilibrium Data on CO2 Hydrate Formation in 20 (wt%) Aqueous Glycerol Solution			
2	Temperature	Experimental Pressure		
3	(K)	(MPa)		
4	270.4	1.502		
5	270.6	1.556		
6	272.3	1.776		
7	273.6	2.096		
8	274.1	2.281		
9	275.5	2.721		
10	276.2	3.001		
11	277.1	3.556		
12		=LN(B4)	=A4	
13	Solution:			
14	ln(P)	T	T^{-2}	=1/(A4*A4)
15	0.406797553	270.4	1.36769E-05	
16	0.442118426	270.6	1.36567E-05	
17	0.574363645	272.3	1.34867E-05	
18	0.740030766	273.6	1.33588E-05	
19	0.824613943	274.1	1.33101E-05	
20	1.00099946	275.5	1.31752E-05	
21	1.098945566	276.2	1.31085E-05	
22	1.268636318	277.1	1.30235E-05	
23				
24	13087245.48	1.402918548	-557.9279543	B,C,A Values
25	2168476.73	0.211819362	86.92736588	Standard errors for the estimated parameter values
26	0.998456887	0.014560548	#N/A	Coefficient of determination and standard error of the v value
27	1617.601643	5	#N/A	F statistic and the number of degrees of freedom (d.f)
28	0.685894021	0.001060048	#N/A	ssreg (regression sum of squares), ssresid (residual sum of squares)
29	**NOTE:** Need to Select Cells A24:C28 & then Write the LINEST function & then click "Control+Shift+Enter"			

Following MATLAB function demonstrates to fit a polynomial for the given experimental data. The code mentioned in the function can also be executed as steps from command prompt itself (instead of executing as a function).

Write the MATLAB function and save the file as CO_2 regression.in content of the file is shown below:

```
function coeff=co2regression(T, Pexp, n)
% This function regress the experimental vapor pressure data and finds
% the polynomial coefficients for the equations:
% Ln P = A+BT+....+CT^n
% T: Temperature in K
% Pexp: Vapor pressure noted from the experimental finding in MPa
% Suggested data: T=[270.4 270.6 272.3 273.6 274.1 275.5 276.2 277.1]
% Suggested data: Pexp=[1.502 1.556 1.776 2.096 2.281 2.721 3.001 3.556]
LnPexp=log(Pexp);
coeff=polyfit(T,LnPexp,n);
% calculate the vaopr pressure using the polynomial coefficients
LnPth=polyval(coeff,T);
Pth=exp(LnPth);
%Calculate percentage error in vapor pressure
errorpc=(Pexp-Pth)*100./Pexp;
plot(T,Pexp,':+', T,Pth, '-*r');
title('Vapor pressure regression Curve');
xlabel('Temperature,K');
ylabel('Vapor pressure, MPa');
```

Execute the function from command prompt to show the results:

```
>> T=[270.4 270.6 272.3 273.6 274.1 275.5 276.2 277.1];
>> Pexp=[1.502 1.556 1.776 2.096 2.281 2.721 3.001 3.556];
>> coeff=co2regression(T, Pexp, 2)

coeff =

    0.0070   -3.7030   490.2598

>> coeff=co2regression(T, Pexp, 1)

coeff =

    0.1246   -33.3143

>>
```

Another Approach to solve the problem from command prompt is shown below (Using the matrices backslash operator):

```
>>  T=[270.4  270.6  272.3  273.6  274.1  275.5  276.2
277.1];

>>  Pexp=[1.502 1.556 1.776 2.096 2.281 2.721 3.001
3.556];

>> X=log(Pexp)'
```

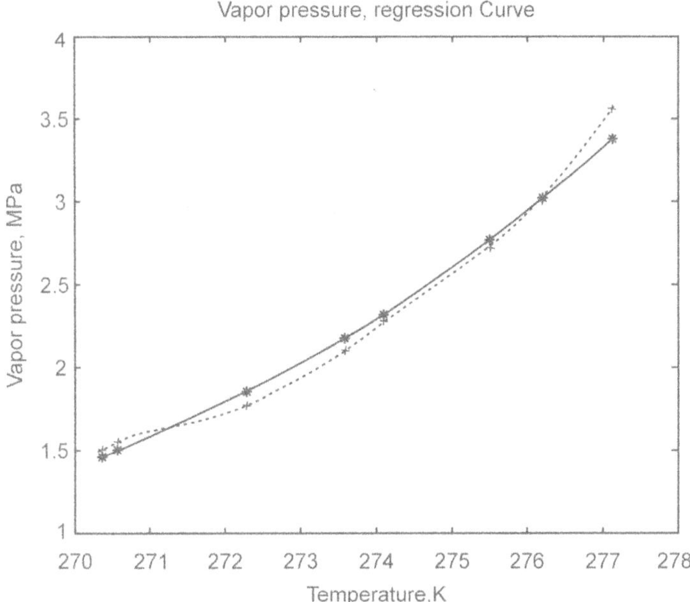

Fig. 15.1

```
>> Y=[ones(arraysize,1), T']

>> coeff=Y\X
coeff =

  -33.3143

    0.1246
```

The second approach can be used for any non-linear equation regression.

Following is the procedure to find coefficients for the equations 15.1 and 15.2.

The MATLAB code demonstrates the procedure

```
>>  T=[270.4  270.6  272.3  273.6  274.1  275.5  276.2
277.1];

>> Pexp=[1.502 1.556 1.776 2.096 2.281 2.721 3.001
3.556];

>> X=log(Pexp)'

>> Y=[ones(arraysize,1), T' [T.^-2']']

>> coeff=Y\X

coeff =
```

```
   1.0e+07  *

   -0.0001

    0.0000

    1.3087

>> format short e

>> coeff=Y\X

coeff =

   -5.5793e+02

    1.4029e+00

    1.3087e+07
```

Hence the equation for the vapor pressure is

Ln P = $-557.93 + 1.4029 \times T + 1.3087 \times 10^7/T^2$

References

Peter Englezos, Nicolas Kalogerakis, Applied Parameter Estimation for Chemical Engineers, Marcel Dekker, Inc.

16

Regress Feature – CO$_2$ Emissions

Objective: Understand curve fitting having sinusoidal waves in up trend.

Problem Description:

A common measurement for studying the effect of burning fossil fuels on weather patterns is the level of carbon dioxide (CO_2) concentration in the atmosphere. Some scientists believe the upward trend in CO_2 levels could cause atmospheric temperatures to rise, polar ice caps to melt and climates of different regions of the earth to change radically.

This example describes how to fit a nonlinear model to available CO_2 data, in parts per million, collected by the U.S. National Oceanic and Atmospheric Administration (NOAA) from 1979 to 1996 at the Mauna Loa Observatory in Hawaii.
http://scrippsco2.ucsd.edu/data/in_situ_co2/monthly_mlo.csv

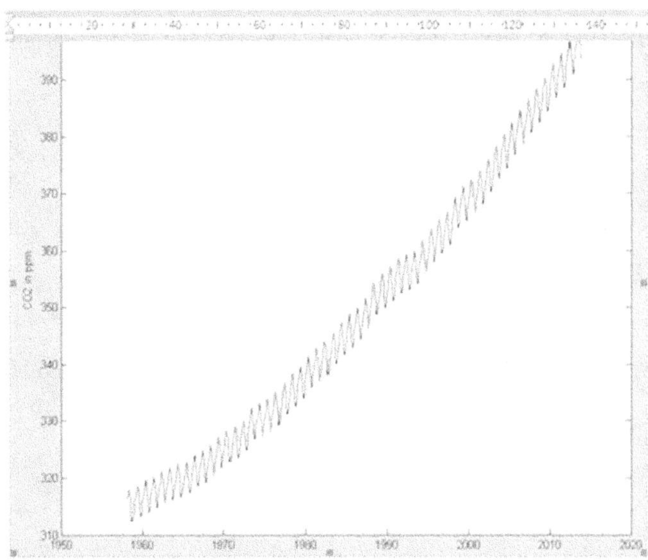

Fig. 16.1 CO$_2$ emissions data from 1979 to 1996

Cyclical phenomena in nature can be modeled by a sinusoid and its higher harmonics, that is,

$$y(t) = \sum_{k=1}^{n} A_k \sin \frac{2\pi k}{T}(t + \phi_k)$$

$A_k T/K$ and Φ_k are the amplitude, period, and phase, respectively, of the k-th harmonic. The sinusoidal terms capture seasonal trends, but the persistent upward trend requires an exponential term. A suggested model for this data, determined by examining records of total world fossil fuel production from the mid 1860s, is

$$y(t) = B + d \exp \alpha t + \sum_{k=1}^{n} A_k \sin \frac{2\pi k}{T}(t + \phi_k)$$

where t (time) is in years, α is believed to be approximately 0.0444/year, and the other parameters are unknown (including the number of sinusoids needed) To build the model, create a custom equation with one sinusoid term.

$$y(t) = B + d \exp \alpha t + A_1 \sin \frac{2\pi k}{T}(t + \phi_1)$$

f(date) = a*sin(2*pi*date/0.192) + d*exp (0.044*date) + b

Coefficients (with 95% confidence bounds):

    a = –0.0004879 (–0.5333, 0.5323)

    d = 3.084e – 037 (3.035e – 037, 3.134e – 037)

    b = 314.3 (313.6, 315)

Goodness of fit:

 SSE: 1.631e+004

 R-square: 0.9576

 Adjusted R-square: 0.9574

 RMSE: 4.956

First, we can determine the coefficients of the exponential term separately. We use the toolbox's built-in exponential model which automatically calculates optimal start points from the data,

$$y(t) = d \exp \alpha t$$

Fitting this model results in coefficients of *d = 348.3* and α = *0.06849* Using these values of *d* and a as start values in the sinusoidal model, we get a new fit that matches the cyclical nature well.

f(x) = d*exp(alpha*t)

where x is normalized by mean 1986 and std 16.06

Coefficients (with 95% confidence bounds):

d = 348.3 (348.1, 348.5)

alpha = 0.06849 (0.06783, 0.06915)

Goodness of fit:

SSE: 6046

R-square: 0.9843

Adjusted R-square: 0.9842

RMSE: 3.015

General model Sin1:

f(x) = a1*sin(b1*x+c1)

where x is normalized by mean 1986 and std 16.06

Coefficients (with 95% confidence bounds):

a_1 = 4951 (−1.8e+005, 1.899e + 005)

b_1 = 0.004811 (−0.1758, 0.1854)

c_1 = 0.07057 (−2.57, 2.711)

SSE is the sum of the squared differences between each observation and its group's mean. It can be used as a measure of variation within a cluster. If all cases within a cluster are identical the SSE would then be equal to 0.

The formula for SSE is:

$$SSE = \sum\nolimits_{i=1}^{n}(x_i - \bar{x})^2$$

Root mean square error (RMSE) is defined mathematically as:

$$RMSE = \sqrt{\frac{1}{N}\sum_{i=1}^{N}(F_1 - 0_i)^2}$$

where

F_i = the forecast values of the parameter in question

0_i = the corresponding verifying value (observed or analysed)

N = the number of verifying points (grid points or observations) in the verification area

A data set has values y_i, each of which has an associated modeled value f_i (also sometimes referred to as $\hat{y}_i$). Here, the values y_i are called the observed values and the modeled values f_i are sometimes called the predicated values.

In what follows $\bar{y}$ is the mean the observed data:

$$\bar{y} = \frac{1}{n}\sum_{i=1}^{n} y_i$$

where n is the number of observations.

The "variability" of the data set is measured through different sums of squares:

$$SS_{tot} = \sum_i (y_i - \bar{y})^2 \text{ , the total sum of squares (proportional to the sample}$$

variance):

$$SS_{reg} = \sum_i (f_i - \bar{y})^2 \text{ , the regression sum of squares, also called the}$$

explained sum of squares.

$$SS_{res} = \sum_i^i (y_i - fi)^2 \text{, the sum of squares of residuals, also called the}$$

residual sum of squares.

The nations SS_R and SS_E should be avoided, since in some texts their meaning is reversed to Residual sum of squares and Explained sum of squares, respectively.

The most general definition of the coefficient of determination is

$$R^2 \equiv 1 - \frac{SS_{res}}{SS_{tot}} \ .$$

The adjusted R^2 is defined as

$$\bar{R}^2 = 1 - (-R^2)\frac{n-1}{n-p-1} = R^2 - (1-R^2)\frac{p}{n-p-1}$$

where p is the total number of regressors in the linear model (not counting the constant term), and n is the sample size.

Solution is provided in the Reference Website.

References

http://in.mathworks.com/company/newsletters/articles/atmospheric-carbon-dioxide-modeling-and-the-curve-fitting-toolbox.htm1.

Mass Transfer

"It was quite natural to suppose that this law for diffusion of a salt in its solvent must be identical with that according to which the diffusion of heat in a conducting body takes place; upon this law Fourier founded his celebrated theory of heat, and it is the same that Ohm applied... to the conduction of electricity ... according to this law, the transfer of salt and water occurring in a unit of time between two elements of space filled with two different solutions of the same salt, must be, ceteris partibus, *directly proportional to the difference of concentrations, and inversely proportional to the distance of the elements from one another"*.

- Thinking about Thomas Graham results, Adolf Fick perceived the deep analogy between diffusion and conduction of heat or electricity, a premonitory intuition.

17

Dynamics of Binary Distillation Column

Objective: Study the response for step change in reflux in binary distillation column.

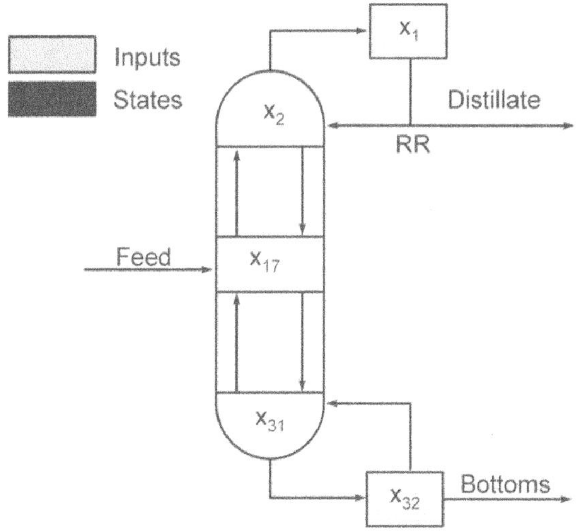

Fig. 17.1 Typical distillation column with condenser, Reboiler stages

Theory

Component continuity

$$\frac{d(M_n x_n)}{dt} = L_{n-1} x_{n-1} - L_n x_n + V y_n - V y_{n+1}$$

We will assume a binary system (two components) with constant relative volatility throughout the column and theoretical (100) percent efficient) trays, i.e., the vapor leaving the tray is in equilibrium with the liquid on the tray. This means the simple vapor-liquid equilibrium relationship can be used.

157

$$y_n = \frac{\alpha x_n}{1 + (a - 1)x_n}$$

where

x_n = liquid composition on the nth tray (mole fraction more volatile component)

y_n = vapor composition on the nth tray (mole fraction more volatile

α = relative volatility

MATLAB Procedure for the Problem Solution:

Content of the file is shown below:

Step 1: Create distill.m file to define derivatives

```
%   t -- time (not used)
%   x --- mole fraction of A at each stage
%   xdot --- state derivatives
function xdot = distill(t,x)
global u
% Inputs (1):
% Reflux Ratio (L/D)
rr=u;
% States (32):
% x(1) - Reflux Drum Liquid Mole Fraction of Component A
% x(2) - Tray 1 - Liquid Mole Fraction of Component A
% x(n) - Tray n-1 - Liquid Mole Fraction of Component A
% x(17) - Tray 16 - Liquid Mole Fraction of Component A (Feed Location)
% x(32) - Reboiler Liquid Mole Fraction of Component A
% Parameters
% Feed Flowrate (mol/min)
Feed =  24.0/60.0;
% Mole Fraction of Feed
x_Feed = 0.5;
% Distillate Flowrate (mol/min)
D=0.5*Feed;
% Flowrate of the Liquid in the Rectification Section (mol/min)
L=rr*D;
% Vapor Flowrate in the Column (mol/min)
V=L+D;
% Flowrate of the Liquid in the Stripping Section (mol/min)
FL=Feed+L;
% Relative Volatility = (yA/xA)/(yB/xB) = KA/KB = alpha(A,B)
vol=1.6;
% Total Molar Holdup in the Condenser
atray=0.25;
% Total Molar Holdup on each Tray
acond=0.5;
% Total Molar Holdup in the Reboiler
areb=1.0;
% Vapor Mole Fractions of Component A
% From the equilibrium assumption and mole balances
% 1) vol = (yA/xA) / (yB/xB)
```

```
% 2) xA + xB = 1
% 3) yA + yB = 1
for i=1:32
    y(i)=x(i)*vol/(1+(vol-1)*x(i));
end
% Compute xdot
xdot(1) = 1/acond*V*(y(2)-x(1));
for j=2:16
    xdot(j) = 1/atray*(L*(x(j-1)-x(j))-V*(y(j)-y(j+1)));
end
xdot(17) = 1/atray*(Feed*x_Feed+L*x(16)-FL*x(17)-V*(y(17)-y(18)));
for j=18:31
    xdot(j) = 1/atray*(L*(x(j-1)-x(j))-V*(y(j)-y(j+1)));
end
xdot(32) = 1/areb*(FL*x(31)-(Feed-D)*x(32)-V*y(32));
xdot = xdot';  % xdot must be a column vector
```

Step 2: From the MATLAB command prompt, solve ODEs

```
>> % Step test for Model - Binary Distillation Column

global u

% Steady State Initial Condition for the Control
u_ss = 3.0;

% Steady State Initial Conditions for the States
x_ss(1:32)=linspace(0.93542,0.0646,32);
x_ss = x_ss';

% Open Loop Step Change
u = 2.7;
% Final Time (min)
tf = 120;

[t,x] = ode15s('distill',[0 tf],x_ss);
>>
```

Step 3: Plot results

```
>> x1=x(:,1);
>> plot(t,x1);
>> xlabel('Time');
>> ylabel('Distillate Composition');
>> title('Dynamics of Binary Distillation Column');
>>
```

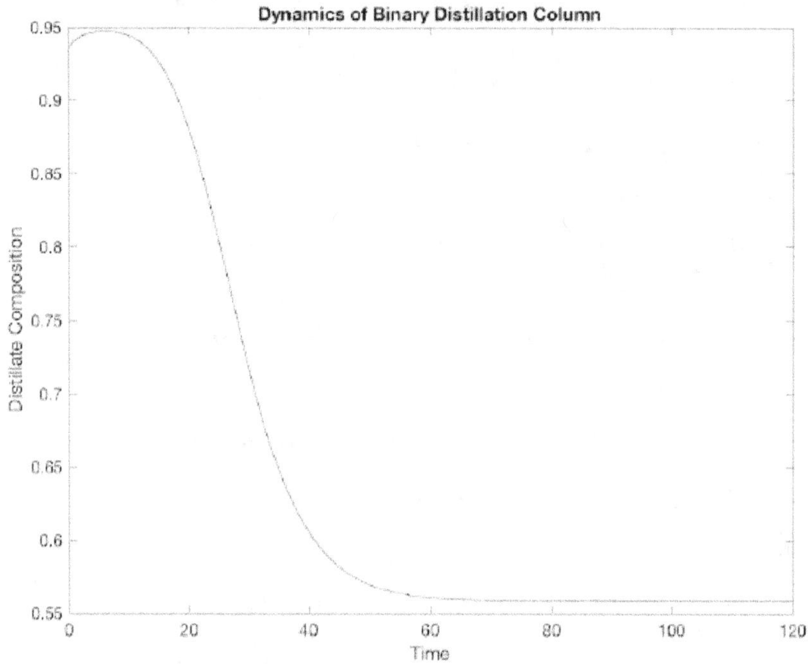

Calculations and Results

Verify the Concentration profile by hand calculations using discretization principle explained in chapter 4.

References

Hahn, J. and T.F. Edgar, An improved method for nonlinear model reduction using balancing of empirical gramians, Computers and Chemical Engineering, 26, pp.1379-1397, (2002).

William. L. Luyben., Process Modeling, Simulation and control for Chemical Engineers, McGraw Hill Publishers.

Chemical Reaction Engineering

"During this long battle over Arrhenius's theory of dissociation tremendous advances were made in chemistry and ever closer links were established between chemistry and physics to the great benefit of both sciences"

- Svante Arrhenius 1903 Nobel presentation speech when he was considered for the physics as well as for the chemistry prize in 1903. He is much better remembered today for the so-called "Arrhenius equation", but Arrhenius's Nobel citation makes no mention of that aspect of his work.

18

Plug Flow Reactor

Objective: Design plug flow reactor and calculate volume

Problem Statement:

Ethyl formate is to be produced from ethanol and formic acid in a continuous flow tubular reactor operated at a constant temperature of 303 K (30° C). The reactants will be fed to the reactor in the proportions 1 mole HCOOH: 5 moles C2H5OH at a combined flowrate of 0.0002 m3/s (0.72m3/h). The reaction will be catalysed by a small amount of sulphuric acid. At the temperature, mole ratio, and catalyst concentration to be used, the rate equation determined from small-scale batch experiments has been found to be:

$$R = kC_F^2$$

R is formic acid reactiong/(kmol/m^{3}s)

C_F is concentration of formic acid kmol/m^3, and

$k = 2.8 \times 10^{-4}$ m^3/kmol s.

The density of the mixture is 820 kg/m3 and this may be assumed constant throughout. Estimate the volume of the reactor required to convert 70 per cent of the formic acid to the ester. If the reactor consists of a pipe of 50 mm i.d. what will be the total length required? Determine also whether the flow will be laminar or turbulent and comment on the significance of this in relation to the estimate of reactor volume. The viscosity of the solution is 1.4 × 10−3 Ns/m2.

Theory

The plug flow reactor is an idealized model of a tubular reactor. Whereas the feed mixture to a CSTR reactor gets instantaneously mixed, the fluid elements entering the plug flow reactor are assumed to be unmixed in the direction of the flow. Since each element of feed spends the same time in the reactor, the plug flow reactor is also a convenient method of modeling a batch reactor (on a spatial basis instead of on a time variable basis).

Analytical Solution

Although the ethanol fed to the reactor is partly consumed in the reaction, the rate equation indicates that the rate of reaction depends only on the concentration of the formic acid.

Thus
$$R = kC_F^2$$

In this liquid phase reaction, it may be assumed that the mass density of the liquid is unaffected by the reaction, allowing the material balance for the tubular reactor to be applied on a volume basis with plug flow.

Thus
$$\frac{V_t}{v} = \int_0^{xf} \frac{dx}{R}$$

In this case: $R = kC_F^2 = k(C_0 - X)^2$

where Cv is the concentration of formic acid in the feed.

Thus,
$$\frac{V_t}{v} = \frac{1}{kC_0}\left[\frac{1}{(1-\alpha f)} - 1\right] = \frac{1}{kC_0}\left(\frac{\alpha f}{1-\alpha f}\right)$$

In terms of fractional conversion $x_f = \alpha_f C_0$, then:

$$\frac{V_t}{v} = \frac{1}{kC_0}\left[\frac{1}{(1-\alpha_f)} - 1\right] = \frac{1}{kC_0}\left(\frac{\alpha_f}{1-\alpha_f}\right)$$

For HCOOH, the molecular mass = 46 kg/kmol.

For C2H5OH, the molecular mass = 46 kg/kmol.

Thus: 1 kmol HCOOH is present in $(1 \times 46) + (5 \times 46) = 276$ kg feed mixture.

or: (276/820) = 0.337 m3 feed mixture.

Thus: C0 = (1/0.337) = 2.97 kmol/m3

The volume of the reactor required to convert a fraction of the feed of 0.7 is given by:

$$Vt = (0.72/3600)(1/(2.8 \times 10{-4} \times 2.97))(0.7/(1 - 0.7)) = 0.561 \text{ m3}$$

The equivalent length of a 50 mm ID pipe is then:

$$0.561/[(\pi/4)0.050^2] = 286 \text{ m}$$

say 29 lengths, each of 10 m length, connected by U-bends.

The mean velocity in the tube, u = $(0.72/3600)/(\pi/4)0.050^2 = 0.102$ m/s and, the

Reynolds Number is then:

$$Re = u\rho d/\mu = (0.102 \times 820 \times 0.050)/(1.4 \times 10{-}3) = 2980$$

confirming turbulent flow and the validity of the assumed plug flow.

CHEMCAD Procedure

- Launch CHEMCAD
- Select Engineering units as "Alt SI" and change the pressure unit to Kg/Cm2G.
- Draw the flow sheet as shown in the diagram (1 Feed, 1 Product Streams 1 kinetic reactor unit).
- Change the mode to "Simulation" and add components both reactants & products.
- Provide the feed stream information as mentioned in the Objective.
- Provide Kinetic Reactor information as given in the objective.
- Run the flow sheet and resolve the errors, if any.

Generate the report to know the volume of PFR.

References

Coulson and Richardson, Chemical Engineering, Vol.3, Butterworth Heinmann Publishers.

19

Batch Reactor with Series Reaction

Objective: Solve simultaneous ODE's of batch reactor and find optimum reaction time for series reaction.

Description: Considered a Batch in which A------$\rightarrow$ B ----$\rightarrow$C, series reaction takes place. Reactant A is consumed in the perfectly mixed batch reactor by a first-order reactions occurring in the liquid phase. Assume that the constant holdup and isothermal conditions.

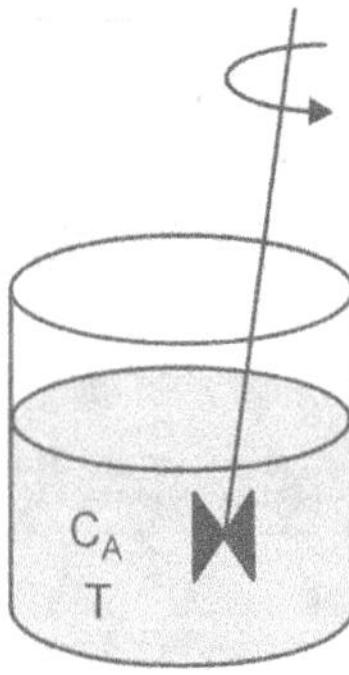

Fig. 19.1 Continuous stirred tank reactor

Modeling Equations

$$\frac{dC_a}{dt} = -k_1 C_a$$

$$\frac{dc_b}{dt} = k_1 C_a - k_2 C_b$$

$$\frac{dC_c}{dt} = k_2 C_b$$

Data:

$$K_1 = 1 \text{ hr-1} \qquad\qquad K_2 = 2 \text{ hr-1}$$

$$\text{At } t = 0 \qquad\qquad C_a = 5 \text{ mol} \qquad C_b = C_c = 0 \text{ mol}$$

MATLAB Solution for the Problem:

Write simple MATLAB function to evaluate the derivatives and save it as batch.m

Content of the function is shown here. C_a, C_b, C_c must be defined within the same array, and C_a represented by 1st element of the array i.e., c(1), C_b is represented as c(2) and C_c represented as as c(3).

```
function dcdt=batch(t,c)
%c(1)=ca, c(2)=cb, c(3)=cc
global k1 k2
dcdt=zeros(3,1);
dcdt=[(-k1*c(1));((k1*c(1))-(k2*c(2)));(k2*c(2))];
```

Execute the function to solve the differential equations from the MATLAB Command prompt by providing the initial values and the name of the function in which the differential equations are defined:

```
>> global k1 k2
>> k1=1;
>> k2=2;
>> tspan=[0 8];
>> c0=[5 0 0];
>>   [t,c]=ode45(@batch,tspan,c0);
>> plot(t,c(:,1),'+',t,c(:,2),'*',t,c(:,3),'x');
>> legend('ca','cb','cc');
>> xlabel('Time(hr)');
>> ylabel('Conc.(mol/hr)');
>> title('Batch Reactor');
>>
```

Typing t at command prompt will display all the time steps at which the derivative values are estimated.

Note that by typing c at MATLAB command prompt will display the concentrations Ca, Cb, Cc in 3 column format for various time steps.

Following graph will be obtained to explain the concentrations of various species in the reactor with time:

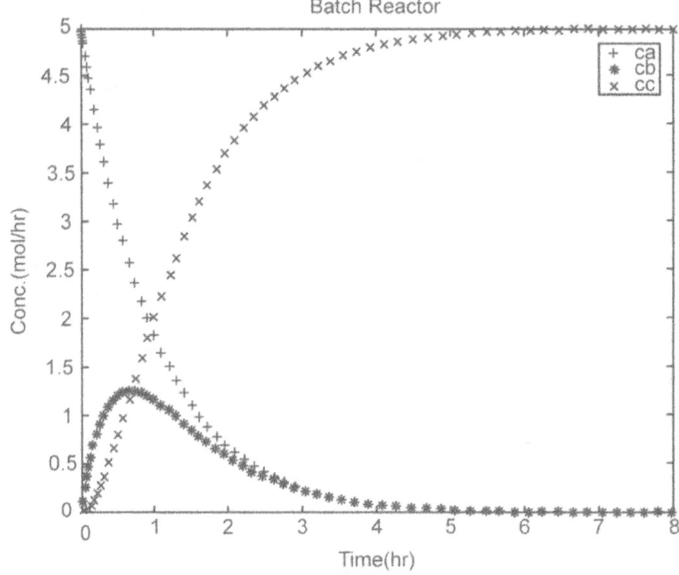

Fig. 19.2 Concentration stirred tank reactor

LineSpec: ':+r' mean the graph will be plotted as dotted line with + symbol for points and in red color.

Specifier	Line Style	Specifier	Marker	Specifier	Color
–	Solid line (default)	o	Circle	y	yellow
– –	Dashed line	+	Plus sign	m	magenta
:	Dotted line	*	Asterisk	c	cyan
– .	Dash-dot line	.	Point	r	red
		x	Cross	g	green
		s	Square	b	blue
		d	Diamond	w	white
		^	Upward-pointing triangle	k	black
		v	Downward-pointing triangle		
		>	Right-pointing triangle		
		<	Left-pointing triangle		
		p	Pentagram		
		h	Hexagram		

Various solvers available in MATLAB for ODEs:

Solver	Problem type	Accuracy	When to use
ode45	Non-stiff	Medium	Most of the time. ode45 should be the first solver you try
ode23	Non-stiff	Low	ode23 can be more efficient than ode45 at problems with crude tolerances, or in the presence of moderate stiffness
ode113	Non-stiff	Low to High	ode113 can be more efficient than ode45 at problems with stringent error tolerances, or when the ODE function is expensive to evaluate
ode15s	Stiff	Low to Medium	Try ode15s when ode45 fails or is inefficient and you suspect that the problem is stiff. Also use ode15s when solving differential algebraic equations (DAEs)
ode23s	Stiff	Low	ode23s can be more efficient than ode15s at problems with crude error tolerances. It can solve some stiff problems for which ode15s is not effective. ode23s computes the Jacobian in each step, so it is beneficial to provide the Jacobian via odeset to maximize efficiency and accuracy. If there is a mass matrix, it must be constant
ode23t	Stiff	Low	Use ode23t if the problem is only moderately stiff and you need a solution without numerical damping. ode23t can solve differential algebraic equations (DAEs)
ode23tb	Stiff	Low	Like ode23s, the ode23tbsolver might be more efficient than ode15s at problems with crude error tolerances
ode15i	Fully implicit	Low	Use ode15i for fully implicit problems $f(t,y,y') = 0$ and for differential algebraic equations (DAEs) of index 1

Verification of the Result in Excel

Concentration profile of Ca for various time intervals can be estimated from the theoretical solution for the mentioned differential equation.

	A	B	C	D	E
8					
9	t	Ca			
10	t	$5*e^{(-t)}$		Formulae Used:	
11	0	5			
12	0.5	3.032653		B12	=5*EXP(-A12)
13	1	1.839397		B13	=5*EXP(-A13)
14	1.5	1.115651		.....	
15	2	0.676676			
16	2.5	0.410425			
17	3	0.248935			
18	3.5	0.150987			
19	4	0.091578			
20	4.5	0.055545			
21	5	0.03369			
22	5.5	0.020434		B22	=5*EXP(-A22)

Note that Formula of B12 can be copied and pasted to B13 to B22 to complete the task. Formulas are always copied with relative positions to the cells.

References

http://ctms.engin.umich.edu/CTMS/index.php

20

Non-isothermal Plug Flow Reactor

Objective

Equation based modeling of a non-isothermal plug flow reactor using COMSOL and MATLAB.

Problem

Consider a reactor oxidizing SO_2 to form SO_3. The equations are given as

$$\frac{dx}{dz} = -50R' \qquad .....(20.1)$$

$$\frac{dT}{dz} = -4.1(T - T_{surr}) + 1.02*10^4 R' \qquad .....(20.2)$$

where the reaction rate R' is

$$R' = \frac{x\left|1 - 0.167(1-x)\right|^{3.5} \, 2.2(1-x)/K_{ey}}{\left|k_1 + k_2(1-x)\right|^2} \qquad .....(20.3)$$

$ln\ k_1 = -14.96 + 11{,}070/T \qquad ln\ k_2 = -1.331 + 2331/T$

$ln\ K_{eq} = -11.02 + 11{,}570/T$

With the parameters: $T_{surr} = 673.2$ K, $T(0) = 673.2$ K, $X(0) = 1$. Here, the variable X is the concentration of SO_2 divided by the inlet concentration, and 1–X is the fractional conversion, and T is the temperature in Kelvin.

Eq.(20.1) is the mole balance on SO_2 and eq.(20.2) is the energy balance. The right hand side of the eq.(20.1) represents cooling at the wall and right hand side of eq.(20.2) represents heat of reaction. Since there are two variables, we need two equations to solve the problem.

Solution using COMSOL

- Follow appendix 1 and select 0D option – mathematics – 'Global ODEs and DAEs (ge)'- study (time dependent) – done.

- (Define equation) under the model node, Right click on 'definitions tab' and select 'Variables'. As shown in the below Table.

Variables	
Name	**Expression**
k1	$\exp(-14.96+11070/T)$
k2	$\exp(-1.331 + 23331/T)$
keq	$\exp(-11.02 + 11570/T)$
num	$x*(1-0.167*(1-\ldots-2.2*(1-x)/keq$
denom	$(k1 + k2*(1-x)\verb|^|2$
Rate	num/denom

Expand 'Global ODEs and DAEs' and click on 'global equations'. Set the two eq.(20.1) and (20.2) as shown in the below Table.

Global Equations			
$f(u,u_t,u_{tt},t) = 0, \ u(t_0) = u_0 \ \ u_t(t_0) = u_{t0}$			
Name	F(u,ut,utt,t)	Initial value (u_0)	Initial…(u_t0)
X	Xt + 50*Rate	1	0
T	Tt+4,1*(T-Tsurr)-Rate *1.02e4	673.2	0

- Expand the study node, choose time dependent and note that the range of solution is from 0 to 1, as desired. Right click on study and click on compute.

- The plot should appear automatically. If not, choose Results under Model Builder; right click to 1D Plot Group. If that already exists, choose it. Right click on 1D Plot Group and choose Global. Then create the expressions to plot as shown in below figure. The results (X, T, Rate, and Keq) are all scaled so that they appear on the graph. The X-Axis Data is changed to Expression, and set to t, so that the description can be changed to length. Unfortunately, in version 4.1 the units of (s) will be displayed. In this problem there are no units in the length, which is dimensionless. It is useful to see what part of the reactor is doing the most work and to see how the equilibrium constant changes with temperature, which changes with axial position.

Expression		
Expression	**Unit**	**Description**
300*mod1.X		State variable X
mod 1.T		State variable T
3000*mod 1.Rate	1	Rate of reaction
mod 1.keq	1	Equilibrium constant

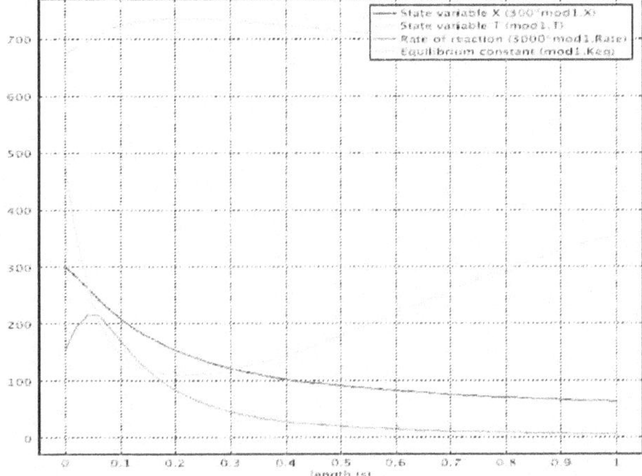

Fig. 20.1 Results for SO_2 reactor

Solution for the Problem in MATLAB:

Write simple MATLAB function to calculate the derivatives of x, T for the non-isothermal reactor and save it as pfreactor.m

Content of the function is shown here.

```
function der=pfreactor(l,var)
% This function estimates the differential equations of non-isothermal plug
% flow reactor
% l: length
% var: Variables for which the derivatives are estimated
%       var(1) is x, SO2 concentration / inlet concentration; 1-x is
%       fractional conversion
%       var(2) is temperature, K
% der: estimated derivative, This will be array- 1st column for time steps,
% 2nd and 3rd columns represented for x and T derivatives
% Data of SO2 oxidation to SO3 is considered for this example
Tsurr=673.2;
x=var(1);
T=var(2);
k1=exp(-14.96+11070/T);
k2=exp(-1.331+2331/T);
```

```
keq=exp(-11.02+11570/T);
num=x*sqrt(1-0.167*(1-x))-2.2*(1-x)/keq;
den=(k1+k2*(1-x))^2;
rate=num/den;
der(1)=-50*rate;
der(2)=-4.1*(T-Tsurr)+1.02e4*rate;
der=der';
```

Execute the function from the MATLAB Command prompt:

```
>> lspan=[0 1];
>> varinit=[1 673.2];
>> [z var]=ode45(@pfreactor, lspan,varinit)

>> plot(z, var(:,1))
>> title('Non-isothermal Plug Flow Reactor conversion study')
>> xlabel('Length');
>> ylabel('SO2 concentration/inlet concentration');
>>
```

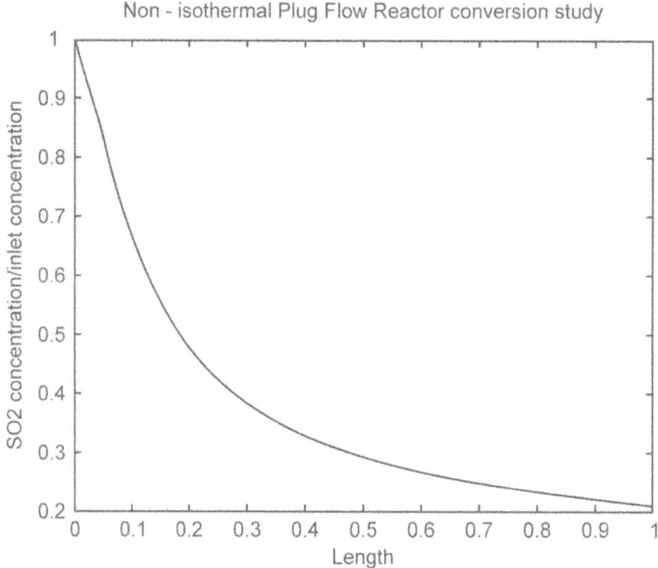

Fig. 20.2 SO2 conversion along the reactor length

```
>> plot(z, var(:,2));
>> title('Non-isothermal Plug Flow Reactor Temperature profile');
>> xlabel('Length');
>> ylabel('Temperature, K');
>>
```

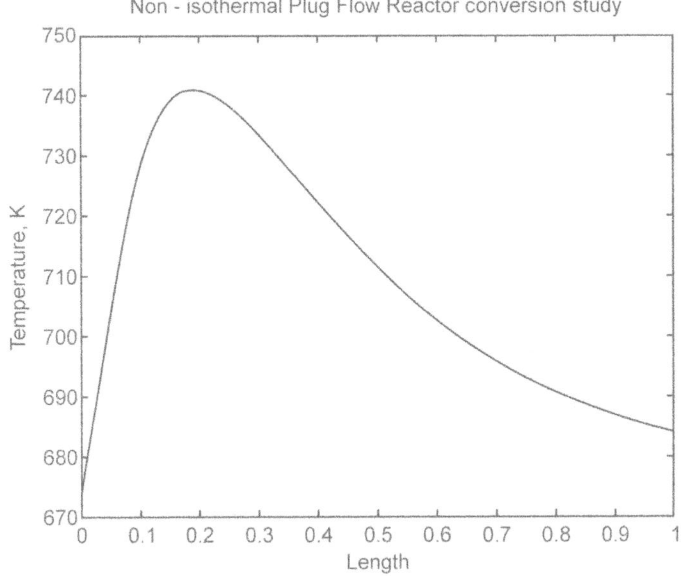

Fig. 20.3 Temperature profile along the reactor length

Verification of Results in Excel sheet:

	A	B	C	D	E	F	G	H	I	J	K	L
3	Tsurr			673.2								
4												
5	Length	x	T	k1	k2	keq	num	den	rate	dx/dz	dT/dz	MATLAB Result
6	0	1.00	673.20	4.41	8.43	8.43	1.00	19.45	0.05	-2.57	524.50	1.0000 673.2000
7	0.0195	0.95	683.43	3.45	8.00	8.00	0.93	14.92	0.06	-3.15	599.78	0.9431 684.3790
8	0.0391	0.89	695.18	2.62	7.55	7.55	0.85	12.01	0.07	-3.53	629.20	0.8763 696.6215
9	0.0586	0.82	707.45	1.99	7.13	7.13	0.75	10.73	0.07	-3.50	573.61	0.8056 708.6819
10	0.0782	0.75	718.70	1.56	6.77	6.77	0.65	10.52	0.06	-3.11	447.52	0.7377 719.2483
11	0.1032	0.67	729.88	1.23	6.44	6.44	0.54	11.12	0.05	-2.44	265.46	0.6628 729.2531
12	0.1282	0.61	736.52	1.07	6.26	6.26	0.46	12.25	0.04	-1.86	119.76	0.6012 735.7154
13	0.1532	0.57	739.51	1.01	6.18	6.18	0.39	13.64	0.03	-1.43	19.84	0.5512 739.3146
14	0.1782	0.53	740.01	1.00	6.17	6.17	0.34	15.19	0.02	-1.12	-45.03	0.5102 740.8091
15	0.2032	0.50	738.88	1.02	6.20	6.20	0.30	16.88	0.02	-0.90	-85.84	0.4759 740.8627
16	0.2282	0.48	736.74	1.07	6.25	6.25	0.27	19.69	0.01	-0.74	-110.51	0.4470 739.8575
17	0.2532	0.46	733.98	1.13	6.33	6.33	0.25	20.62	0.01	-0.61	-124.35	0.4224 738.1216
18	0.2782	0.45	730.87	1.20	6.41	6.41	0.23	22.66	0.01	-0.52	-130.91	0.4013 735.8932
19	0.3032	0.43	727.59	1.28	6.51	6.51	0.22	24.80	0.01	-0.44	-132.58	0.3829 733.3404
20	0.3282	0.42	724.28	1.38	6.60	6.60	0.21	27.05	0.01	-0.38	-130.97	0.3689 730.5935
21	0.3532	0.41	721.01	1.48	6.70	6.70	0.20	29.39	0.01	-0.34	-127.21	0.3527 727.7481
22	0.3782	0.40	717.82	1.59	6.80	6.80	0.19	31.80	0.01	-0.30	-122.05	0.3401 724.8751
23	0.4032	0.40	714.77	1.69	6.89	6.89	0.18	34.29	0.01	-0.27	-116.05	0.3288 722.0242
24	0.4282	0.39	711.87	1.80	6.98	6.98	0.18	36.82	0.00	-0.24	-109.60	0.3187 719.2320
25	0.4532	0.38	709.13	1.92	7.07	7.07	0.17	39.40	0.00	-0.22	-102.94	0.3095 716.5248
26	0.4782	0.38	706.56	2.03	7.16	7.16	0.17	41.99	0.00	-0.20	-96.28	0.3012 713.9210
27	0.5032	0.37	704.15	2.14	7.24	7.24	0.16	44.60	0.00	-0.18	-89.75	0.2935 711.4321
28	0.5282	0.37	701.91	2.25	7.32	7.32	0.16	47.20	0.00	-0.17	-83.42	0.2865 709.0651
29	0.5532	0.36	699.82	2.36	7.39	7.39	0.16	49.78	0.00	-0.16	-77.36	0.2801 706.8234
30	0.5782	0.36	697.89	2.46	7.46	7.46	0.15	52.33	0.00	-0.15	-71.61	0.2741 704.7081
31	0.6032	0.36	696.10	2.57	7.52	7.52	0.15	54.83	0.00	-0.14	-66.18	0.2685 702.7178
32	0.6282	0.35	694.44	2.67	7.58	7.58	0.15	57.28	0.00	-0.13	-61.08	0.2633 700.8497
33	0.6532	0.35	692.92	2.76	7.64	7.64	0.14	59.87	0.00	-0.12	-56.31	0.2584 699.1001
34	0.6782	0.35	691.51	2.85	7.69	7.69	0.14	61.98	0.00	-0.11	-51.87	0.2538 697.4646
35	0.7032	0.34	690.21	2.94	7.74	7.74	0.14	64.22	0.00	-0.11	-47.74	0.2495 695.9379
36	0.7282	0.34	689.02	3.02	7.78	7.78	0.14	66.38	0.00	-0.10	-43.92	0.2454 694.5148
37	0.7532	0.34	687.92	3.10	7.83	7.83	0.13	68.46	0.00	-0.10	-40.38	0.2416 693.1898
38	0.7782	0.34	686.91	3.18	7.87	7.87	0.13	70.45	0.00	-0.09	-37.12	0.2379 691.9574
39	0.8032	0.33	685.98	3.25	7.90	7.90	0.13	72.35	0.00	-0.09	-34.11	0.2344 690.8119
40	0.8282	0.33	685.13	3.31	7.93	7.93	0.13	74.16	0.00	-0.09	-31.34	0.2310 689.7480
41	0.8532	0.33	684.35	3.37	7.97	7.97	0.13	75.89	0.00	-0.08	-28.79	0.2278 688.7606
42	0.8782	0.33	683.63	3.43	7.99	7.99	0.12	77.53	0.00	-0.08	-26.45	0.2247 687.8445
43	0.9032	0.33	682.97	3.49	8.02	8.02	0.12	79.09	0.00	-0.08	-24.30	0.2217 686.9950
44	0.9282	0.32	682.36	3.54	8.05	8.05	0.12	80.56	0.00	-0.07	-22.33	0.2188 686.2074
45	0.9532	0.32	681.80	3.58	8.07	8.07	0.12	81.96	0.00	-0.07	-20.53	0.2161 685.4774
46	0.9782	0.32	681.29	3.63	8.09	8.09	0.12	83.29	0.00	-0.07	-18.87	0.2134 684.8008
47	0.9836	0.32	681.19	3.64	8.09	8.09	0.12	83.86	0.00	-0.07	-18.55	0.2128 684.6598
48	0.9891	0.32	681.08	3.65	8.10	8.10	0.12	83.83	0.00	-0.07	-18.22	0.2122 684.5211
49	0.9945	0.32	680.99	3.65	8.10	8.10	0.12	84.09	0.00	-0.07	-17.91	0.2117 684.3846
50	1	0.32	680.89	3.66	8.10	8.10	0.12	84.35	0.00	-0.07	-17.59	0.2111 684.2505

176   Computational Simulation Tools in Engineering

Note the deviations in the x and T values as compared to the MATLAB result. Excel spread sheet was using Explicit Euler method to estimate x and T where as ode45 uses R-K 4th order method (Refer Ch 32. Simulate the CSTR's in series for detailed procedure).

Formulae used for the excel calculations are shown below:

	A Length	B x	C T	D k1	E k2	F keq
6	0	1.00	673.20	=EXP(-14.96+11070/C6)	=EXP(-1.331+2331/C6)	=EXP(-1.331+2331/C6)
7	0.0195	=B6+J6*(A7-A6)	=C6+K6*(A7-A6)	==>		
8	0.0391	=B7+J7*(A8-A7)	=C7+K7*(A8-A7)	=EXP(-14.96+11070/C8)	=EXP(-1.331+2331/C8)	=EXP(-1.331+2331/C8)
47	0.9836					
48	0.9891					
49	0.9945	=B48+J48*(A49-A48)	=C48+K48*(A49-A48)	=EXP(-14.96+11070/C49)	=EXP(-1.331+2331/C49)	=EXP(-1.331+2331/C49)

	G num	H den	I rate	J dx/dz	K dT/dz
6	=B6*SQRT(1-0.167*(1-B6))-2.2*(1-B6)/F6	=(D6+E6*(1-B6))^2	=G6/H6	=-50*I6	=-4.1*(C6-673.2)+10200*I6
7					
8	=B8*SQRT(1-0.167*(1-B8))-2.2*(1-B8)/F8	=(D8+E8*(1-B8))^2	=G8/H8	=-50*I8	=-4.1*(C8-673.2)+10200*I8
47					
48					
49	=B49*SQRT(1-0.167*(1-B49))-2.2*(1-B49)/F49	=(D49+E49*(1-B49))^2	=G49/H49	=-50*I49	=-4.1*(C49-673.2)+10200*I49

References

COMSOL User Guide.

21

Non-Isothermal CSTR

Objective: Emphasizes the following principle:

Understand the step response for the flow and jacket conditions are changed.

Problem Statement: Model and solve the step change response of CSTRs in series with configuration mentioned below.

Data:

Steady state values:

Fs= 40ft3/h	V= 48ft3	C_{A0}= 0.5lb.molA/ft3
C_A = 0.2455 lb.molA/ft3	T = 600^{0}R	T_J = 594.6^{0}R
F_J= 49.9ft3/min	T_0= 530^{0}R	

Parameter values:

V_J= 3.85ft3	α = 7.08x1010 h-1	E = 30,000Btu/lbmol
R= 1.99Btu/lb.mol.^{0}R	U= 150 Btu/ h Ft^{20}R	A_H = 250ft^2
T_{J0} = 5300R	λ = –30,000Btu/ lb.mol	
Cp= 0.75 Btu/lbm)^{0}R	C_J= 1.0 Btu/ lbm.^{0}R	ρ = 50 lbm/ft^3
ρ_J= 62.3 lbm / ft3	Kc= 4 (ft3/h)/0R	T_{set}= 600^{0}R

Consider a system in which temperature can change with time. An irreversible, exothermic reaction is carried out in a single perfectly mixed CSTR as shown in Fig. 21.1 in which A→B, the reaction is nth-order in reactant A and has a heat of reaction A (Btu/lb. mol of A reacted). Negligible heat losses and constant densities are assumed. To remove the heat of reaction, a cooling jacket surrounds the reactor. Cooling water is added to the jacket at a volumetric flow rate F, and with an inlet temperature of I", . The volume of water in the jacket V, is constant. The mass of the metal walls is assumed negligible so the "thermal inertia" of the metal need not be considered.

A hydraulic- between reactor holdup and the flow out of the reactor is also needed. A level controller is assumed to change the outflow as the volume in the tank rises or falls: the higher the volume, the larger the outflow. The outflow is shut off completely when the volume drops to a minimum value Vmin

$$F = Kv(V - V\min)$$

If the reactor is a flat-bottomed vertical cylinder with diameter D and if the jacket is only around the outside, not around the bottom.

$$A_H = \frac{4}{D}V$$

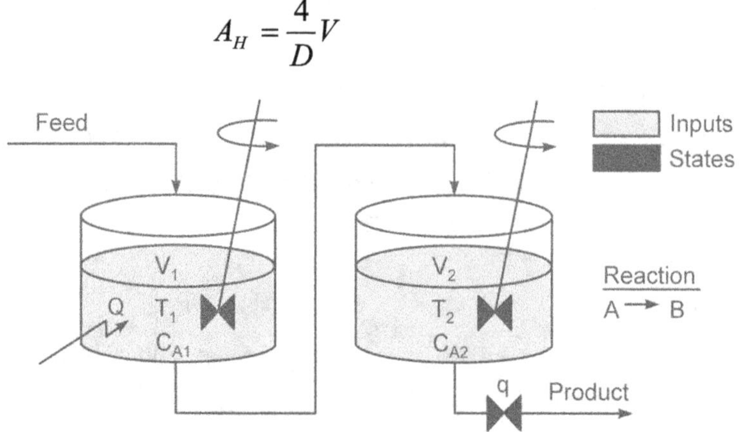

Fig. 21.1 Two CSTR's in series model

Modeling Equations:

$$\frac{dV}{dt} = F_0 - F$$

$$\frac{d(VC_A)}{dt} = F_0 C_{A0} - FC_A = V(C_A)^n \alpha e^{-E/RT}$$

$$pC_p \frac{d(VT)}{dt} = pC_p(F_0 T_0 - FT) - \lambda V(C_A)^n \alpha e^{-E/RT} - UA_E(T - T_1)$$

$$p_s V_J C_s \frac{dT_1}{dt} = F_J p_j C_J(T_{jo} - T_1) + UA_H(T - T_j)$$

MATLAB solution for the problem with following concent:

Create cstr4.m file

```matlab
% Nonlinear model of two CSTRs in series from
%
% Hahn, J. and T.F. Edgar, An improved method for nonlinear model reduction
%  using balancing of empirical gramians, Computers and Chemical
%  Engineering, 26, pp. 1379-1397, (2002)
%  t -- time (not used)
%  x -- state value vector
%  xdot -- set to the vector of state derivatives

function xdot = cstr4(t,x)

global u

% Inputs (2):
% Valve position at the outlet of the second reactor
u1 = u(1,1);
% Heat transferred to the first reactor (J/min)
u2 = u(2,1);

% States (6):
% Volume of Reactor #1 (L)
V1 = x(1,1);
% Concentration of A in Reactor #1 (mol/L)
Ca1 = x(2,1);
% Temperature of Reactor #1 (K)
T1 = x(3,1);
% Volume of Reactor #2 (L)
V2 = x(4,1);
% Concentration of A in Reactor #2 (mol/L)
Ca2 = x(5,1);
% Temperature of Reactor #2 (K)
T2 = x(6,1);

% Parameters
% Constant 1
c1 = 10;
% Constant 2
c2 = 48.1909;
% Feed Volumetric Flowrate (L/min)
qf = 100.0 ;
% Feed Concentration of A (mol/L)
Caf = 1.0;
% Feed Temperature (K)
Tf  = 350.0 ;
% Flowrate out of Reactor #1 (L/min)
q1 = c1*sqrt(V1-V2);
% Flowrate out of Reactor #2 (L/min)
q2 = c1*sqrt(V2)*u1;
% Energy Flow from Reactor #1 - Cooling (J/min)
Qc  = - c2 * u2;
% Pre-exponential Factor for A->B Arrhenius Equation
k0 = 7.2e10;
% EoverR - E/R (K) - Activation Energy (J/mol) / Gas Constant (J/mol-K)
EoverR = 1e4;
% Density of Fluid (g/L)
rho = 1000;
% Heat Capacity of Fluid (J/g-K)
Cp = 0.239;

% Heat of Reaction (J/mol)
dH = 4.78e4;

%   Dynamic Balances

%   dV1/dt
xdot(1,1) = qf - q1;
%   dCa1/dt
xdot(2,1) = qf*Caf/V1 - k0*Ca1*exp(-EoverR/T1) - q1*Ca1/V1 - Ca1/V1*xdot(1,1);
%   dT1/dt
xdot(3,1) = qf*Tf/V1 + (dH*k0/rho/Cp)*Ca1*exp(-EoverR/T1) - q1*T1/V1 + Qc - T1/V1*xdot
(1,1);
```

```
%   dV2/dt
xdot(4,1) = q1 - q2;
%   dCa2/dt
xdot(5,1) = q1*Ca1/V2 - k0*Ca2*exp(-EoverR/T2) - q2*Ca2/V2 - Ca2/V2*xdot(4,1);
%   dT2/dt
xdot(6,1)  = q1*T1/V2 + (dH*k0/rho/Cp)*Ca2*exp(-EoverR/T2) - q2*T2/V2 - T2/V2*xdot(4,1);
```

Main Program

```
% Step test for Model 4 - Two CSTRs in Series

global u

% Steady State Initial Conditions for the Controls
% Valve position at the outlet of the second reactor
u1_ss = 1.0;
% Heat transferred from the first reactor (Cooling)
u2_ss = 1.0;

% Steady State Initial Conditions for the States
V1_ss = 200.0;
CA1_ss = 0.0357;
T1_ss = 446.471;
V2_ss = 100.0;
CA2_ss = 0.0018;
T2_ss = 453.2585;
x_ss = [V1_ss;CA1_ss;T1_ss;V2_ss;CA2_ss;T2_ss];

% Open Loop Step Change
u = [2.0; 0.0];

% Final Time (min)
tf = 10;

[t,x] = ode15s('cstr4',[0 tf],x_ss)

% Volume of Reactors 1 & 2
V1 = x(:,1);
V2 = x(:,4);

% Temperature of Reactors 1 & 2
T1 = x(:,3);
T2 = x(:,6);

% Concentration of Reactors 1 & 2
C1 = x(:,2);
C2 = x(:,5);

% Plot the results
figure(1);
plot(t,V1,t,V2);
ylabel('Volume (L)');
xlabel('Time (min)');
legend('Reactor 1','Reactor 2');

figure(2)
plot(t,T1,t,T2);
ylabel('Temperature (K)');
xlabel('Time (min)');
legend('Reactor 1','Reactor 2');

figure(3)
plot(t,C1,t,C2);
ylabel('Concentration (mol/L)');
xlabel('Time (min)');
legend('Reactor 1','Reactor 2');
```

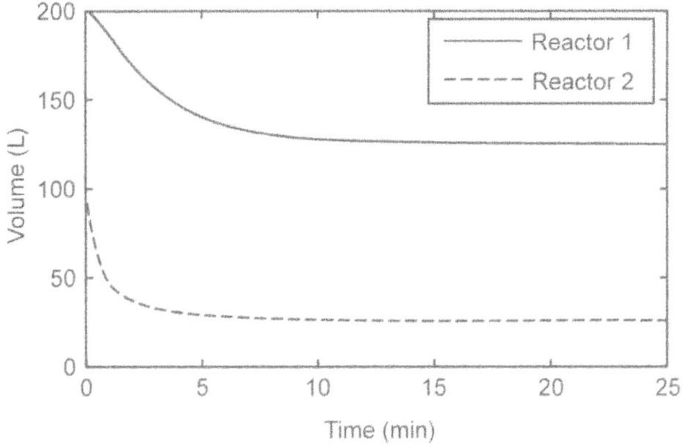

Fig. 21.2 Volume profile with time

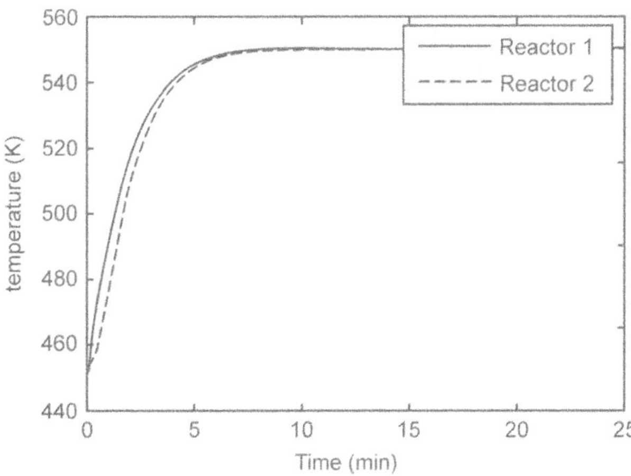

Fig. 21.3 Temperature profile with time

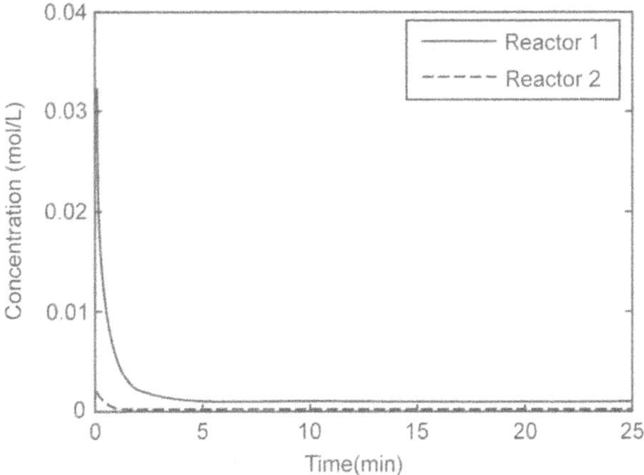

Fig. 21.4 Concentration profile with time

References

Hahn, J. and T.F. Edgar, An improved method for nonlinear model reduction using balancing of empirical gramians, Computers and Chemical Engineering, 26, pp. 1379-1397, (2002).

22

Fluidized Bed Reactor

Objective: Simulation of Fluidized Bed Oxychlorination of Ethylene Process

Process Details: The oxychlorination of ethylene is a two-step process. In the first reactor, ethylene is passed through a bed of cupric chloride particles in which ethylene is converted to ethylene dichloride. In the second reactor, the cuprous chloride generated in the first reactor is converted back into cupric chloride by reacting with a mixture of HCl and oxygen gases.

A three-phase fluidization has been assumed in which bubble phase, cloud-wake phase and emulsion phase exist. A first order reaction between cupric chloride and ethylene has been assumed. The model predicts the outlet concentration of ethylene, Ethylene Dichloride (EDC) and percentage conversion of ethylene.

$$C_2 H_4 + 2CuCl_2 \rightarrow C_2H_4Cl_2 + Cu_2 Cl_2$$

$$Cu_2 Cl_2 + 2HCl + 1/2O_2 \rightarrow 2CuCl_2 + H_2O$$

$$C_2H_4 + 1/2O_2 + 2HCl \rightarrow C_2H_4Cl_2 + H_2O \quad D H_{298}^0 = -295 \text{ kJ/mol}$$

The hydrodynamic and transport property correlations are listed in table.

Table 22.1 Hydrodynamic Parameters

Parameters	Theoretical or Empirical Correlation
Minimum Fluidization velocity (Wen and Yu, 1960)	$u_{mf} = \left(\dfrac{\mu}{d_p\rho_s}\right)\left[\left\{\left((33.7)^2 + \dfrac{0.040d_P^3\rho_g\left(\rho_s - \rho_g\right)g}{}\right)\right\}^{\frac{1}{2}} - 33.7\right]$
Viscosity of Fluidizing Gas	$\mu_g = 1.4(10^{-5})\,(Tb)^{1/2}$
Gas Density	$\rho_g = 353.2(10^{-3})/T_b$

Table 22.1 *contd...*

Parameters	Theoretical or Empirical Correlation
Average Equivalent Bubble Diameter (D_1)	$D_b = 0.43(u_o - u_{mf})0.4 \left(Z + 4\sqrt{A_0}\right)^{0.8} g^{-0.2}$
Bubble velocity	$u_b = u_0 - u_{mf} + u_{br}$, $u_{br} = 0.711 \sqrt{gD_b}$
Bubble Properties (Ratio of Cloud volume to Bubble Volume)	$f_c = \dfrac{3U_{mf}}{\epsilon_{mf}\, u_{br} - u_{mf}}$
Volume Ratio of the Cloud-Wake Phase to Bubble Phase (f_m)	$Fm = f_w + f_c$
Volume Fraction of the Bubble Phase	$\varepsilon_b = U_b / u_b$
Gas Velocity through the Cloud-Wake Phase	$U_{cw} = [(u_0 - u_{mf})/(1 + f_{cw}\,\varepsilon_{mf})]f_{cw}\,\varepsilon_{mf}$
Bed Expansion above the Height at Minimum Fluidization	$H_{if} = H_{if}/(1 - \varepsilon_b)$
Gas Interchange Coefficients	$1/k_{bc} = 1/k_{bc} + 1/k_{ce}$ $k_{bc} = 4.5U_{mf}/D_b + 5.85D_g^{1/2}\, g^{1/4}/D_b^{5/4}$ $k_{ce} = 6.78\left(\varepsilon_{mf}D_g\, u_b/D_b^3\right)^{1/2}$

The reaction exhibited first-order kinetics with regard to the concentration of cupric chloride, the dependency on ethylene concentration was interpreted by a Langmuir-Hinshelwood mechanism. However, the reaction sequence probably proceeds via chlorination of ethylene by cupric chloride. HCl and oxygen then regenerate the copper salt. The reaction rate has the following form:

$$r = \frac{KK_a C_E C_C}{1 + K_a C_E}$$

where k = 269 exp(–37.8/RT), K_c = 0.63

The mass balance for the reactant ethylene and product EDC over a differential element of height (dz) for each phase is shown in Fig.22.1.

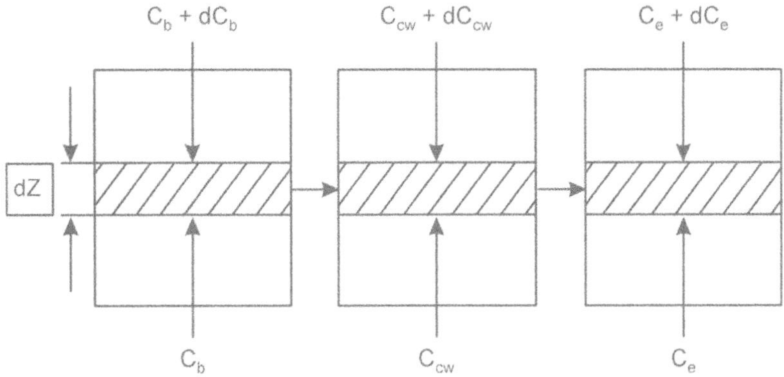

Fig. 22.1 Schematic Representation of Mass Balance

Bubble Phase

$$u_b C_{bEthylene} - U_b(C_{bEthylene} + dC_{bEthylene}) - (K_{bc})_b\, \varepsilon_b\, (C_{bEthylene} - C_{cwEthylene})\, dz = 0$$

After rearranging, it can be written as

$$\frac{dC_{bEthylene}}{dZ} + \frac{(K_{bc})_b\, \varepsilon_b}{u_b} C_{bEthylene} = \frac{(K_{bc})_b\, \varepsilon_b}{u_b} C_{cwEthylene}$$

Cloud-Wake Phase

$$u_{cw} C_{cwethylene} - u_{cw}(C_{cwethylene} + dC_{cwEthylene}) + (K_{bc})_b\, \varepsilon_b\, (C_{bEthylene} - C_{cwethylene})dz$$
$$-\varepsilon_b (K_{ce})b\, (C_{cwEthylene} - C_{eEthylene})\, dZ - Kf_{cw}\, \varepsilon_b\, C_{cwEthylene}\, dz = 0$$

After rearranging, it can be written as:

$$\frac{dC_{cwEthylene}}{dZ} + \frac{\left[(K_{bc})_b + (K_{ce})b + Kf_{cw}\right]\varepsilon_b}{u_{cw}} C_{cwEthylene}$$

$$= \frac{(K_{bc})_b\, \varepsilon_b}{u_{cw}} C_{bEthylene} + \frac{(K_{ce})_b\, \varepsilon_b}{u_{cw}} C_{bEthylene}$$

Here, K is the reaction rate constant based on the unit volume of the dense phase, i.e., emulsion phase and cloud-wake phase.

Emulsion Phase

$$U_{mf}\, C_{eEthylene} - u_{mf}(C_{eEthylene} + dC_{eEthylene}) + (K_{ce})_b\, \varepsilon_b\, (C_{cwEthylene} - C_{eEthylene})dz$$
$$-K[1 - \varepsilon_b\, (1 + f_{cw})]\, C_{eEthylene}\, dz = 0$$

After rearranging, the above equation can be written as:

$$\frac{dC_{eEthylene}}{dZ} + \frac{\left[(K_{ce})_b \, \epsilon_b + K\{1 - \epsilon_b(1 + f_{cw})\}\right]}{u_{mf}} C_{eEthylene} = \frac{(K_{ce})_b \, \epsilon_b}{u_{mf}} C_{bEthylene}$$

At the bottom of bed, i.e., at $Z = 0$, the concentration of reactants fed to each phase is same as that of the incoming feed. Hence the appropriate boundary conditions at $Z = 0$ are as follows:

$$C_{bEthylene} = C_{0Ethylene} \quad C_{cwEthylene} = C_{0Ethylene} \text{ and } C_{eEthylene} = C_{0Ethylene}$$

Bubble Phase

$$u_b C_{bEDC} - u_b(C_{bEDC} + dC_{bEDC}) - (K_{bc})b \, \epsilon_b(C_{bEDC} - C_{cwEDC}) \, dz = 0$$

After rearranging, the above equation can be written as

$$\frac{dC_{bEDC}}{dZ} + \frac{K_{bc}}{U_b} C_{bEDC} = \frac{(K_{bc})_b \, \epsilon_b}{u_b} C_{cwEDC}$$

Cloud-Wake Phase

$$U_{cw} C_{cwEDC} - u_{cw}(C_{cwEDC} + dC_{cwEDC}) + (K_{bc})_b \, \epsilon_b (C_{bEDC} - C_{cwEDC})dz$$

$$- eb(K_{ce})b \, (C_{cwEDC} - C_{eEDC}) \, dz - kf_{cw} \, \epsilon_b \, [K_d C_E C_c/(1 + K_a C_E)]dz = 0$$

After rearranging, the above equation can be written as

$$\frac{dC_{cwEDC}}{dZ} + \frac{\left[(K_{bc})_b + (K_{ce})_b\right]\epsilon_b}{u_{cw}} C_{cwEDC}$$

$$= \frac{(K_{bc})_b \, \epsilon_b}{u_b} C_{bEDC} + \frac{(K_{ce})_b \, \epsilon_b}{u_{cw}} C_{eEDC} + \frac{Kf_{cw} \epsilon_b}{u_{cw}} \left[\frac{K_a C_E C_c}{1 + K_a C_E}\right]$$

Here, k_d is adsorption equilibrium constant, cm^3/mol

Emulsion Phase

$$u_{mf} C_{eEDC} - u_{mf}(C_{eEDC} + dC_{eEDC}) + (K_{ce})_b \, \epsilon_b (C_{cwEDC} - C_{eEDC})dz$$

$$+ K[1 - \epsilon_b(1 + f_{cw})] [K_d C_E C_c/(1 + K_d C_E)] \, dz = 0$$

After rearranging the above equation can be written as

$$\frac{dC_{eEDC}}{dZ} + \frac{(K_{ce})_b \, \epsilon_b}{u_{mf}} C_{eEDC} = \frac{(K_{ce})_b \, \epsilon_b}{u_{mf}} C_{cwEDC} + \frac{K[1 - \epsilon_b(1 + f_{cw})]}{u_{mf}} \left[\frac{K_a C_E C_c}{1 + K_d C_E}\right]$$

$$C_{bEDC} = C_{cEDC} \quad C_{cwEDC} = C_{DEDC} \text{ and } CeEDC = C_{DEDC}$$

Table 22.2 Data used for Simulation

Superficial velocity (u_c)	300 cm/s
Bed temperature (T_t)	240-250 °C
Expanded bed height	72 cm
Bed height at minimum fluidizing condition (H_{mf})	36 cm
Distributor Area (A_c)	10.159 cm^2
Miminum void fraction (e_{mf})	0.55
Diameter of particles (d_p)	0.29 cm
Density of particles (ρ_c)	3.29 g/cm^3
Height above the distributor (z)	18 cm
Reaction rate constant (K)	105 cm^{-1}

Solution provided in CD/Publisher Website.

References

T. Bala Narsaiah, G. Karunya and G. Venkat Reddy, Modeling and Simulation of Fluidized Bed Oxychlorination of Ethylene Process, The IUP Journal of Chemical Engineering, Vol. IV, No. 1, 2012.

Equipment Design

"We may suppose all substances to be measured by weight or mass, [although] convenience may dictate the use of chemical equivalents"

- Gibbs 1876-1878.

23

Three Phase Separator

Objective: Design Gas – Oil – Water Separator

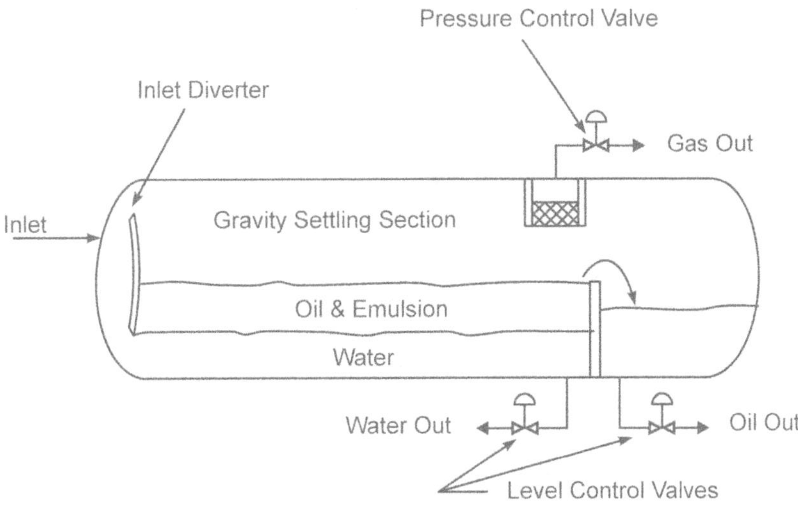

Fig. 23.1 Schematic diagram of Horizontal three phase separator

Problem Description

Determine the diameter and seam-to-seam length of a three-phase horizontal separator for the following operating conditions:

Oil production rate (Q_D):	8000 BPD
Water production rate (Q_w):	3000 BPD
Gas-oil ratio:	1000 SCF/bbl
Oil viscosity:	20 cP
Oil specific gravity:	0.89
Water specific gravity:	1.04

Gas specific gravity:	0.65
Gas compressibility:	0.89
Operating pressure:	250 psia
Operating temperature (T):	95oF
Oil retention time (t_0):	15 min
Water retention time (t_u):	10 min

Useful Formulae for Design:

the maximum allowable oil pad thickness, $H_{o,max}$ expressed as follows:

$$H0,max = \frac{1.28 \times 10^{-3} t_o (\Delta y) d_m^2}{\mu_o} \text{ in.} \qquad ....(23.1)$$

$$\frac{A_w}{A} = 0.5 \frac{Q_w t_w}{Q_o t_o + Q_w t_w} \qquad .....(23.2)$$

the maximum vessel diameter associated with the maximum oil pad height according to Eq. (12):

$$D_{max} = \frac{H_{o,max}}{H_o / D} \qquad .....(23.3)$$

The equation provides a relationship between the separator diameter and effective length as follows:

$$LD = 422 \left(\frac{Q_g TZ}{P} \right) \left[\left(\frac{\rho_g}{\rho_o - \rho_g} \right) \left(\frac{C_d}{d_m} \right) \right]^{1/2} \qquad .....(23.4)$$

where D is the separator internal diameter (in), L is the effective length of the separator (ft), T is the operating temperature $(°R)$, Z is the gas compressibility at operating pressure and temperature, P is the operating pressure ρ_g and ρ_o are the gas and oil densities, respectively (lb/ft^3), C_d is the drag coefficient, and d_m is the minimum oil droplet size to be separated from gas (μm). As d_m is normally taken as 100 μm and C_d determined by the iterative procedure described there.

$$D^2 L = 1.429(Q_o t_o + Q_w t_w) \text{ in.}^2 \text{ ft} \qquad .....(23.5)$$

Procedure

The procedure for determining the diameter and length of a three-phase horizontal separator can, therefore, be summarized in the following steps:

1. Determine the value of A_w A from Eq. (23.2).

2. Use Fig.23.1 to determine the value of H_o/D for the calculated value of A_w/A.

3. Determine the maximum oil pad thickness, $H_{o,\,max}$ from Eq.(23.1) with dm equal to 500 µm.

4. Determine D_{max} from Eq. (23.3).

5. For diameters smaller than D_{max}. Determine the combinations of D and L that satisfy the gas capacity constraint from Eq. (23.4), substituting 100 µm for d_m.

6. For diameters smaller than Dmax, determine the combinations of D and L that satisfy the retention time constraint from Eq. (23.5).

7. Compare the results obtained in steps 5 and 6 and determine whether the gas capacity or retention time (liquid capacity) governs the separator design.

8. If the gas capacity governs the design, determine the seam-to-seam length of the separator, L_s, from

$$L_s = L + \frac{D}{12} \qquad \qquad .....(23.6)$$

If the liquid retention time (liquid capacity) governs the design, determine Ls from

$$L_s = 4 + \frac{L}{3} \qquad \qquad ....(23.7)$$

9. Recommend a reasonable diameter and length with a slenderness ratio in the range of 3–5. In making the final selection, considerations such as cost and availability will be important. It should be mentioned that, in some cases, the slenderness ratio might be different from the range of 3–5. In such cases, especially when the slenderness ration is larger than 5, internal baffles should be installed to act as wave breakers in order to stabilize the gas-liquid interface.

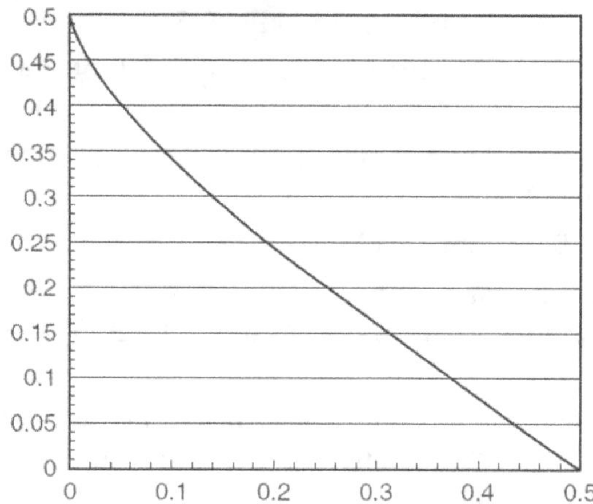

Fig. 23.2 H_o/D as a function of A_w/A

Excel solution for the given problem:

	A	B	C	D	E
1	**Input Data**				
2	Oil production rate	8000	bpd		
3	Water production rate, Qw	3000	bpd		
4	Gas–oil ratio	1000	SCF/bbl		
5	Oil viscosity, μ_o	20	cP		
6	Oil specific gravity, γ_o	0.89			
7	Water specific gravity, γ_w	1.04			
8	Gas specific gravity, γ_g	0.65			
9	Gas compressibility, Z	0.89			

10 Operating pressure — 250 psia

11 Operating temperature — 95 F

12 Oil retention time, to — 15 min

13 Water retention time, tw — 10 min

14

15 dm oil — 500 μ

16 Temperature, T — 555 R

17 Cd — 0.65

18 Water density, γW — 62.4 lb/ft3

19 dm gas — 100 μ

21 Calculations:

23 HoMax — 36 inch

$$H_{o,\max} = \frac{1.28 \times 10^{-3}\, t_o (\Delta \gamma) d_m^2}{\mu_o} \quad \text{in.}$$

26 Aw/A — 0.1

$$\frac{A_w}{A} = \frac{0.5\, Q_w t_w}{Q_o t_o + Q_w t_w}$$

x	y
Aw/A	Ho/D
0	0.5
0.1	0.338
0.2	0.225
0.3	0.155
0.4	0.08
0.5	0

30 Ho/D — 0.337

32 Dmax — 106.91 inch

$$D_{\max} = \frac{H_{o,\max}}{H_o D}$$

35 gas density — 0.888 lb/ft3

$$\rho_g = \frac{2.7\, \gamma_g P}{TZ}$$

38 oil density — 55.54 lb/ft3

$$\rho_l = \rho_w \gamma_l$$

40 Gas Production rate, Qg — 8 SCF/day

41 DL — 68.24

$$DL = 420 \left(\frac{TZQ_g}{P} \right) \left(\frac{\rho_g C_d}{d_m (\rho_o - \rho_g)} \right)^{0.5}$$

Ho/D

46 D2L

$$D^2 L = 1.429 (Q_o t_o + Q_w t_w) \quad \text{in.}^2 \text{ ft}$$

$y = -0.8935x^3 + 3.8736x^2 - 1.9824x + 0.5001$

$R^2 = 0.9956$

D^2/L	214350
thicness of cast steel	0.006
density of carbon steel	7850
unit cost /kg	200

D (in.)	D(m)	L(ft)	L(m)	Ls (=4l/3)	L/(d/12)	Volume of	ft-metre
66	1.6764	49.207989	1.24988292	65.61065	8.946907	0.039476	18649
72	1.8288	41.3483796	1.05024884	55.13117	6.891397	0.036186	17095
78	1.9812	35.2317554	0.89488659	46.97567	5.42027	0.033402	15780
84	2.1336	30.3784014	0.77161139	40.50454	4.339772	0.031016	14653
90	2.286	26.462963	0.67215926	35.28395	3.528395	0.028949	13676
96	2.4384	23.2584635	0.59076497	31.01128	2.907308	0.027139	12821
102	2.5908	20.6026528	0.52330738	27.4702	2.423842	0.025543	12067

water flow rate — 0.004140294 m^3/sec

velocity — 3 m/sec

cross sectional area — 0.001380098 m^2

Matlab Solution for the Problem:

Write MATLAB function with following content and save the file as design H Separator.m

```
function [length, dia]=designHSeparator(oilrate, waterrate,GOR, oiltime, watertime)
%  [length, dia]=designHSeparator(8000, 3000,1000, 15,  10)
spgrwater=1.04;spgroil=0.89;spgrgas=0.65;
dm=500;muoil=20;
Hmax=1.28e-3*oiltime*(spgrwater-spgroil)*dm^2/muoil;
AwAratio=0.5*waterrate*watertime/(oilrate*oiltime+waterrate*watertime);
HoDratio=-2.8333*AwAratio^3 + 3.1786*AwAratio^2 - 1.8681*AwAratio + 0.5012;
Dmax=Hmax/HoDratio;
gasrate=oilrate/GOR;
T=555;z=0.89;P=250;rhowater=62.4;Cd=0.65;
rhogas=2.7*spgrgas*P/(T*z);rhooil=rhowater*spgroil;
DL=422*(gasrate*T*z/P)*((rhogas/(rhooil-rhogas))*(Cd/dm))^0.5;
D2L=1.429*(oilrate*oiltime+waterrate*watertime);
D=D2L/DL;
dia=min(D,Dmax);
L=D2L/dia^2;
%check volumes
length=4*L/3;
```

Execution Procedure

```
>> [length, dia]=designHSeparator(8000, 3000,1000, 15,  10)

length =

   25.9964


dia =

   104.8515

>>
```

References

Abdel-Aal, Md Aggour and Fahim, Petroleum and Gas Feild Processing, Chapter 4, Marcel-Dekker Inc.

24

Control Valve Sizing

Objective: Learn various valves, sizing the valve for various scenarios.

Valve Sizing and Selection

Step 1: Define the System

The system is pumping water from one tank to another through a piping system with a total pressure drop of 150 psi. The fluid is water at 70 °F. Design (maximum) flow rate of 150 gpm, operating flow rate of 110 gpm, and a minimum flow rate of 25 gpm. The pipe diameter is 3 inches. At 70 °F, water has a specific gravity of 1.0.

Key Variables: Total pressure drop, design flow, operating flow, minimum flow, pipe diameter, specific gravity

Step 2: Define a maximum allowable pressure drop for the valve

When defining the allowable pressure drop across the valve, you should first investigate the pump. What is its maximum available head? Remember that the system pressure drop is limited by the pump. Essentially the Net Positive Suction Head Available (NPSHA) minus the Net Positive Suction Head Required (NPSHR) is the maximum available pressure drop for the valve to use and this must not be exceeded or another pump will be needed. It's important to remember the trade off, larger pressure drops increase the pumping cost (operating) and smaller pressure drops increase the valve cost because a larger valve is required (capital cost). The usual rule of thumb is that a valve should be designed to use 10-15% of the total pressure drop or 10 psi, whichever is greater. For our system, 10% of the total pressure drop is 15 psi which is what we'll use as our allowable pressure drop when the valve is wide open (the pump is our system is easily capable of the additional pressure drop).

Step 3: Calculate the valve characteristic

$$C_v = Q\sqrt{\frac{G}{\Delta P}}$$

where

Q = design flowrate (gpm)

G = specific gravity relative to water

ΔP = allowable pressure drop across wide open valve

For our system:

$$C_v = 150\sqrt{\frac{1}{15}} = 38. \cong 39$$

At this point, some people would be tempted to go to the valve charts or characteristic curves and select a valve. Don't make this mistake, instead, proceed to Step #4!

Step 4: Preliminary Valve Selection

Don't make the mistake of trying to match a valve with your calculated Cv value. The Cv value should be used as a guide in the valve selection, not a hard and fast rule. Some other considerations are,

(a) Never use a valve that is less than half the pipe size

(b) Avoid using the lower 10% and upper 20% of the valve stroke. The valve is much easier to control in the 10-80% stroke range.

Before a valve can be selected, you have to decide what type of valve will be used. For our case, we'll assume we're using an equal percentage, globe valve (equal percentage will be explained later). The valve chart for this type of valve is shown below. This is a typical chart that will be supplied by the manufacturer (as a matter of fact, it was!)

FLOW CHARAC-TERISTIC	VALVE SIZE		MAXI-MUM TRAVEL	PORT DIA.	DESIGNS ED AND ET (FLOW DOWN)					DESIGN ES (FLOW UP)				
					Valve Opening, Percent of Total Travel									
					10	30	70	100	100	10	30	70	100	100
	DIN	Inches	mm	mm	C_v				F_L	C_v				F_L
	DN 25	1, 1-1/4	19	33.3	.783	2.20	7.83	17.2	.88	.783	1.86	9.54	17.4	.95
	DN 40	1-1/2	19	47.6	1.52	3.87	17.4	35.8	.84	1.54	3.57	17.2	33.4	.94
	DN 50	2	29	58.7	1.66	4.66	25.4	59.7	.85	1.74	4.72	25.0	56.2	.92
	DN 65	2-1/2	38	73.0	3.43	10.8	49.2	99.4	.84	4.05	10.6	45.5	82.7	.93
	DN 80	3	38	87.3	4.32	10.9	66.0	136	.82	4.05	10.0	59.0	121	.89
	DN 100	4	51	111.1	5.85	18.3	125	224	.82	6.56	17.3	103	200	.91
	DN 150	6	51	177.8	12.9	43.3	239	394	.85	13.2	41.1	223	357	.86
Equal Percentage	DN 200	8	76	203.2	27.0	105	605	818	.96	25.9	97.8	618	808	.85
					X_T				...	X_T				...
	DN 25	1, 1-1/4	19	33.3	.766	.587	.743	.667	...	.754	.763	.830	.721	...
	DN 40	1-1/2	19	47.6	.780	.716	.690	.679	...	.674	.694	.698	.793	...
	DN 50	2	29	58.7	.827	.774	.702	.687	...	.863	.849	.792	.848	...
	DN 65	2-1/2	38	73.0	.778	.678	.661	.660	...	.747	.745	.783	.878	...
	DN 80	3	38	87.3	.774	.682	.663	.675	...	.768	.761	.754	.757	...
	DN 100	4	51	111.1	.731	.643	.672	.716	...	.722	.739	.718	.822	...
	DN 150	6	51	177.8	.688	.682	.736	.778	...	.723	.767	.808	.816	...
	DN 200	8	76	203.2	.844	.636	.725	.807	...	.825	.681	.735	.827	...

For our case, it appears the 2 inch valve will work well for our Cv valve at about 80-85% of the stroke range. Notice that that we're not trying to squeeze our Cv into the 11/2 valve which would need to be at 100% stroke to handle our maximum flow. If this valve were use, two consequences would be experienced: the pressure drop would be a little higher than 15 psi at our design (max) flow and the valve would be difficult to control at maximum flow. Also, there would be no room for error with this valve, but the valve we've chosen will allow for flow surges beyond the 150 gpm range with severe headaches! So we've selected a valve...but are we ready to order? Not yet, there are still some characteristics to consider.

Step 5: Check the Cv and stroke percentage at the minimum flow

If the stoke percentage falls below 10% at our minimum flow, a smaller valve may have to be used in some cases. Judgments plays role in many cases. For example, isyour system more likely to oerate closer to the maximum flowrates more often than the minimum flowrates? Or is it more likely to operate near the minimum flowrate for extended periods of time. It's difficult to find the perfect valve, but you should find one that operates well most of the time. Let's check the valve we've selected for our system.

$$C_v = 25\sqrt{\frac{1}{15}} = 65$$

Referring back to our valve chart, we see that a Cv of 6.5 would correspond to a stroke percentage of around 35-40% which is certainly acceptable. Notice that we used the maximum pressure drop of 15 psi once again in our calculation. Although the pressure drop across the valve will be lower at smaller flowrates, using the maximum valve gives us a "worst case" scenario. If our Cv at the minimum flow would have been around 1.5, there would not really be a problem because the valve has a Cv or 1.66 at 10% stroke and since we use the maximum pressure drop, our estimate is conservative. Essentially, at lower pressure drops, Cv would only increase which in this case would be advantageous.

Step 6: Check the gain across applicable flowrates

Gain is defined as:

$$\text{Gain} = \frac{\Delta \text{ Flow}}{\Delta \text{ Stro ker Travel}}$$

Now, at our three flowrate:

Q min = 25 gpm

Q op = 110 gpm

Q des = 150 gpm

We have corresponding Cv values of 6.5, 28, and 39. The corresponding stroke percentages are 35%, and 85% respectively. Now we construct the following table:

Flow (gpm)	Stroke (%)	Change in flow (gpm)	Change in Stroke (%)
25	35	110-25 = 85	73-35 = 38
110	73		
150	85	150-110 = 40	85-73 =12

Q_{min} = 25 gpm

Q_{op} = 110 gpm

Q_{des} = 150 gpm

We have corresponding Cv values of 6.5, 28, and 39. The corresponding stroke percentages are 35%, 73%, and 85% respectively. Now we construct the following table:

Flow (gpm)	Stroke (%)	Change in flow (gpm)	Change in Stroke (%)
25	35	110-25 = 85	73-35 = 38
110	73		
150	85	150-110 = 40	85-73 = 12

Gain #1 = 85/38 = 2.2

Gain #2 = 40/12 = 3.3

The difference between these values should be less than 50% of the higher value.

0.5 (3.3) = 1.65 and 3.3 – 2.2 = 1.10. Since 1.10 is less than 1.65, there should be no problem in controlling the valve. Also note that the gain should never be less than 0.50. So for our case, I believe our selected valve will do nicely!

Other Notes

Another valve characteristic that can be examined is called the choked flow. The relation uses the FL value found on the valve chart. I recommend checking the choked flow for vastly different maximum and minimum flowrates. For example if the difference between the maximum and minimum flows is above 90% of the maximum flow, you may want to check the choked flow. Usually, the rule of thumb for determining the maximum pressure drop across the valve also helps to avoid choking flow.

Selecting a Valve Type

When speaking of valves, it's easy to get lost in the terminology. Valve types are used to describe the mechanical characteristics and geometry (Ex/ gate, ball, globe valves). We'll use valve control to refer to how the valve travel or stroke (openness) relates to the flow:

1. Equal Percentage: equal increments of valve travel produce an equal percentage in flow change

2. Linear: valve travel is directly proportional to the valve stoke

3. Quick opening: large increase in flow with a small change in valve stroke. So how do you decide which valve control to use? Here are some rules of thumb for each one:

   1. Equal Percentage (most commonly used valve control)

      (a) Used in processes where large changes in pressure drop are expected

      (b) Used in processes where a small percentage of the total pressure drop is permitted by the valve

      (c) Used in temperature and pressure control loops

   2.   Linear

      (a) Used in liquid level or flow loops

      (b) Used in systems where the pressure drop across the valve is expected to remain fairly constant (ie. steady state systems)

   3. Quick Opening

      (a) Used for frequent on-off service

      (b) Used for processes where "instantly" large flow is needed (i.e. safety systems or cooling water systems)

      Now that we've covered the various types of valve control, we'll take a look at the most common valve types.

Gate Valves

Best Suited Control: Quick Opening
Recommended Uses:

1. Fully open/closed, non-throttling

2. Infrequent operation

3. Minimal fluid trapping in line

Applications: Oil, gas, air, slurries, heavy liquids, steam, noncondensing gases, and corrosive liquids

Advantages

1. High capacity
2. Tight shutoff
3. Low cost
4. Little resistance to flow

Disadvantages

1. Poor control
2. Cavitate at low pressure drops
3. Cannot be used for throttling

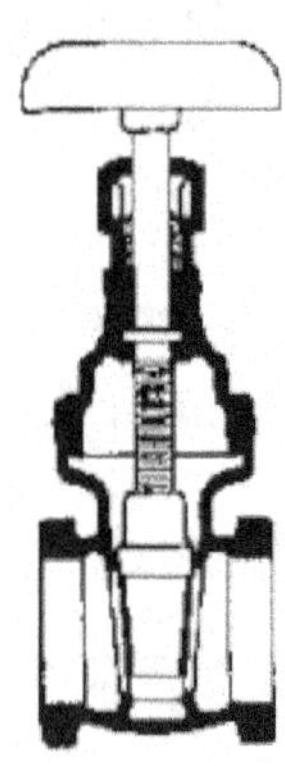

Fig. 24.1 Gate valve

Globe Valves

Best Suited Control: Linear and Equal percentage
Recommended Uses:

1. Throttling service/flow regulation
2. Frequent operation

Applications: Liquids, vapors, gases, corrosive substances, slurries

Advantages

1. Efficient throttling
2. Accurate flow control
3. Available in multiple ports

Disadvantages

1. High pressure drop
2. More expensive than other valves

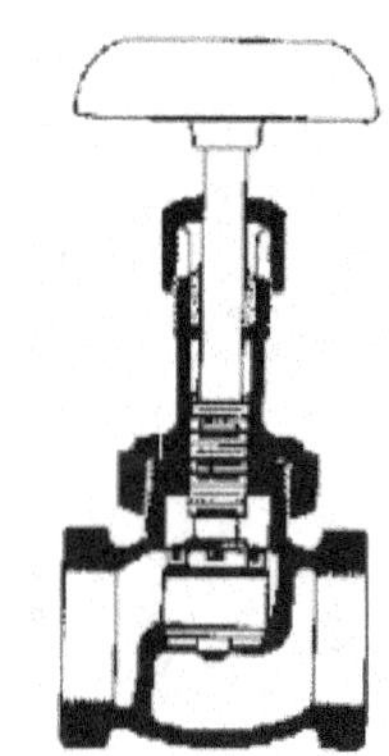

Fig. 24.2 Globe valve

Ball Valves

Best Suited Control: Quick opening, linear

Recommended Uses

1. Fully open/closed, limited-throttling
2. Higher temperature fluids

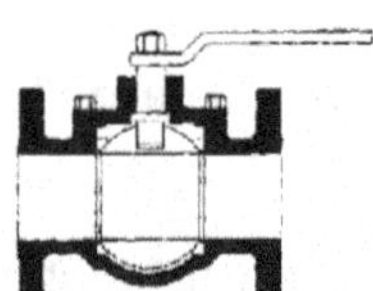

Fig. 24.3 Ball valves

Applications: Most liquids, high temperatures, slurries

Advantages

1. Low cost
2. High capacity
3. Low leakage and maintenance
4. Tight sealing with low torque

Disadvantages

1. Poor throttling characteristics
2. Prone to cavitation

Butterfly Valves

Best Suited Control: Linear, Equal percentage
Recommended uses:

1. Fully open/closed or throttling services
2. Frequent operation
3. Minimal fluid trapping in line

Applications: Liquids, gases, slurries, liquids with suspended solids

Fig. 24.4 Butterfly
valve

Advantages

1. Low cost and maintenance
2. High capacity
3. Good flow control
4. Low pressure drop

Disadvantages

1. High torque required for control
2. Prone to cavitation at lower flows

Other Valves

Another type of valve commonly used in conjunction with other valves is called a check valve. Check valves are designed to restrict the flow to one direction. If the flow reverses direction, the check valve closes. Relief valves are used to regulate the operating pressure of incompressible flow. Safety valves are used to release excess pressure in gases or compressible fluids.

References

Rosaler, Robert C., Standard Handbook of Plant Engineering, McGraw-Hill, New York, 1995, pages 10-110 through 10-122.

Purcell, Michael K., "Easily Select and Size Control Valves", Chemical Engineering Progress, March 1999, pages 45-50.

25

Single Effect Evaporator

Objective: Estimate capacity, economy of evaporator

Evaporator

Performance or Tubular Evaporator

Capacity is defined as the number of kilograms of water vaporized per hour.

Economy is the number of kilograms vaporized net kilogram or steam fed to the unit.

Enthalpy Balance for Single-Effect Evaporator

$$q_s = \dot{m}_g \left(H_s - H_c \right) = \dot{m}_s \lambda_s$$

The enthalpy balance for the steam side is

where

q_s = rate of heat transfer through heating surface from steam
H_v = specific enthalpy of steam
H_c = specific enthalpy of condensate
λ_s = latent heat of condensation of steam
m_s = rate of flow of steam

The enthalpy balance for the liquor side is $q = (\dot{m}_f - \dot{m})H_v - \dot{m}_f H_s = \dot{m}_s H$

where

q = rate of heat transfer from heating surface to liquid
H_v = specific enthalpy of vapour
H_f = specific enthalpy of thin liquor
H = specific enthalpy of thick liquor

When the heat of dilution is negligible $q = \dot{m}_f c_{pf} \left(T - T_f \right) + (\dot{m}_f - \dot{m})\lambda$

where

  q = rate of heat transfer from heating surface to liquid

  H_v = specific enthalpy of vapour

  H_f = specific enthalpy of thin liquor

  H = specific enthalpy of thick liquor

When the heat of dilution is negligible $q = \dot{m}_f c_{pf}\left(T - T_f\right) + (\dot{m}_f - \dot{m})\lambda$

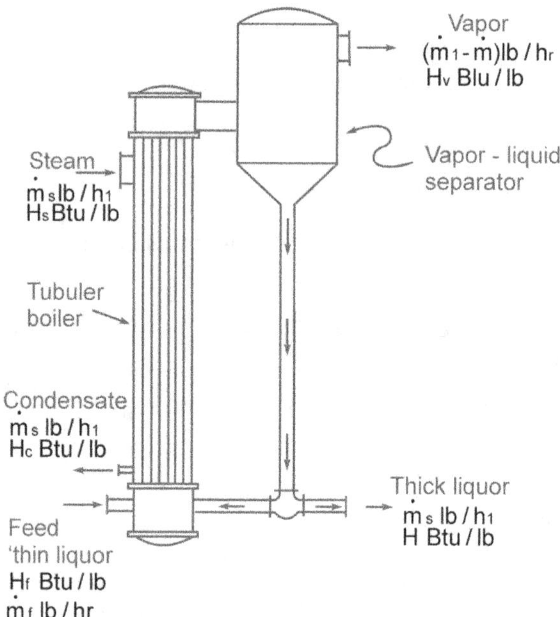

Fig. 25.1 Single effect evaporator

Solution

Mass Balance

Total feed rate, m_f = 7.5 kg/s

Total feed in the feed = 7.5 × 0.05 (as feed concentration by wt: 5%)

  = 0.375 kg/s

Total product   = 0.375/0.2 (as product conc. By weight is: 20%)

  = 1.875 kg/s

Total evaporation   = 5.625 kg/s

Energy Balance

Heat of vaporization@0.13 bar = 2200 kJ/kg

Therefore, Total heat to be supplied by steam = 5.625×2200

$$= 12375 \text{ KJ/s}$$

Latent Heat of vaporisation of Steam, λ_s $\quad = 2000 \text{ KJ/kg}$

Therefore, Stream Required $\quad = 12375/2000$

$$= 6.1875 \text{ Kg/s}$$

Steam Economy

$$= 5.625/6.1875$$

Heat transfer Area required

$$= 12375/(4750 \times (107 - 52))$$

$$= 0.0474 \text{ m}^2$$

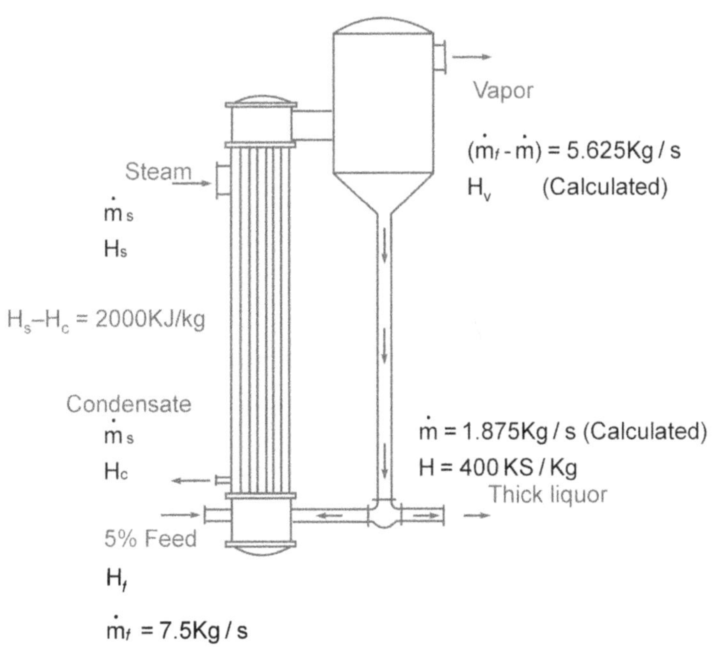

Fig. 25.2 Evaporator case study

References

McCabe, Smith and Hariott, Unit Operations of Chemical Engineering, McGraw Hill Publishers.

26

Column Design using FUG Method

Problem Statement: Design debutanizer using FUG method.

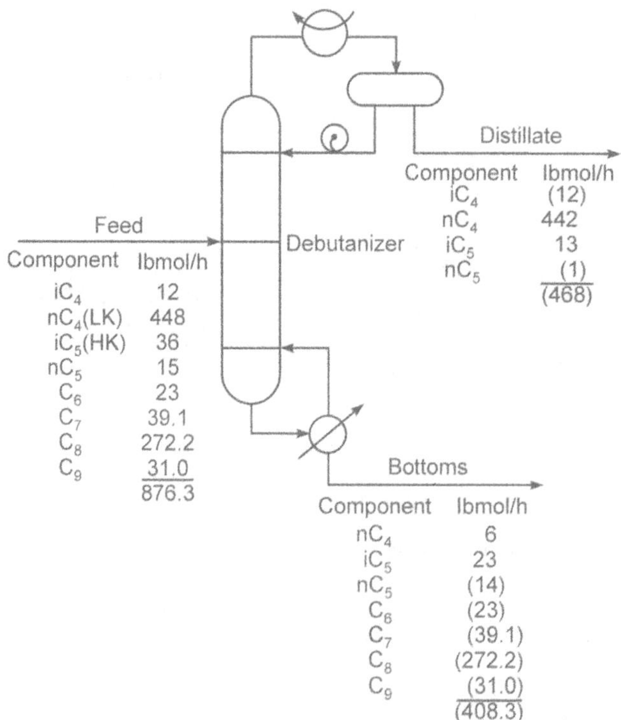

Fig. 26.1 Debutanizer flow diagram

Theory

Fenske Equation for Minimum Equilibrium Stages

For a specified separation between two key compartments of a multicomponent mixture, an exact expression is easily developed for the required minimum number of equilibrium stages, which corresponds to total reflux. This condition can be achieved in practice by changing the column with feedstock and operating it with no further input of feed and no withdrawal of distillate or bottoms, as illustrated in Fig. 26.1. To facilitate derivation of the Fenske equation, stages are numbered from the bottom up. All vapour leaving stage N is condensed and returned to stage N as reflux. All liquid leaving stage 1 is vaporized and returned to stage 1 as boilup. For steady-state operation within the column, heat input to the reboiler and heat output from the condenser are made equal (assuming no heat losses). Then by a material balance, vapour and liquid streams passing between any pair of stages have equal flow rates and compositions, for example $V_{N-1} = L_N$ and $y_{i,N-1} = x_{i,N}$. However, molar vapour and liquid flow rates will change from stage to stage unless the assumption of constant molar overflow is valid.

Derivation of an exact equation for the minimum number of equilibrium stages involves only the definition of the K-value and the mole-fraction equality between stages. For component i at stage 1 in Fig. 26.1.

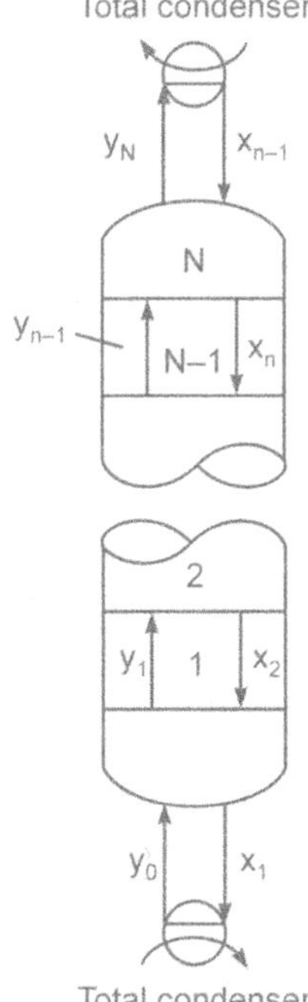

Fig. 26.2 Stage numbering convention

$$y_i l = K_{i,i} x_i, l \qquad \qquad \dots (26.1)$$

But for passing streams

$$y_i l = x_{i,2} \qquad \qquad \dots (26.2)$$

Combining these two equations,

$$x_{i,2} = K_{i,l} x_i, l \qquad \qquad \dots (26.3)$$

Similarly, for stage 2

$$y_{i,2} = K_{i,2}x_i, 2 \qquad\qquad .....(26.4)$$

Combining (26.3) and (26.4), we have

$$y_{i,2} = K_{i,2}K_{i,2}\, K_{i,1}x_{i,1} \qquad\qquad .....(26.5)$$

Eq. (26.5) is readily extended in this fashion to give

$$y_{i,N} = K_{i,N}K_{i,N-1}...K_{i,2}K_{i,1}\, x_{i,1} \qquad\qquad .....(26.6)$$

Similarly, for component j,

$$y_j,N = K_{j,N}\, K_j, N-1...\, K_{j,2}K_{j,1}x_{j,1} \qquad .....(26.7)$$

Combining (26.6) and (26.7), we find that

$$\frac{y_i,N}{y_j,N} = \alpha_N\alpha_{N-1}...\alpha_2\alpha_1\left(\frac{x_i,1}{x_j,1}\right) \qquad .....(26.8)$$

or

$$\left(\frac{x_i,N+1}{x_i,1}\right)\left(\frac{x_j,1}{x_j,N+1}\right) = \prod_{k=1}^{N_{min}} \alpha_k \qquad .....(26.9)$$

where $a_k = K_{i,k}/K_{j,k}$, the relative volatility between components i and j, Eq. (26.9) relates the relative enrichments of any two components i and j over a cascade of N the two components. Although (26.9) is exact, it is rarely used in practice because the conditions of each stage must be known to compute the set of relative volatilities. However, if the relative volatility is assumed constant, (26.9) simplifies to

$$\left(\frac{x_i,N+1}{x_i,1}\right)\left(\frac{x_j,1}{x_j,N+1}\right) = \alpha^N \qquad .....(26.10)$$

or

$$N_{min} = \frac{\log\left\{\left[(x_i,N+1)/x_i 1\right]\left[x_j,N+1\right]\right\}}{\log\alpha_i,j} .....(26.11)$$

Eq.(26.11) is extremely useful. It is referred to as the Fenske eq.(26.5). When i = the light key (LK) and j = heavy key (HK), the minimum number of equilibrium stages is influenced by the nonkey components only by their effect (if any) on the value of the relative volatility between the key components.

Eq.(26.11) permits a rapid estimation of minimum equilibrium stages. A more convenient form of (26.11) is obtained by replacing the product of the mole-fraction ratios by the equivalent product of mole-distribution ratios in terms of component distillate and bottoms flow rates d and b, respectively, and by replacing the relative volatility by a geometric mean of the top-stage and bottom-stage values. Thus,

$$N_{min} = \frac{\log\left[\left(d_i / d_j\right)\left(b_j / b_i\right)\right]}{\log \alpha_m} \qquad .....(26.12)$$

where the mean relative volatility is approximated by

$$\alpha_m = [(a_{i,j})N(\alpha_{i,j})_1]^{1/2} \qquad .....(26.13)$$

Thus, the minimum number of equilibrium stages depends on the degree of separation of the two key components and their relative volatility, but is independent of fee-phase condition. Eq. (26.12) in combination with (26.13) is exact for two minimum stages. For one stage, it is equivalent to the equilibrium-flash equation. In practice, distillation columns are designed for separations corresponding to as many as 150 minimum equilibrium stages.

The two key components are n-butane and isopentane. Distillate and bottoms conditions based on the estimated product distributions for nonkey components in Fig.(26.3) are

Component	$x_{N+1} = x_D$	$x_1 = x_B$
iC_4	0.0256	-0
$_nC_4$ (LK)	0.9445	0.0147
$_iC_5$ (HK)	0.0278	0.0563
$_nC_5$	0.0021	0.0343
$_nC_6$	-0	0.0563
$_nC_7$	-0	0.0958
$_nC_8$	-0	0.6667
$_nC_9$	-0	0.759
	1.0000	1.0000

From Fig.(26.2), at 123 °F, the assumed top-stage temperature is

$$(\alpha_n C_4 / C_5)_t = 5.20/3.60 = 1.44$$

From (26.13),

$$\alpha_m = [(2.08)(1.44)]^{1/2} = 1.73$$

Noting that $(d_i/d_j) = (x_{Di}/x_{Dj})$ and $(b_i/b_j) = (x_{Bi}/x_{Bj})$, (26.12) becomes

$$N_{min} = \frac{\log[(0.9445/0.0278)(0.0563/0.0147)]}{\log 1.73} = 8.88 \text{ stages}$$

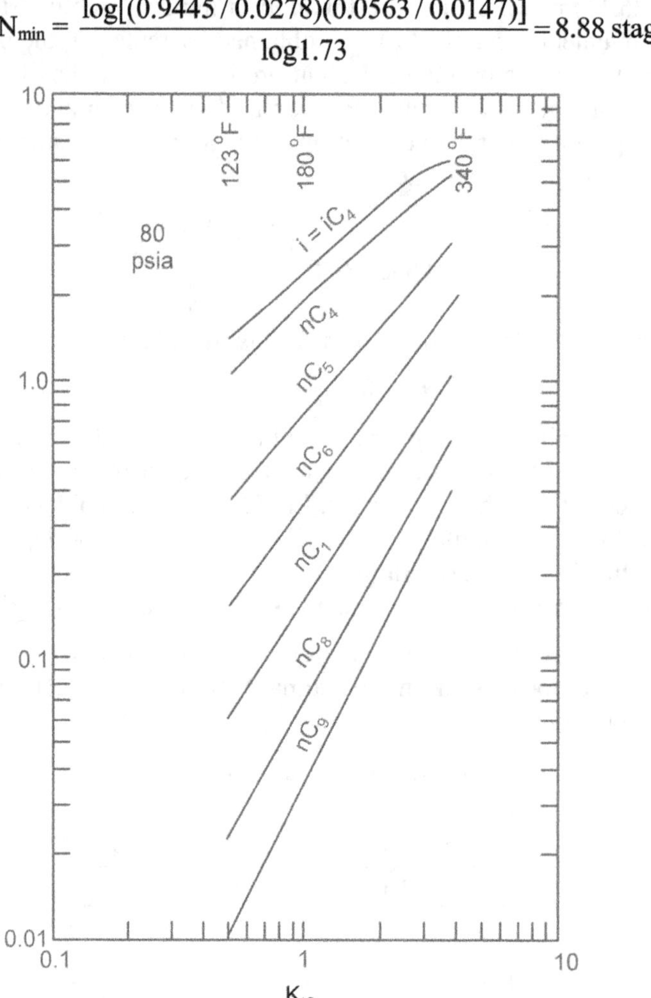

Fig. 26.3 Ideal K-values for hydrocarbons at 80 psia

The line drawn through the data represents the equation developed by Molokanov et al. (26.17).

$$Y = \frac{N - N_{min}}{N+1} = 1 - \exp\left[\left(\frac{1+54.4X}{11+117.2X}\right)\left(\frac{X-1}{X^{0.5}}\right)\right] \qquad .....(26.14)$$

where

$$X = \frac{R - R_{min}}{R+1}$$

This equation satisfies the end points $(Y = 0, X = 1)$ and $(Y = 1, X = 0)$. At a value of R/R_{min} near the optimum of 1.3, Fig. 26.10 predicts an optimal ratio for N/N_{min} of approximately 2. The value of N includes one stage for a partial reboiler and one stage for a partial condenser, if any.

Use the Gilliland correlation to estimate the theoretical-stage requirements for the debutanizer of examples 26.1, 26.2 and 26.5 for an external reflux of 379.6 lbmol/h (30% greater than the exact value of the minimum-reflux rate from bachelor).

Solution

From the examples cited, values of R_{min} and $[(R - R_{min})]/(R + 1)]$ are obtained using a distillate rate from example 9.5 of 469.56 lbmol/h. Thus, $R = 379.6/469.56 = 0.808$. With $N_{min} = 8.88$.

$$R_{min} = 0.479 \text{ and } X = \frac{R - R_{min}}{R + 1} = X = 0.182$$

From (26.34),

$$\frac{N - N_{min}}{N + 1} = 1 - \exp\left[\left(\frac{1 + 54.4(0.182)}{11 + 117.2(0.182)}\right)\left(\frac{0.182 - 1}{0.182^{0.5}}\right)\right]$$

$$= 0.476$$

$$N = \frac{8.88 + 0.476}{1 - 0.476} = 17.85$$

$$N - 1 = 16.85$$

where $N-1$ corresponds to the equilibrium stages in the tower allowing one theoretical stage for the reboiler, but no stage for the total condenser.

It should be kept in mind that, had the exact value of R_{min} not been known and a value of R equal to 1.3 times R_{min} from the underwood method been used, the value of R would have been 292 lbmol/h. But this, by coincidence, is only the true minimum reflux. Therefore the desired separation would not be achieved.

Solution provided in CD/Publisher Website

References

Seader and Henley, Separation process principles, 2nd Edition, John Wieley and Sons, Inc.

27

Shell and Tube Heat Exchanger Design using Kern Method

Problem Statement

Design an exchanger to sub-cool condensate from a methanol condenser from 95 °C to 40 °C. Flow-rate of methanol 100,000 kg/h. Brackish water will be used as the coolant with a temperature rise from 25° to 40°C.

Shell and Tube Heat Exchangers

Fluid allocation guidelines: more corrosive fluid, The fluid that has the greatest tendency to foul the surface, fluid temperatures are high enough to require the use of special alloys, higher pressure stream, fluid with the lowest allowable pressure drop, low viscous fluid, higher flow rate fluid should be taken on tube side.

Tube wall conditions affecting overall heat transfer and associated temperature profile

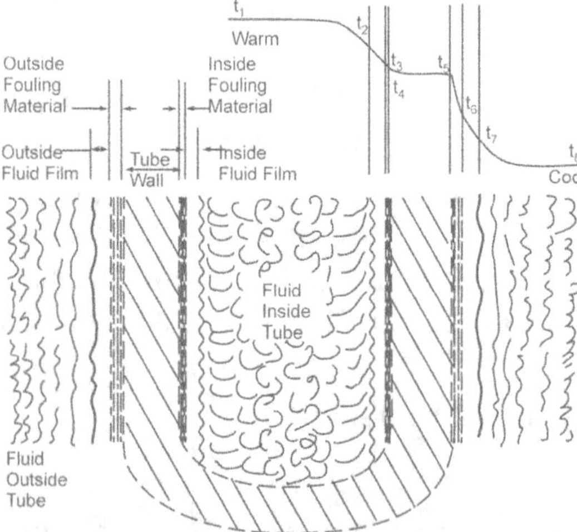

Fig. 27.1 Fluid flows through two tube passes: part of flow is parallel to shell-side fluid, and part is counterflow

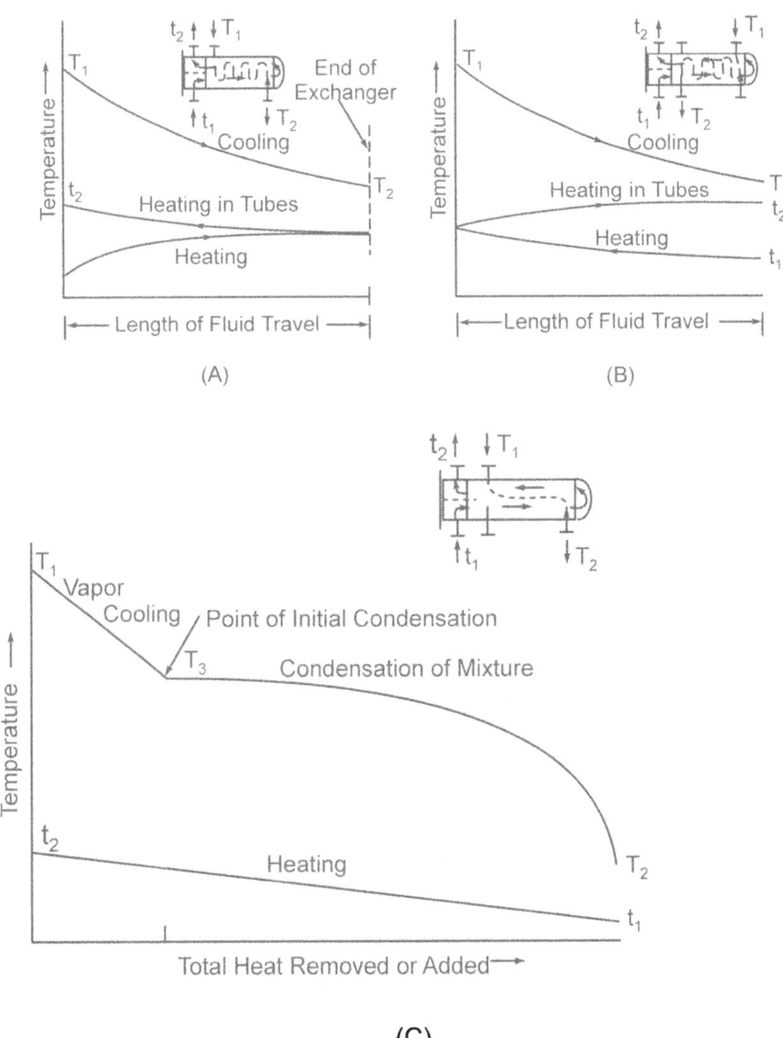

Fig. 27.2

Nomenclature for Heat Exchanger Components. *Standards of Tubular Exchanger Manufacturers Association*, 7th Ed., Fig. N—1.2, © 1988.

Heat Exchangers
Selection Guide Heat Exchanger Types

Type Designation	Significant Feature	Applications Best Suited	Limitations	Approximate Relative Cost in Carbon Steel Construction
Fixed Tube Sheet	Both tube sheets fixed to shell.	Condensers: liquid-liquid: gas-gas: gas-liquid: cooling and heating horizontal or vertical, reboiling.	Temperature difference at extremes of about 200oF due to differential expansion.	1.0
Floating Head or Tube sheet (removable and no removable bundles)	One tube sheet "floats" in shell or with shell, tube bundle may or may not be removable from shell, but back cover can be removed to expose tube ends:	High temperature differentials, above about 200oF extremes: dirty fluids requiring cleaning of inside as well as outside of shell, horizontal or vertical	Internal gaskets offer danger of leaking. Corrosiveness of fluids on shell-side floating parts. Usually confined to horizontal units.	1.28
U-Tube: U-Bundle	Only one tube sheet required. Tubes bent in U-shape. Bundle is removable.	High temperature differentials, which might require provision for expansion in fixed tube units. Clean service or easily cleaned conditions on both tube side and shell side. Horizontal or vertical.	Bends must be carefully made, or mechanical damage and danger of rupture can result. Tube side velocities can cause erosion of inside of bends. Fluid shout be free of suspended particles.	0.9–1.1
vKettle	Tube bundle removable as U-type of Floating head. Shell enlarged to allow boiling and vapor disengaging.	Boiling fluid on shell side, as refrigerant, or process fluid being vaporized. Chilling or cooling of rube-side fluid in refrigerant evaporating on shell side.	For horizontal installation. Physically large for other applications.	1.2-14

Table *Contd...*

Type Designation	Significant Feature	Applications Best Suited	Limitations	Approximate Relative Cost in Carbon Steel Construction
Double Pipe	Each tube has own shell forming annular space for shell-side fluid. Usually use externally finned tube.	Relatively small transfer area service or in banks for larger applications. Especially suited for high pressures in tube (greater than 400 psig).	Services suitable for finned tube. Piping-up a large number often requires cost and space.	0.8-1.4
Pipe Coil	Pipe coal for submersion in coil-box of water or sprayed with water is simplest type of exchanger.	Condensing, or relatively low heat loads on sensible transfer.	Transfer coefficient is low, requires relatively large space if heat load is high.	0.5-0.7
Open Tube Sections (water cooled)	Tubes require no shell, only end headers, usually long water sprays over surface, sheds scales on outside tubes by expansion and contraction. Can also be used in water box.	Condensing, relatively low heat loads on sensible transfer.	Transfer coefficient is low, takes up less space than pipe coil.	0.8-11
Open Tube Section (air cooled): Plain or Finned Tubes	No shell required, only end headers similar to water units.	Condensing, high-level heat transfer.	Transfer coefficient is low, if natural convection circulation, but is improved with forced air flow across tubes.	0.8-1.8
Plate and Frame	Composed of metal-formed thin plates separated by gaskets. Compact, easy to clean.	Viscous fluids, corrosive fluids slurries, high heat transfer.	Now well suited for boiling or condensing limit 350-500°F by gaskets. Used for liquid-liquid only: not gas-gas	0.8-1.5
Small-tube Teflon	Chemical resistance of tubes: no tube fouling.	Clean fluids, condensing, cross-exchange.	Low heat transfer coefficient.	2.0-4.0
Spiral	Compact, concentric plates; no bypassing high turbulence.	Cross-flow, condensing, heating.	Process corrosion, suspended materials.	0.8-1.5

Design Procedure

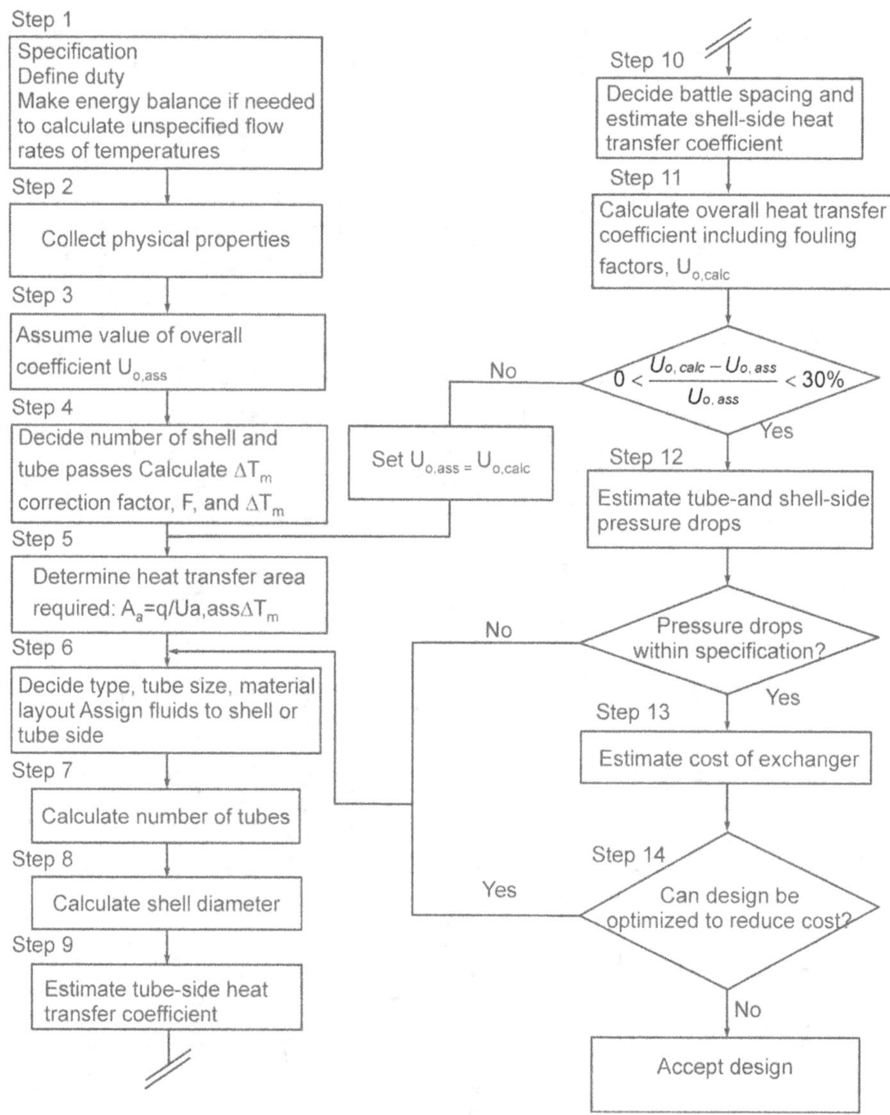

Fig. 27.3 Heat exchanger design flow chart

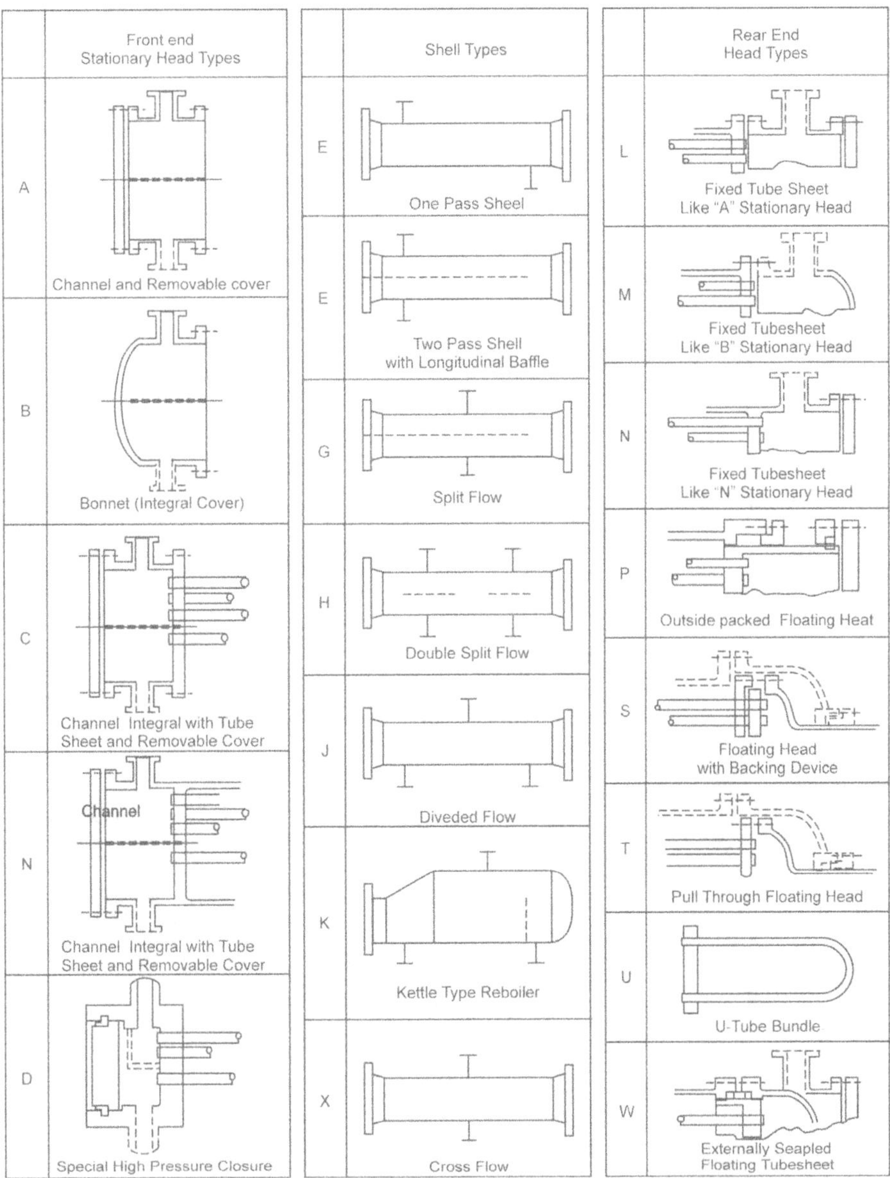

Fig. 27.4 Heat exchanger with floating head (two tube-pass, one shell-pass)

LMTD

$$\Delta T = \frac{(T_2 - t_1) - (T_1 - t_2)}{\ln\left(\dfrac{T_2 - t_1}{T_1 - t_2}\right)} = \frac{GTD - LTD}{\ln\dfrac{GTD}{LTD}}$$

where

GTD _ Greater Terminal Temperature Difference, °F

LTD _ Lesser Terminal Temperature Difference, °F

LMTD _ Logarithmic Mean Temperature Difference, °F

T_1 _ Inlet temperature of hot fluid, °F

T_2 _ Outlet temperature of hot fluid, °F

t_1 _ Inlet temperature of cold fluid, °F

t_2 _ Outlet temperature of cold fluid, °F

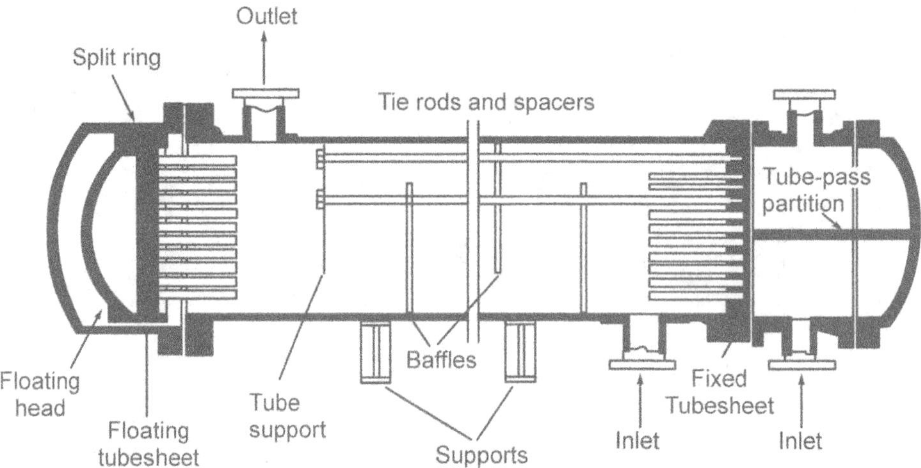

Fig. 27.5 Typical shell and Tube Heat Exchanger

Design Procedure

Coolant is corrosive, so assign to tube-side

Heat capacity methanol = 2.84 kJ/kg °C

$$\text{Heat load} = \frac{100.000}{3600} \times 2.84(95 - 40) = 4340 \text{ kw}$$

Heat capacity water = 4.2 kJ/kg °C

$$\text{Cooling water flow} = \frac{4340}{4.2(40-25)} = 68.9 \text{ kg/s}$$

$$\Delta T_{lm} = \frac{(95-40)-(40-25)}{ln\dfrac{(95-40)}{(40-25)}} = 31 \text{ °C} \qquad .....(27.1)$$

Use one shell pass and two tube passes

$$R = \frac{95-40}{40-25} = 3.67 \qquad .....(27.2)$$

$$S = \frac{40-25}{95-25} = 0.21 \qquad .....(27.3)$$

Assume $\quad F_t = 0.85$

$$\Delta T_m = 0.85 \times 31 = 26°C$$

$$U = 600 \text{ W/m}^2 \text{ °C} \quad \text{(Assumed)}$$

Provisional area

$$A = \frac{4340 \times 10^3}{26 \times 600} = 278 \text{ m}^2 \qquad .....(27.4)$$

Choose 20 mm o.d., 16 mm i.d., 4.88 m long tubes ($\frac{3}{4}$ in. 16 ft), cupro-nickel. Allowing for tube-sheet thickness, take

$$L = 4.83 \text{ m}$$

Area of one tube $= 4.83 \times 20 \times 10^{-3} \pi = 0.303 \text{ m}^2$

$$\text{Number of tubes} = \frac{278}{0.303} = 918$$

As the shell-side fluid is relatively clean use 1.25 triangular pitch

$$\text{Bundle diameter } D_b = 20\left(\frac{918}{0.249}\right)^{1/2.207} = 8.26 \text{ mm} \qquad .....(27.5)$$

Use a split-ring floating head type

bundle diametrical clearance = 68 mm (Assumed)

Shell diameter, $D_s = 826 + 68 = 894$ mm

(**Note:** nearest standard pipe sizes ae 863.6 or 914.4 mm)

Tube-side coefficient

$$\text{Mean water temperature} = \frac{40+25}{2} = 33\ ^\circ\text{C}$$

$$\text{Tube cross-sectional area} = \frac{\pi}{4} \times 16^2 = 201\ \text{mm}^2$$

$$\text{Tube per pass} = \frac{918}{2} = 459$$

$$\text{Total flow area} = 459 \times 201 \times 10^{-6} = 0.092\ \text{m}^2$$

$$\text{Water mass velocity} = \frac{68.9}{0.092} = 749\ \text{kg/s m}^2$$

$$\text{Density water} = 995\ \text{kg/m}^3$$

$$\text{Water linear velocity} = \frac{749}{995} = 0.75\ \text{m/s}$$

$$h_i = \frac{4200(1.35 + 0.02 \times 33)0.75^{0.8}}{16^{0.2}} = 3852\ \text{W/m}^2\ ^\circ\text{C} \quad .....(27.6)$$

The coefficient can also be calculated using the following equation

$$\frac{h_i d_i}{k_f} = j_h\, \text{Re}\, \text{Pr}^{0.33}\left(\frac{\mu}{\mu_w}\right)^{0.14}$$

Viscosity of water $= 0.8\ \text{mNs/m}^2$

Thermal conductivity $= 0.59\ \text{W/m}\ ^\circ\text{C}$

$$R_c = \frac{\rho u d_i}{\mu} = \frac{995 \times 0.75 \times 16 \times 10^{-3}}{0.8 \times 10^{-3}} = 14.925$$

$$P_r = \frac{C_p \mu}{k_f} = \frac{4.2 \times 10^3 \times 0.8 \times 10^{-3}}{0.59} = 5.7$$

$$\text{Neglect}\left(\frac{\mu}{\mu_w}\right)$$

$$\frac{L}{d_i} = \frac{4.83 \times 10^3}{16} = 302$$

$$j_h = 3.9 \times 10^{-3}\ \text{(Assumed)}$$

$$h_i = \frac{0.59}{16 \times 10^{-3}} \times 3.9 \times 10^{-3} \times 14.925 \times 5.7^{0.33} = 3812 \text{ W/m}^2 \, ^\circ C$$

Shell-side coefficient

Choose baffle spacing $= \dfrac{D_s}{5} = \dfrac{894}{5} = 178$ mm

Tube pitch $= 1.25 \times 20 = 25$ mm

Cross-flow area $A_s = \dfrac{(25-20)}{25} \, 894 \times 178 \times 10^{-6} = 0.032 \text{ m}^2$ $\qquad$ .....(27.7)

Max velocity, $G_s = \dfrac{100.000}{3600} \times \dfrac{1}{0.032} = 868 \text{ kg/s m}^2$

Equivalent diameter $d_e = \dfrac{1.1}{20} \, (25^2 - 0.917 \times 20^2) = 14.4$ mm $\qquad$ ......(27.8)

Mean shell side temperature $= \dfrac{95+40}{2} = 68 \, ^\circ C$

Methanol density $= 750 \text{ kg/m}^3$

Viscosity $= 0.34 \text{ mNs/m}^2$

Heat capacity $= 2.84 \text{ kJ/kg } ^\circ C$

Thermal conductivity $= 0.19 \text{ W/m } ^\circ C$

$$R_c = \frac{G_s d_e}{\mu} = \frac{868 \times 14.4 \times 10^{-3}}{0.34 \times 10^{-3}} \qquad .....(27.9)$$

$$P_r = \frac{C_p \mu}{k_f} = \frac{2.84 \times 10^3 \times 0.34 \times 10^{-3}}{0.19} = 5.1$$

Choose 25 percent baffle cut,

$\quad J_h = 3.3 \times 10^{-3}$ (Assumed)

Without the viscosity correction term

$$h_s = \frac{0.19}{14.4 \times 10^{-3}} \times 3.3 \times 10^{-3} \times 36.762 \times 5.1^{1/3} = 2740 \text{ W/m}^2 \, ^\circ C$$

Estimate wall temperature

Mean temperature differences $= 68 - 33 = 35 \, ^\circ C$

Across all resistances

Across methanol film $= \dfrac{U}{h_o} \times \Delta T = \dfrac{600}{2740} \times 35\ = 8\ ^{\circ}C$

Mean wall temperature $= 68 - 8 = 60\ ^{\circ}C$

$\mu_w = 0.37\ \text{mNs/m}^2$

$\left(\dfrac{\mu}{\mu_w}\right)^{0.14} = 0.99$

which shows that the correction for a low-viscosity fluid is not significant

Overall Coefficient

Thermal conductivity of cupro-nickel alloys $= 50$ W/m $^{\circ}$C

Take the fouling coefficients from for methanol (light organic) as $5000\ \text{Wm}^{-2}\ ^{\circ}\text{C}^{-1}$, brackish water (sea water), take as highest value $3000\ \text{Wm}^{-2}\ ^{\circ}\text{C}^{-1}$

$$\frac{1}{U_o} = \frac{1}{2740} + \frac{1}{5000} + \frac{20 \times 10^{-3} \ln\left(\dfrac{20}{16}\right)}{2 \times 50} + \frac{20}{16} \times \frac{1}{3000} + \frac{20}{16} \times \frac{1}{3812} \quad .....(27.10)$$

$U_o = 738$ W/m^2 $^{\circ}$C

well above assumed value of 600 W/m^2 $^{\circ}$C

Pressure drop

Tube-side

From Fig. 12.24, for Re $= 14,925$

$J_f = 4.3 \times 10^{-3}$

Neglecting the viscosity correction term

$$\Delta P = 2\left(8 \times 4.3 \times 10^{-3}\left(\frac{4.83 \times 10^3}{16}\right) + 2.5\right)\frac{995 \times 0.75^2}{2} \quad .....(27.11)$$

$= 7211 \ N/m^2 = 7.2 \ kPs \ (1.1 \ Psi)$

low, could consider increasing the number of tube passes

Shell Side

$$\text{Linear velocity} = \frac{G_s}{\rho} = \frac{868}{750} = 1.16 \ m/s$$

at Re = 36.762

$$J_f = 4 \times 10^{-2}$$

Neglect viscosity correction

$$\Delta p_s = 8 \times 4 \times 10^{-2} \left(\frac{894}{14.4} \right) \left(\frac{4.83 \times 10^3}{178} \right) \frac{750 \times 1.16^2}{2} \qquad \qquad .....(27.12)$$

$$= 272.019 \ N/m^2$$

$$= 272 \ kPa \ (39 \ psi) \ \text{too high}$$

could be reduced by increasing the baffle pitch. Doubling the pitch halves the shell-side velocity, which reduces the pressure drop by a factor of approximately $(1/2)^2$.

$$\Delta p_s = \frac{272}{4} = 68 \ kPa \ (10 \ psi), \ \text{acceptable}$$

This will reduce the shell-side heat-transfer coefficient by a factor of $(1/2)^{0.8}$ $(h_e \ \alpha \ Re^{0.8} \ \alpha \ u_s^{0.8})$

$$h_o = 2740 \times \left(\frac{1}{2} \right)^{0.8} = 1573 \ W/m^2 \ ^{\circ}C$$

This gives an overall coefficient of 615 $W/m^2 \ ^{\circ}C$ still above assumed value of 600 $W/m^2 \ ^{\circ}C$.

Excel Solution for the problem

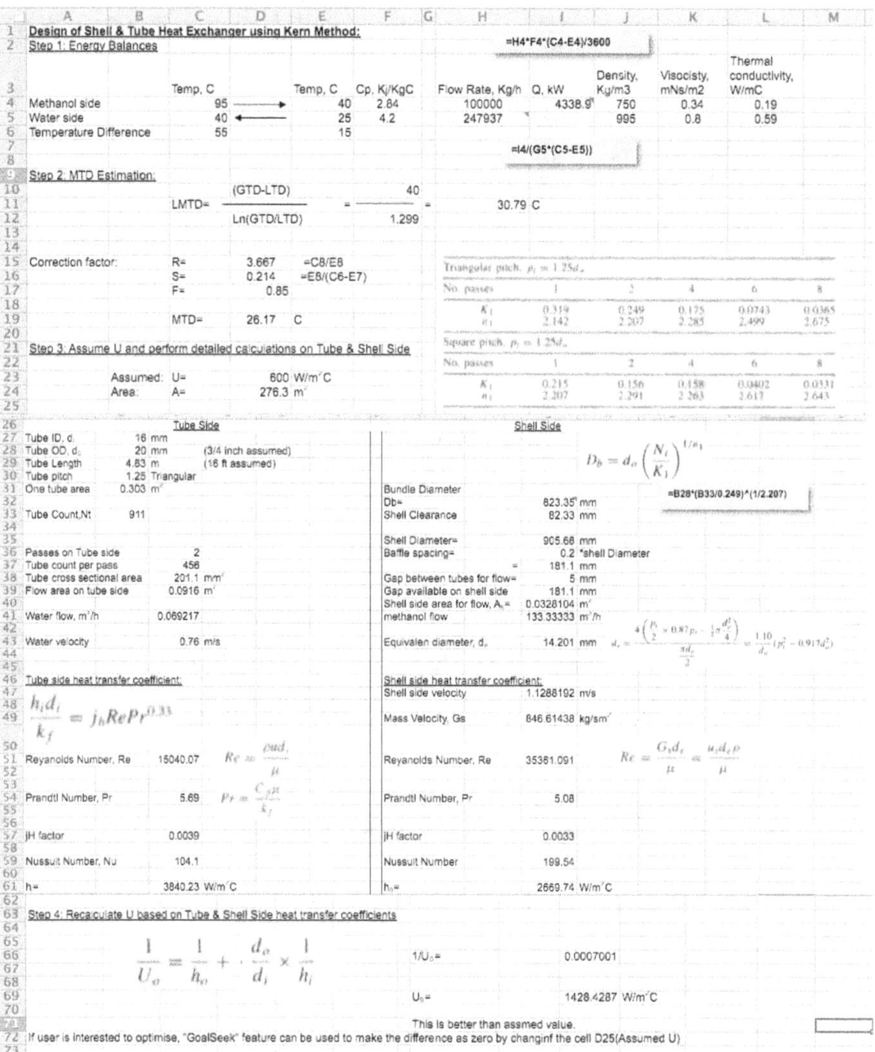

Design of Shell & Tube Heat Exchanger using Kern Method:

Step 1: Energy Balances

=H4*F4*(C4-E4)/3600

	Temp, C		Temp, C	Cp, Kj/KgC	Flow Rate, Kg/h	Q, kW	Density, Kg/m3	Viscosity, mNs/m2	Thermal conductivity, W/mC
Methanol side	95 →		40	2.84	100000	4338.9	750	0.34	0.19
Water side	40 ←		25	4.2	247937		995	0.8	0.59
Temperature Difference	55		15						

=I4/(G5*(C5-E5))

Step 2: MTD Estimation:

$$LMTD = \frac{(GTD-LTD)}{Ln(GTD/LTD)} = \frac{40}{1.299} = 30.79 \; C$$

Correction factor:

R=	3.667	=C8/E8	
S=	0.214	=E8/(C6-E7)	
F=	0.85		
MTD=	26.17	C	

Triangular pitch, $p_t = 1.25d_o$

No. passes	1	2	4	6	8
K_1	0.319	0.249	0.175	0.0743	0.0365
n_1	2.142	2.207	2.285	2.499	2.675

Square pitch, $p_t = 1.25d_o$

No. passes	1	2	4	6	8
K_1	0.215	0.156	0.158	0.0402	0.0331
n_1	2.207	2.291	2.263	2.617	2.643

Step 3: Assume U and perform detailed calculations on Tube & Shell Side

Assumed:	U=	600	W/m²C
Area:	A=	276.3	m²

Tube Side

Tube ID, d_i	16 mm	
Tube OD, d_o	20 mm	(3/4 inch assumed)
Tube Length	4.83 m	(16 ft assumed)
Tube pitch	1.25 Triangular	
One tube area	0.303 m²	
Tube Count, Nt	911	
Passes on Tube side	2	
Tube count per pass	456	
Tube cross sectional area	201.1 mm²	
Flow area on tube side	0.0916 m²	
Water flow, m³/h	0.069217	
Water velocity	0.76 m/s	

Shell Side

$$D_b = d_o \left(\frac{N_t}{K_1}\right)^{1/n_1}$$

=B28*(B33/0.249)^(1/2.207)

Bundle Diameter Db=	823.35 mm
Shell Clearance	82.33 mm
Shell Diameter=	905.68 mm
Baffle spacing=	0.2 *shell Diameter
	181.1 mm
Gap between tubes for flow=	5 mm
Gap available on shell side	181.1 mm
Shell side area for flow, A_s=	0.0328104 m²
methanol flow	133.33333 m³/h
Equivalen diameter, d_e	14.201 mm

$$d_e = \frac{4\left(\frac{P_t}{2} \times 0.87 p_t - \frac{1}{2}\pi\frac{d_o^2}{4}\right)}{\frac{\pi d_o}{2}} = \frac{1.10}{d_o}(p_t^2 - 0.917 d_o^2)$$

Tube side heat transfer coefficient:

$$\frac{h_i d_i}{k_f} = j_h Re Pr^{0.33}$$

Reyanolds Number, Re	15040.07	$Re = \dfrac{\rho u d_i}{\mu}$
Prandtl Number, Pr	5.69	$Pr = \dfrac{C_p \mu}{k_f}$
jH factor	0.0039	
Nussult Number, Nu	104.1	
h=	3840.23 W/m²C	

Shell side heat transfer coefficient:

Shell side velocity	1.1288192 m/s
Mass Velocity, Gs	846.61438 kg/sm²
Reyanolds Number, Re	35361.091
Prandtl Number, Pr	5.08
jH factor	0.0033
Nussult Number	199.54
h_o=	2669.74 W/m²C

$$Re = \frac{G_s d_e}{\mu} = \frac{u_s d_e \rho}{\mu}$$

Step 4: Recalculate U based on Tube & Shell Side heat transfer coefficients

$$\frac{1}{U_o} = \frac{1}{h_o} + \frac{d_o}{d_i} \times \frac{1}{h_i}$$

$1/U_o$=	0.0007001
U_o=	1428.4287 W/m²C

This is better than assmed value.

If user is interested to optimise, "GoalSeek" feature can be used to make the difference as zero by changinf the cell D25(Assumed U)

Additional Practice: Design a Heat Exchanger with following Data:

	Shellside	Tubeside
Fluid	Crude oil	Heavy gas oil circulation reflux
Flow rate, kg/h	399.831	277,200
Temperature in/out, oC	277/249	302/275
Operating pressure drop, kg/cm2	28.3	13.0

	Shellside	Tubeside
Allowable pressure drop, kg/cm2	1.2	0.7
Fouling resistance, h*m2*°C/kcal	0.0007	0.0006
Heat duty, MM kcal/h	5.4945	5.4945
Viscosity in/out, cP	0.664/0.563	0.32/0.389
Design pressure, kg/cm2(gage)	44.0	17.0
Line size, mm (nominal)	300	300
Material of construction	Carbon steel	Tubes Type 410 stainless steel Other. 5Cr1/2Mo

References

Applied Process Design - Vol 3, Ernest E. Ludwig, Gulf Professional Publishing.

Chemical Engineering, Vol 1, Fluid Flow, Heat Transfer and Mass Transfer J. M. Coulson and J. F. Richardson with J. R. Buckhurst and J. H. Harker.

28

Centrifuge Design

Objective: Process design of centrifuge

Description:

When an aqueous slurry is filtered in a plate and frame press, fitted with two 50mm thick frames each 150mm square, operating with a pressure difference of 350 kN/m^2, the frames are filled in 3600 × (1 h). How long will it take to produce the same volume of filtrate as is obtained from a single cycle when using a centrifuge with a perforated basket, 300 mm diameter and 200 mm deep? The radius of the inner surface of the slurry is maintained constant at 75 mm and the speed of rotation is 65 Hz (3900 rpm).

It may be assumed that the filter cake is incompressible that the resistance of the cloth is equivalent to 3 mm of cake in both cases, and that the liquid in the slurry has the same density as water.

Solution

In the filter press

Noting that V = 0 when t = 0, then,

$$V^2 + 2\,\frac{AL}{v}v = \frac{2(-\Delta P)A^2 t}{r\mu v}$$

and

$$v = \frac{1A}{v}$$

Thus

$$\frac{l^2 A^2}{v^2} + \left(\frac{2AL}{v}\right)\left(\frac{lA}{v}\right) = \frac{2(-\Delta P)A^2 t}{r\mu v}$$

or

$$l^2 + 2Ll = \frac{2(-\Delta P)vt}{r\mu}$$

For one cycle

$l = 25$ mm = 0.025 m; L = 3 mm = 0.003 m

$-\Delta P = 350$ kN/m^2 = 3.5 × 10^5 N/m^2

t = 3600 s

$$\therefore \quad 0.025^2 + (0.003 \times 0.025) = 2 \times 3.5 \times 10^5 \times 3600 \times \left(\frac{v}{r\mu}\right)$$

and $\quad \left(\dfrac{rm}{v}\right) = 3.25 \times 10^{12}$

In the centrifuge

$$(R^2 - r^2)\left(1 + 2\frac{L}{R}\right) + 2r^2 \, ln\left(1 + 2\frac{L}{R}\right)\frac{r'}{R} = \frac{2vt\rho w^2}{r\mu}\left(R^2 - r_0^2\right)$$

R = 0.15 m, H = 0.20 m and the volume of cake = $2 \times 0.050 \times 0.15^2$ = 0.00225 m^3

Thus $\qquad \pi(R^2 - r^2) \times 0.20 = 0.00225$

and $\qquad (R^2 - r^2) = 0.00358$

Thus $\qquad r^2 = (0.15^2 - 0.00358) = 0.0189$ m^2

and $\qquad r' = 0.138$ m

$\qquad\qquad r_0 = 65 \times 2\pi = 408.4$ rad/s

and $\qquad \omega = 65 \times 2\pi = 408.4$ rad/s

The time taken to produce the same volume of filtrate or cake as in one cycle of the filter press is therefore given by

$$(0.15^2 - 0.138)^2 \, (1 + 2 \times 0.003/0.15) + 2(0.0189) \, ln \, (0.138/0.15)$$

$$= \frac{2 \times t \times 1000 \times 408.4^2}{3.25 \times 10^{12}} (0.15^2 - 0.075^2)$$

or $0.00359 - 0.00315 = 1.732 \times 10^{-6}$t

from which $\qquad t = \dfrac{4.4 \times 10^{-4}}{1.732 \times 10^{-6}}$

$$= 254 \text{ s or } 4.25 \text{ min}$$

Solution provided in CD/Publisher Website

References

Coulson and Richardson's Chemical Engineering. Vol.6, 3rd Edition, Butterworth Heinmann.

Process Control

1st Law: *The best control system is the simplest one that will do the job*

2nd Law: *You must understand the process before you can control it.*

3rd Law: *Liquid Levels must always be controlled.*

29

Basics of Process Control

MATLAB for process Control

1. Laplace transform calculation

Calculate Laplace Transform of $X^3+3Cos(2X)$:

```
>> syms x
>> f=laplace(x^3+3*cos(2*x))
ans =
6/s^4+3*s/(s^2+4)
```

2. Inverse Laplace transform calculation

```
>> ilaplace(f)
 ans =
 t^3+3*cos(2*t)
>> ilaplace(f,'x')
ans =
x^3+3*cos(2*x)
```

3. Transfer function calculation

$$G(s) = \frac{2x+1}{s^2+4s+3}$$

```
>> num=[2 1];
>> den=[1 4 3];
>> G=tf(num,den)
```

Transfer function:

$$\frac{2s+1}{s^2+4s+3}$$

Alternatively, the entry could have been done in one expression by typing:-

```
>> G=tf([2 1],[1 4 3])
```

The roots of the polynomial can be calculated by typing:-

```
>> roots(den)

ans =

    -3

    -1
```

Alternatively the transfer function can be entered in zero, pole, gain form where the command is in the form G=*zpk*(zeroes, poles, gain)

```
>> G=zpk([-0.5],[-1 -3],2)

Zero/pole/gain:
```

$$\frac{2(s+0.5)}{(s+1)(s+3)}$$

A state space model or object formed from known A,B,C,D matrices, often denoted by (A,B,C,D), can be entered with the command G=*ss*(A,B,C,D).

```
>> A=[0 1;-3 -4];

>> B=[0;1];

>> C=[1,2];

>> D=0;

>> G=ss(A,B,C,D);

>> tf(G)

Transfer function:
```

$$\frac{2s+1}{s^2+4s+3}$$

4. Transient (Step) response for Transfer function

$$G(s) = \frac{1+sT}{(s+1)(s+2)}$$

It is easy to show mathematically that the response will have an overshoot for T>1. The response for T=0.5, T=1 and T=2 are obtained using following commands:

```
>> G0=tf([1],[1 3 2]);
>> step(G0);
>> hold on
>> G1=tf([0.5 1],[1 3 2]);
>> G2=tf([1 1],[1 3 2]);
>> G3=tf([2 1],[1 3 2]);
>> step(G1)
>> step(G2)
>> step(G3)
>>   legend('Step    Response:    T=0','Step
Response:       T=0.5','Step       Response:
T=1','Step
Response: T=2');
```

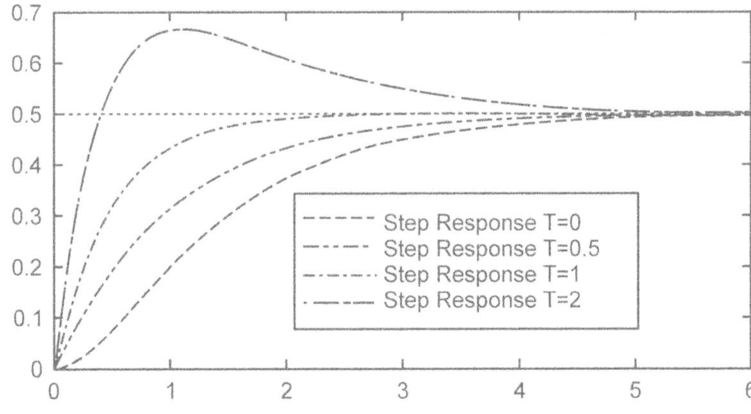

Fig. 29.1 Step response for second order system for different ρ

5. Bode plot for given Transfer function

Consider again the one zero, three pole transfer function

$$G(s) = \frac{4(s+1)}{(s+2)(s^2+s+1)}$$

Dividing the numerator and denominator by 2, it can be written in the form:

$$G(s) = \frac{2(1+s)}{(1+0.5s)(s^2+s+1)}$$

For plotting the Bode diagram it can be thought of as 4 transfer functions A constant gain of 2

A single zero with a break point of 1

A single pole with a break point of 2

A quadratic pole with natural

frequency θ and damping ratio, $\zeta = 0.5$

The instruction in MATLAB to obtain the Bode plot of a transfer function G is bode (G). The resultant Bode magnitude plot is shown.

For plotting the Bode diagram it can be thought of as 4 transfer functions

A constant gain of 2

A single pole with a break point of 2

A quadratic pole with natural frequency 1 and damping ratio, $\zeta = 0.5$.

The instruction in MATLAB to obtain the Bode plot of a transfer function G is bode (G). The resultant Bode magnitude plot is shown.

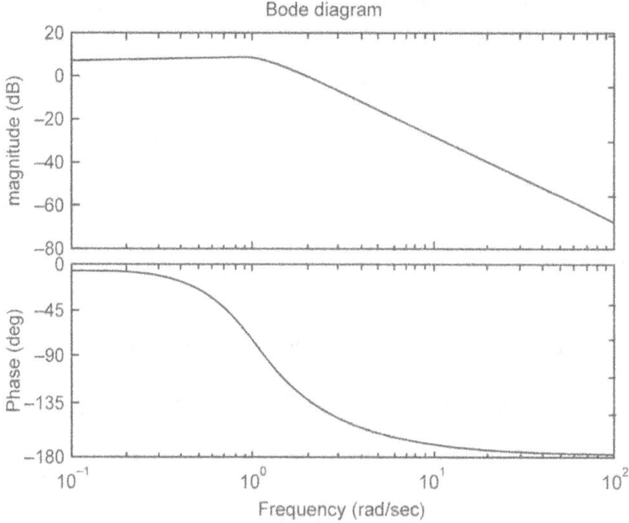

Fig. 29.2 Bode magnitude plot

For plotting the Bode diagram it can be thought of as 4 transfer functions

A constant gain of 2

A single zero with a break point of 1

A single pole with a break point of 2

A quadratic pole with natural frequency 1 and damping ratio,$\zeta = 0.5$.

The instruction in MATLAB to obtain the Bode plot of a transfer function G is *bode*(G). The resultant Bode magnitude plot is shown.

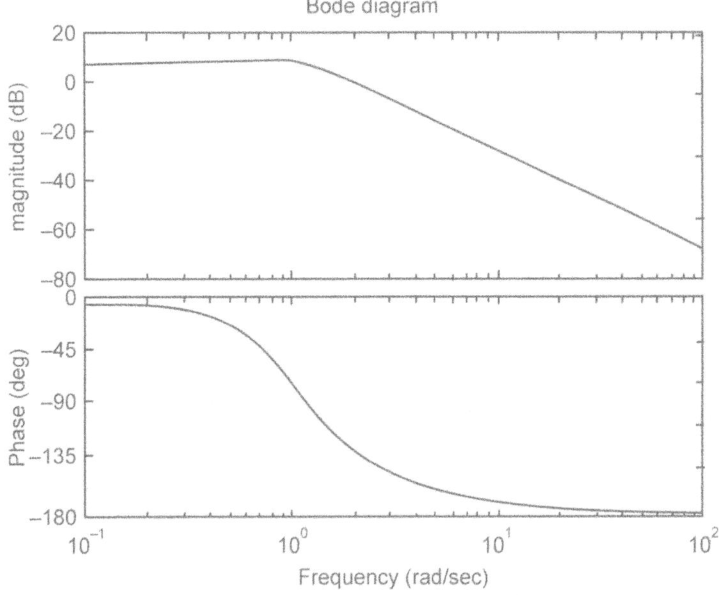

Fig. 29.3 Bode magnitude and phase curve

6. Stability from the characteristic equation (denominator of closed loop transfer function)

Characteristic equation is the denominator of closed loop transfer function.

$$F(s) = f_n s^n + f_{(n-1)} s^{(n-1)} + \dots f_1 s + f_0 \text{ with } f_0 > 0$$

Roots can easily found using command *roots*(poly), where *poly* is entered as a string of coefficients with highest power os s first.

```
>> roots([1 6 11 6])
ans =
-3.0000
-2.0000
-1.0000
```

If there are positive roots, the system is unstable.

7. Root Locus

Design of a single control loop may sometimes just involve the choice of a suitable gain, K, in which case the characteristic equation will be

$$1 + KG_{ol}(s) = 0$$

MATLAB plots a root locus with the command ***rlocus*(G)**

Consider Transfer function

$$\frac{Y(s)}{U(s)} = \frac{B(s)}{A(s)} G(s)$$

B(s) and A(s) are polynomials of order m and n respectively.

$$A(s) = s^n + a_{n-1}s^{n-1} + a_{n-2}s^{n-2} \ldots\ldots\ldots a_1 s + a_0$$

$$B(s) = s^m + b_{m-1}s^{m-1} + b_{m-2}s^{m-2} \ldots\ldots\ldots b_1 s + b_0$$

Since a and b are coefficients of the polynomials are real numbers. The roots of the polynomial are either real or complex pairs. The transfer function is zero for those values of s which are the roots of B(s), hence called them as *zeroes*. The transfer function will be infinite at the roots of denominator, which are called as *poles*. The general transfer function has m zeroes and n poles and is said to have a relative degree of n-m, which can be shown from physical realization considerations cannot be negative.

Simple example for root locus:

```
>> rlocus(tf([1],[1
```

Simple example for root locus:

```
>> rlocus(tf([1],[1 3 2 0]))
```

```
3 2 0]))
```

Some simple rules which enable quick check of a root locus:

1. The number of root locus paths will be n, assuming n>=m

2. The loci start at the poles of the open loop transfer function with K=0

3. The loci finish at the zeros of the open loop transfer function as k→∞

4. A number of loci equal to the relative degree, n-m, or the so-called number of zeros at infinity, of the open loop transfer function will tend to infinity as K tends to infinity.

5. Loci exist on the real axis to the left of an odd number of singularities (poles+zeros).

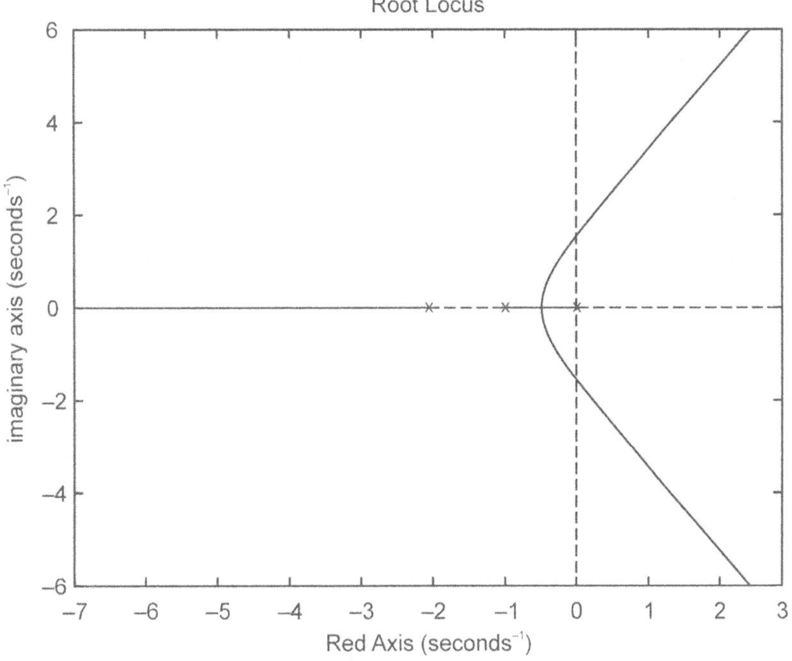

Fig. 29.4 Root Locus plot

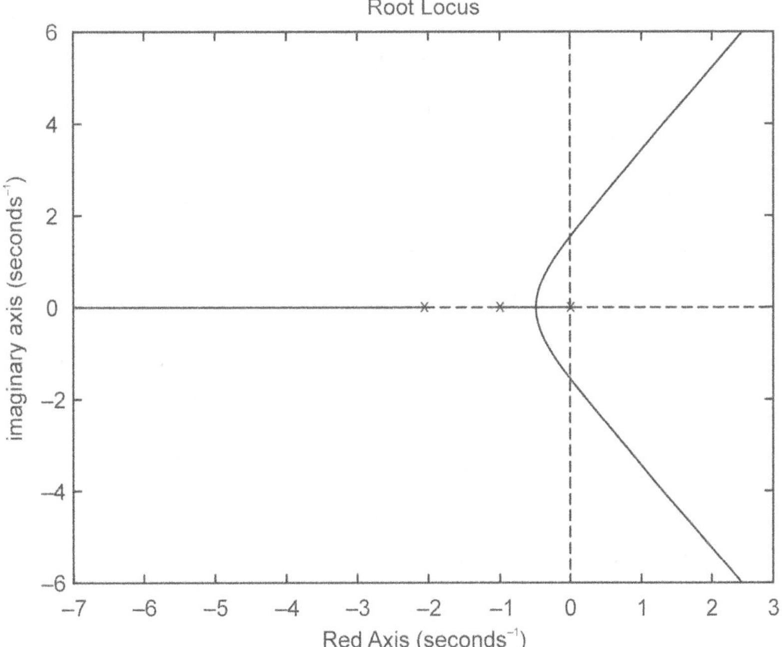

Fig. 29.5 Root locus diagram

8. Nyquist plot for given Transfer function

simple example of this point consider the transfer function

$$G(s) = \frac{1}{s(s+1)^2}$$

then putting $s = j\omega$, and writing $G(j\omega)$ in the form $X(\omega) + jY(\omega)$ gives

$$X(\omega) = \frac{-2}{(1+\omega^2)^2} \text{ and } Y(\omega) = \frac{-(-1\omega^2)}{(1+\omega^2)^2}. \text{ Clearly as } \omega \to 0, X(\omega)$$

$\to -2,$

not the imaginary axis, although the phase does tend to-90°. This will always be the case that the locus for a transfer function with one or more integrator will tend to an asymptote which in principle can be calculated. The Nyquist plot of this transfer function is obtained with the instruction defining the transfer function of G in the Nyquist statement:-

$$\gg \text{nyquist(tf(1,[1 2 1 0]))}$$

Information about where a Nyquist plot cuts the axes can be obtained from the facts that the real axis is cut when $Y(\omega) = 0$ or arg $G(j\omega) = 0°$ or 180°, and the imaginary axis when $X(\omega) = 0$ or arg $G(j\omega) = -90°$ or $+ 90°$. Which are the easiest calculations can depend on the transfer function. For the above example it is easily seen from $Y(\omega)$ that the real axis is cut when $\omega = 1$ and the imaginary axis is only reached as ω tends to infinity. However for $G(s) = 1/(1 + s)^6$ then where it cuts the axes is best obtained using arg $G(j\omega)$, which is simply equal to $-6 \tan^{-1}\omega$.

Three further comments must be made here about the plot of Fig. 4.4:

1. For reasons to be explained later the graph is drawn for both positive and negative frequencies. The labeling of these has been added to the plot afterwards.

2. It can be shown for all transfer functions that $X(\omega)$ is an even function and $Y(\omega)$ an old function of ω. Thus the negative frequency part of the plot is a reflection of the positive frequency plot in the real axis.

3. MATLAB does not label the frequencies automatically on the plot but they can be selected by use of the cursor as has been done to obtain one frequency point on this plot.

4. The frequency response plot instructions *bode(G) and nyquist(G)* in MATLAB automatically select the frequency range. This can be done by the user by selecting a vector w, typically on a log scale using the instruction $\omega = \log space(a,b,n)$, which generates n points on a log scale between 10^2 and 10^3. If n is omitted the default is 50 points. The plot instructions are then *bode(G,ω)* and *nyquist(G,ω)*.

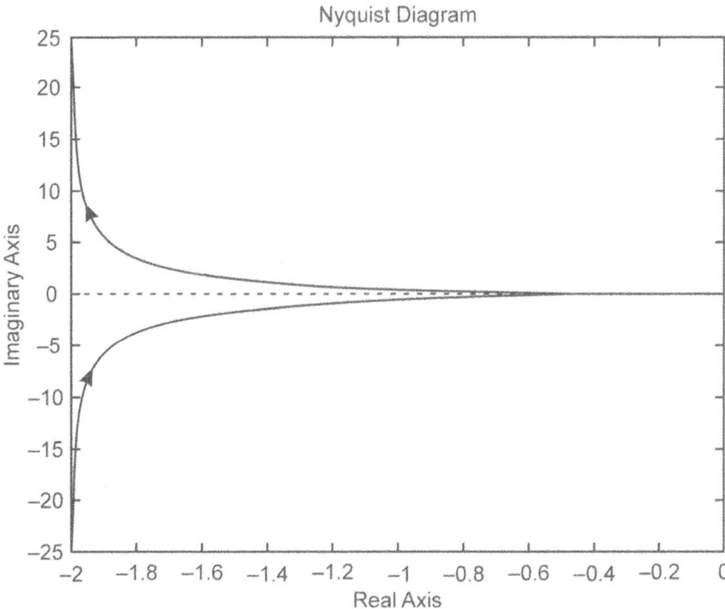

Fig. 29.6 Nyquist plot for $1/s(s+1)^2$

Oscillatory Behaviour

Suppose a variable y (a tank liquid temperature) oscillates regularly in time:

$$y(t) = A\sin\left(2\pi\frac{t}{p} + \varphi\right) = A\sin(2\pi ft + \varphi) = A\sin(\omega t + \varphi)$$

where   A is the amplitude (dimension of y)

       p is the period (dimension of time)

       f is the cyclical frequency (dimension of time^{-1})

ω is the radian frequency (dimension of radians time^{-1})

φ is the phase angle (dimension of radians)

Inserting 2π into the argument is necessary because we are attempting to describe physical behavior (something varying in time) by an abstract math function that doesn't care what our time units are. View $\omega = 2\pi/p = 2\pi f$ as the conversion factor between time units and radians.

Generate Unit Sin wave:

```
>> t=0:pi/10:2*pi;
```

```
>> plot(t,sin(t))
```

Sine wave of Amplitude = 3	Sine wave of Amplitude = 3 Radian frequency = 0.9	Sine wave of Amplitude = 3 Radian frequency = 0.9 Phase Angle = $-\pi/2$
>> plot(t,3*sin(t))	>> plot(t,3*sin(0.9*t),'r')	>> plot(t,3*sin(0.9*t-pi/2),'k')

Practice

1. Temperature of Hyderabad is following a sine wave trend with max. temperature at noon as 33C and lowest temperature in the early morning is 27C. Draw this wave.

2. The Open loop transfer function for a three capacity process is

$$G = 2/ [(10s+1)(5s+1)(2s+1)]$$

Where the time constants are in minutes. (a). If a proportional controller with a gain of 3.0 is used, will the system be stable? (b). If reset action is added and a reset time of 5 min is used, will the system be stable for a controller gain of 3.0? [Use Nyquist plots].

References

Derek Atherton, Control Engineering – An Introduction with use of MATLAB.

30

Simulation of Temperature Control of Water Tank

Simulate the Temperature Control of Water Tank

Objective: Emphasizes the following principle control the water tank temperature as constant using P, PI, PID controllers when inlet temperature suddenly changes.

Problem Statement: Water tank was operated by a heater of duty 2500 kcal/h for heating water fed at 20 °C. If the set point of water tank changed to 80 °C, simulate and calculate the outlet temperatures and heat duty for P, PI, PID controlled scenarios.

Theory

A liquid stream at a temperature T_i enters an insulated, tank at a constant flow rate. It is desired to maintain (or control) the temperature in the tank at T_R by means of the controller. If the measured temperature T_m differs from the desired temperature T_R, the controller senses the difference or error, $\varepsilon = T_R - T_m$, and changes the heat input in such a way to reduce the magnitude of ε.

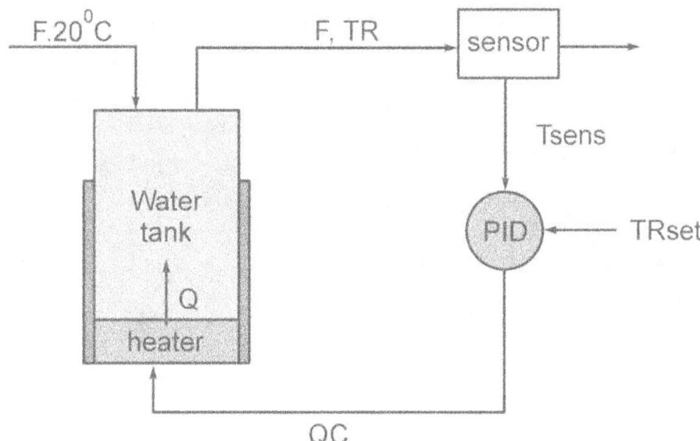

Fig. 30.1 Heating tank control diagram

Modeling Equations

If the System is controlled by a Proportional Controller

$$\frac{dT1}{dt} = \frac{1}{V\rho Cp}[F\rho Cp(T0 - T1) + Q]$$

$$\frac{dQ}{dt} = (Qc - Q)/\tau_{tank}$$

$$\frac{dT3}{dt} = \frac{1}{\tau_{sensor}}(T1 - T3)$$

Create `Heatingcontrol.m` **to test controllers such as P, PI, PID with following code:**

```matlab
% Control of a water tank temperature
% Input arguments:   t    - time
%                    x(1) - Tank temp; x(2) - Duty;
%                    x(3) - Thermocople reading Temperature
% Outputs:           xdot - Derivatives for T1,Q,T3
function xdot=Heatingcontrol(t,x)
global PID_Controller;
V = 100;
F = 10;
rho = 1;
cp = 4.19;
Q0 = 2500;
TRset = 80;
T0 = 20;
TauSens = 2;
Kp = 300;
TauQ = 1;
TauI=9;
TauD=0.8;
switch (PID_Controller)
    case 0 %proportional
        QC= Q0 + Kp*(TRset - x(3));
    case 1 %proportional Integral
        QC= Q0 + Kp*(TRset - x(3))+ (Kp/TauI) ;
    case 2 %proportional Integral Derivative
      QC=Q0+ Kp*(TRset - x(3))+(Kp/TauI) -(Kp*TauD/TauSens*(x(1)-x(3)));
    otherwise
        error('This control is impossible')
  QC=Q0;
end
xdot(1) = 1/(V*rho*cp)*(F*rho*cp*(T0-x(1))+x(2));
xdot(2) = (QC-x(2))/TauQ;
xdot(3) = 1/TauSens*(x(1)-x(3));
xdot=xdot';
```

Write Main Program and save the file as controltanktemp.m content of the file is shown below:

```matlab
% Control of a water tank temperature
% Case of Proportional Control
% Main program calls function Heatingcontrol
% and plot results: sensor and tank temperatures and heat input to the water tank
global PID_Controller;
PID_Controller=0;
tf = 100;
x0=[20 2500 20];
[t,x] = ode15s('Heatingcontrol',[0 tf],x0);
x1=x(:,1);
x2=x(:,2);
x3=x(:,3);
figure(1);
hold on;
plot(t,x1,'r');
plot(t,x3,'b');
hold off;
figure(2);
plot(t,x2,'c');
```

Execution Procedure

```
>> controltanktemp()
>>
```

Result: P –System:

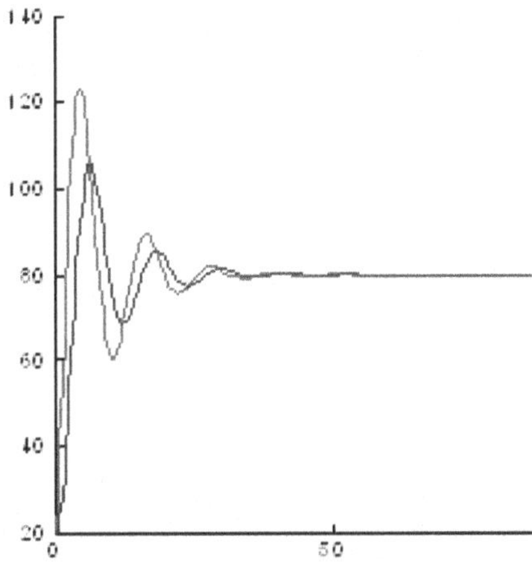

Fig. 30.2 Response curve for P-control

Main Program – PI System

```
% Case of Proportional-Integral Control
```

```
PID_Controller=1;
```

Set in the main function controltanktemp.m file and execute the same function from command prompt to get the response of PI-control.

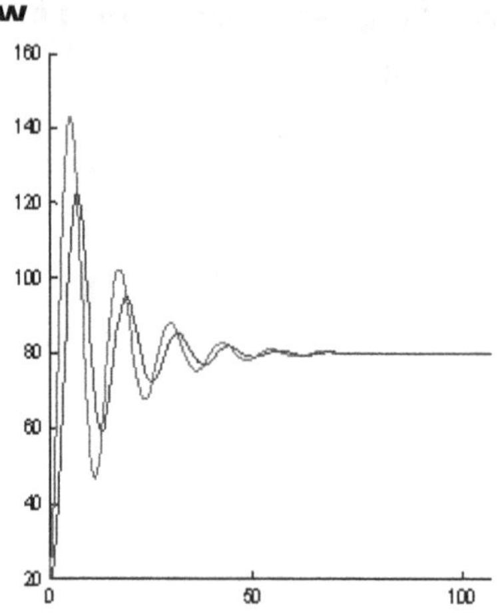

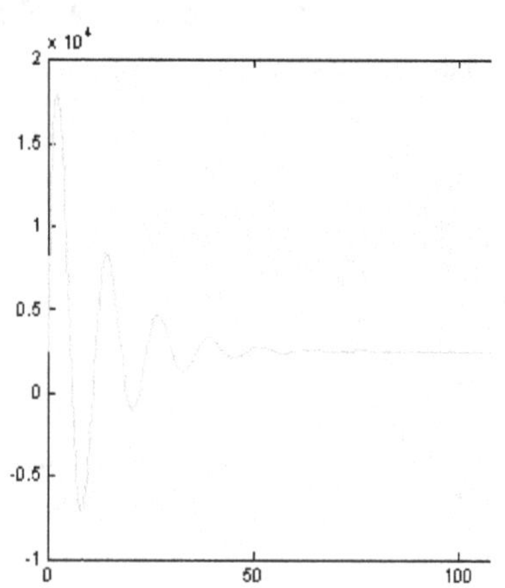

Fig. 30.3 Response of PI control   **Fig. 30.4** Response of PI Control for 'Q'

Remaining commands are same as Proportional controller case except set PID controller = z

Result PID System

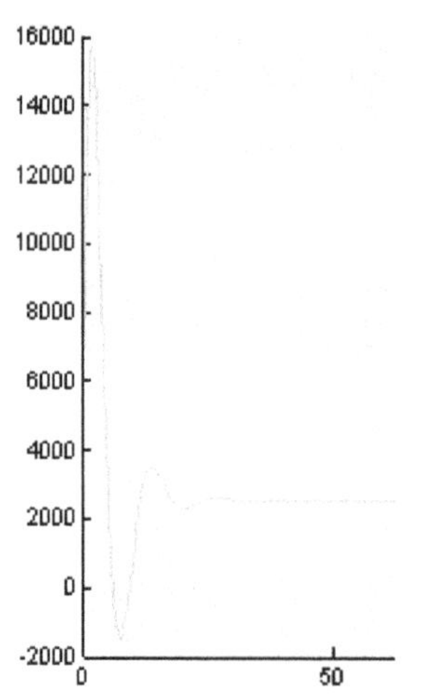

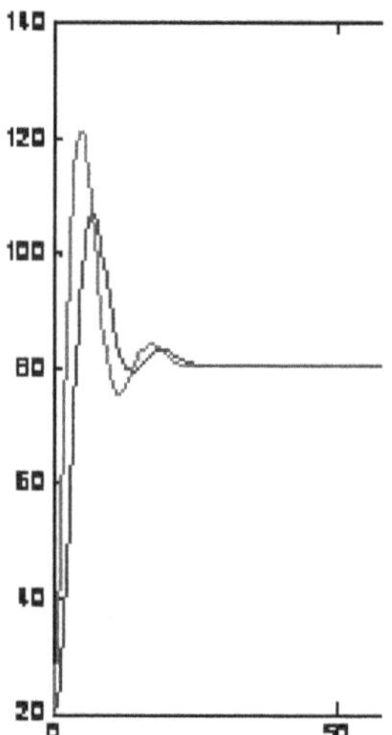

Fig. 30.5 Response of PID control for Q   **Fig. 30.6** Response of PID controller
                                                                    for temperatures

References

Process System Analysis and Control by Coughanower, McGraw-Hill Publishers

web.iku.edu.tr/courses/ee/ee670/ee670/lecture%20notes/ch01.pdf

31

PID Control of Three CSTR's in Series

Theory

Considered a three CSTR's in series in which A → B first order reaction takes place. Product B is produced and reactant A is consumed in each of the three perfectly mixed reactors by a first-order reaction occurring in the liquid. For the moment let us assume that the temperatures and holdups (volumes) of the three tanks can be different, but both temperatures and the liquid volumes are assumed to be constant (isothermal and constant holdup). Density is assumed constant throughout the system, which is a binary mixture of A and B.

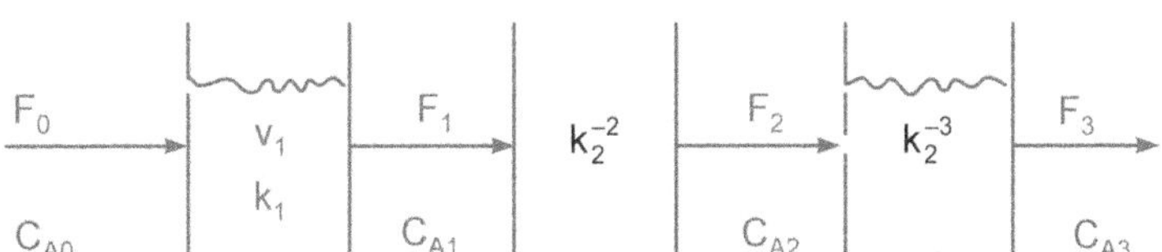

Fig. 31.1 Three CSTRs in series Flow diagram

Modeling Equations

$$V_1 \frac{dC_{A1}}{dt} = F(C_{A0} - C_{A1}) - V_1 k_1 C_{A1}$$

$$V_2 \frac{dC_{A2}}{dt} = F(C_{A1} - C_{A2}) - v_2 k_2 C_{A2}$$

$$V_3 \frac{dC_{A3}}{dt} = F(C_{A2} - C_{A3}) - v_3 k_3 C_{A3}$$

Data

Steady state conditions

Time = 0 $C_{A0} = 0.8$ $C_{A1} = 0.4$ $C_{A2} = 0.2$ $C_{A3} = 0.1$ kmol/m^3

Parameters

Tau = 2.0 min   Rate constant K = 0.5 min^{-1}

Study the behaviour if the inlet concentration is changed to 1.8 kmol/m^3

Matlab Program: Create cstrs3.m file

```
function uprime=cstrs3(t,u)
uprime = zeros(3,1);
uprime(1)= (1.8-u(1))/2-0.5* u(1);
uprime(2) = (u(1)-u(2))/ 2-0.5 * u(2);
uprime(3) = (u(2)-u(3))/ 2-0.5 * u(3);
```

Execute the program from MATLAB Command prompt:

```
>> u0 = [0.4 0.2 0.1];
>> [t,u] = ode45(@cstrs3,[0,15],u0)

>> plot(t,u(:,1),'+',t,u(:,2),t,u(:,3),'*');
>> legend('CA1','CA2','CA3');
>> xlabel('Time');
>> ylabel('Concentration');
>> title('CSTRs in series');
>>
```

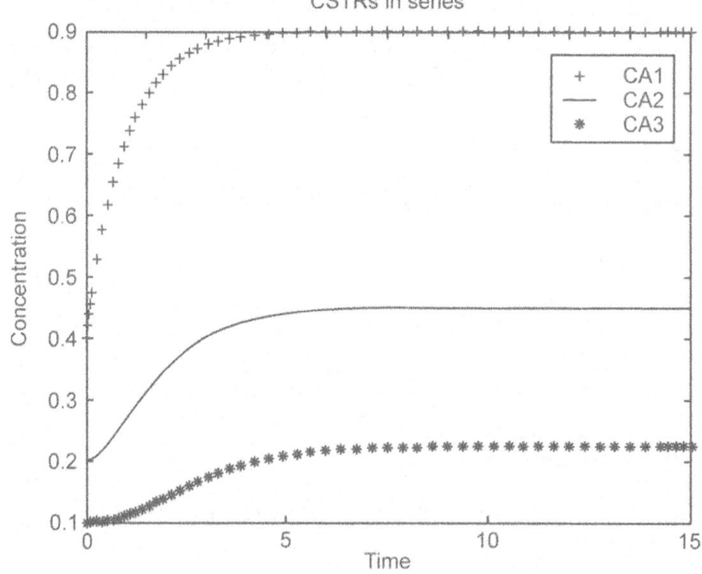

Fig.31.2 Concentration profiles of CSTR outlets

Verification of MATLAB calculations with 4th Order Method in Excel sheet for 1st reactor was performed. Following the result from the excel sheet:

MATLAB Result:

t =	u =			dt
0	0.4000	0.2000	0.1000	
0.04	0.4197	0.2002	0.1000	0.0402
0.08	0.4386	0.2008	0.1000	0.0402
0.121	0.4558	0.2017	0.1000	0.0402
0.161	0.4743	0.2029	0.1001	0.0402
0.297	0.5288	0.2091	0.1005	0.1366
0.434	0.5760	0.2177	0.1013	0.1365
0.571	0.6174	0.2281	0.1026	0.1366
0.707	0.6535	0.2396	0.1044	0.1366
0.846	0.6854	0.2519	0.1068	0.1365
0.984	0.7131	0.2648	0.1097	0.1366
1.123	0.7373	0.2773	0.1130	0.1365
1.261	0.7583	0.2899	0.1167	0.1365
1.433	0.7806	0.3048	0.1210	0.1713
1.604	0.7994	0.3190	0.1272	0.1712
1.775	0.8153	0.3324	0.1328	0.1713
1.946	0.8286	0.3448	0.1386	0.1712
2.164	0.8426	0.3591	0.1456	0.2174
2.361	0.8538	0.3718	0.1532	0.2175
2.599	0.8628	0.3831	0.1601	0.2174
2.816	0.8701	0.3929	0.1658	0.2174
3.082	0.8771	0.4032	0.1743	0.2659
3.348	0.8825	0.4118	0.1812	0.2658
3.614	0.8865	0.4188	0.1874	0.2659
3.879	0.8896	0.4248	0.1930	0.2658
4.224	0.8927	0.4309	0.1991	0.3447
4.569	0.8949	0.4355	0.2042	0.3446
4.913	0.8963	0.4391	0.2088	0.3447
5.258	0.8974	0.4419	0.2119	0.3447
5.663	0.8960	0.4442	0.2151	0.4046

Cell annotation: `=((SK$2-H5)/2)-(0.6*H6)`

3.7 RANGE-KUTTA (FOURTH-ORDER)

Column K annotation: `Uprime(1)=(xxxxxxxx-u(1))/2 - 0.5* u(1);`
Column N annotation: `=(O6-G9)*K5`
Column P annotation: `=(((SK$2-(H5+N5/2))/2)-(0.5*(H5+N6/2)))*(O6-G5)`

t =	u(1)	Uprime(1)	k1	k2	k3	k4	u(1) next
0	0.4	1.4500	0.0583	0.0571	0.0571	0.0560	0.4571
0.0402	0.4571	1.3929	0.0560	0.0549	0.0549	0.0538	0.5120
0.0804	0.5120	1.3380	0.0538	0.0527	0.0527	0.0517	0.5647
0.1206	0.5647	1.2853	0.0517	0.0506	0.0506	0.0496	0.6154
0.1608	0.6154	1.2346	0.1686	0.1571	0.1579	0.1471	0.7730
0.2974	0.7730	1.0770	0.1470	0.1370	0.1377	0.1282	0.9104
0.4339	0.9104	0.9398	0.1283	0.1196	0.1202	0.1119	1.0304
0.5705	1.0304	0.8198	0.1120	0.1043	0.1049	0.0976	1.1350
0.7071	1.1350	0.7150	0.0990	0.0922	0.0928	0.0862	1.2275
0.8456	1.2275	0.6225	0.0863	0.0803	0.0807	0.0751	1.3081
0.9842	1.3081	0.5419	0.0751	0.0699	0.0702	0.0653	1.3782
1.1227	1.3782	0.4718	0.0653	0.0608	0.0611	0.0569	1.4392
1.2612	1.4392	0.4106	0.0704	0.0643	0.0649	0.0593	1.5039
1.4325	1.5039	0.3461	0.0593	0.0542	0.0548	0.0499	1.5583
1.6037	1.5583	0.2917	0.0500	0.0457	0.0461	0.0421	1.6042
1.775	1.6042	0.2456	0.0421	0.0385	0.0388	0.0354	1.6429
1.9462	1.6429	0.2071	0.0450	0.0401	0.0407	0.0362	1.6834
2.1838	1.6834	0.1968	0.0362	0.0320	0.0327	0.0291	1.7159
2.3611	1.7159	0.1341	0.0291	0.0260	0.0263	0.0234	1.7421
2.5985	1.7421	0.1079	0.0234	0.0209	0.0212	0.0189	1.7632
2.8159	1.7632	0.0866	0.0231	0.0200	0.0204	0.0176	1.7835
3.0818	1.7835	0.0865	0.0177	0.0153	0.0156	0.0135	1.7990
3.3476	1.7990	0.0510	0.0138	0.0118	0.0120	0.0104	1.8109
3.6135	1.8109	0.0391	0.0104	0.0090	0.0092	0.0079	1.8200
3.8793	1.8200	0.0300	0.0103	0.0085	0.0089	0.0073	1.8288
4.224	1.8288	0.0212	0.0073	0.0061	0.0063	0.0052	1.8350
4.5688	1.8350	0.0150	0.0052	0.0043	0.0044	0.0037	1.8393
4.9133	1.8393	0.0107	0.0037	0.0030	0.0031	0.0026	1.8424
5.258	1.8424	0.0076	0.0031	0.0024	0.0026	0.0020	1.8450
5.8828	1.8450	0.0050	0.0020	0.0018	0.0017	0.0013	1.8468

Cell annotations:
`=H5`
`=((O6-G9)*((SK$2-(H5+O9/2))/2)-(0.5*(H5+O9/2)))`
`=(O6-G9)*(((SK$2-(H5+P9))/2)-(0.5*(H5+P9)))`
`=H5+(N5+2*O9+2*P5+Q5)/6`

Runge-kutta (Fourth-Order). The fourth-order Runge-Kutta algorithm is widely used in chemical engineering. For the ODE the Runge-Kutta algorithm is

$$k_1 = \Delta t \, f(x_n, t_n)$$

$$k_2 = \Delta t \, f\!\left(x_n + \frac{1}{2}k_1, \, t_n + \frac{1}{2}\Delta t\right)$$

$$k_3 = \Delta t \, f\!\left(x_n + \frac{1}{2}k_2, \, t_n + \frac{1}{2}\Delta r\right)$$

$$k_4 = \Delta t \, f(x_n + k_3, \, t_n + \Delta r)$$

$$x_{n+1} = x_n + \frac{1}{6}(k_1 + 2k_2 + 2k_3 + k_4)$$

For numerically integrating two first-order ODEs

$$\frac{dx_1}{dt} = f_1(x_1, x_2\, t)$$

$$\frac{dx_2}{dt} = f_2(x_1, x_2\, t)$$

Formulae Used

Row/Column	C	E	F	G	H
6	=A6–A5	0	0.4	=((1.8-F5)/2)-(0.5*F5)	=(E6-E5)*G5
7	=A7–A6	0.0402	=L5	==>	
8	.	0.0804	=L6	=((1.8-F7)/2)-(0.5*F7)	=(E8-E7)*G7
		.	.		
		.	.		
		.	.		
		.	.		
60	=A60-A59	14.8145	=L59	=((1.8-F60)/2)-(0.5*F60)	=(E61-E60)*G60

Row/Column	I	J
6	=(((1.8-(F5+H5/2))/2)-(0.5*(F5+H5/2)))*(E6-E5)	=((E6-E5))*(((1.8-(F5+I5/2))/2)-(0.5*(F5+I5/2)))
7		
8	=(((1.8-(F7+H7/2))/2)-(0.5*(F7+H7/2)))*(E8-E7)	=((E8-E7))*(((1.8-(F7+I7/2))/2)-(0.5*(F7+I7/2)))
60	=(((1.8-(F60+H60/2))/2)-(0.5*(F60+H60/2)))*(E61-E60)	=((E61-E60))*(((1.8-(F60+I60/2))/2)-(0.5*(F60+I60/2)))

Row/Column	K	L
6	=(E6-E5)*(((1.8-(F5+J5))/2)-(0.5*(F5+J5)))	=F5+(H5+2*I5+2*J5+K5)/6
7		
8	=(E8-E7)*(((1.8-(F7+J7))/2)-(0.5*(F7+J7)))	=F7+(H7+2*I7+2*J7+K7)/6
60	=(E61-E60)*(((1.8-(F60+J60))/2)-(0.5*(F60+J60)))	=F60+(H60+2*I60+2*J60+K60)/6

NOTE: User need not write the formulas for all rows and columns; Complete the formulas for 1st row using RK method as guideline; complete the 2nd row also; then copy the cells of 2nd row i.e., C7 to L7 and paste the same to remaining rows (this copies the formulas with relative positions).

Controlled Scenario

Now, Add a feedback controller to the third tank. the controller looks the product concentration leaving the tank C_{A3} and makes adjustments in the inlet concentration to the first reactor C_{A0} in order to keep C_{A3} near its desired setpoint C_{A3}^{Set}.

The variable C_{AD} is the disturbance variable and C_{AM} is the manipulated variable. Now initial concentration to the tank 1 is $C_{A0} = C_{AM} + C_{AD}$

Control Loop Diagram

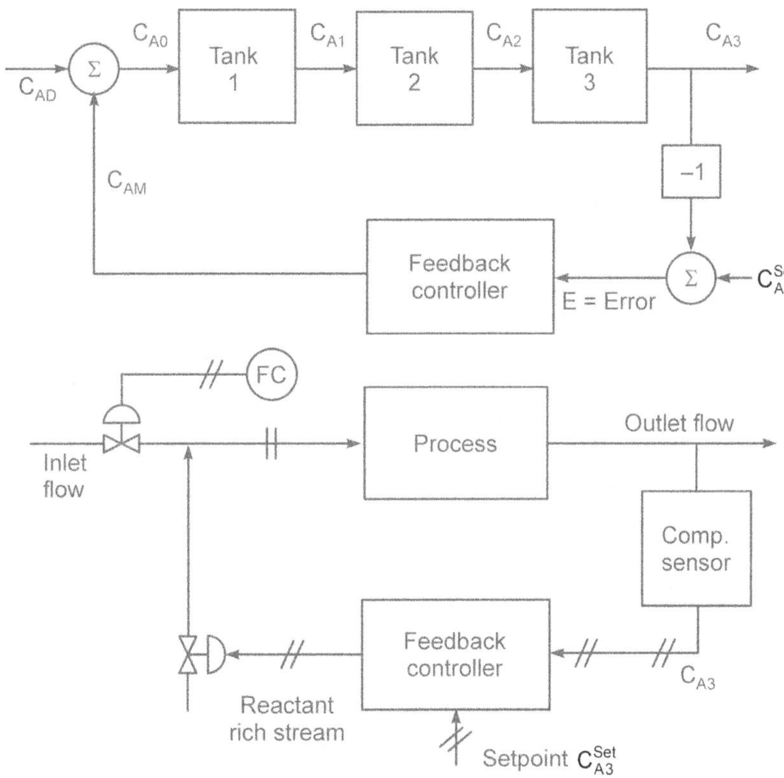

Fig. 31.3 Control loop diagram

The feedback controller has PI controller action It changes CAM based on the magnitude of the error. The equation is

$$C_{AM} = 0.8 + k_c \left(E + \frac{1}{\pi_1} \int E_{(t)} dt \right)$$

Data

Steady state conditions

Time = 0	$C_{A0} = 0.8$	$C_{A1} = 0.4$	$C_{A2} = 0.2$	$C_{A3}=0.1$
	C_{A3}	Set = 0.1	Distrubance	$C_{AD} = 0.2$

Parameters

Tau = 2.0   Rate constant   K = 0.5   Controller gain = 30   taui = 5.0

MATLAB   Program: Open   model_3CSTRsInSeries_PID.mdl   from MATLAB

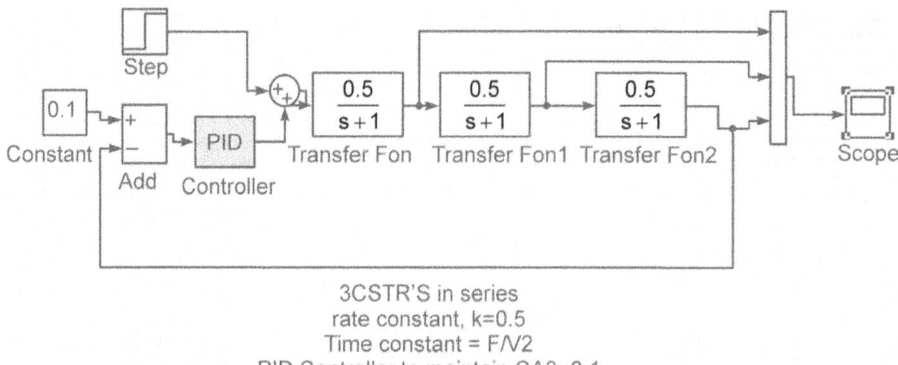

Fig. 31.4 Block diagram for 3 CSTRs in series

Increase Kc till the response is on the verge of instability. Note the Kc and Time period.

Analyse and find the PI tuning parameters.

Reference Example 9.5 from Luebyn:

<table>
<tr><td align="center">Model Equations</td><td align="center">Transfer Functions</td></tr>
</table>

$$\frac{dC_{A1}}{dt} + \left(k_1 + \frac{1}{\tau_1}\right)C_{A1} = \frac{1}{\tau_1}C_{A0} \qquad G_{1(A)} = \frac{C_{A1(s)}}{C_{A0(s)}} = \frac{1/\tau_1}{s + k_1 + 1/\tau_1}$$

$$\frac{dC_{A2}}{dt} + \left(k_2 + \frac{1}{\tau_2}\right)C_{A2} = \frac{1}{\tau_2}C_{A1} \qquad G_{2(s)} = \frac{C_{A2(s)}}{C_{A1(s)}} = \frac{1/\tau_2}{s + k_2 + 1/\tau_2}$$

$$\frac{dC_{A3}}{dt} + \left(k_3 + \frac{1}{\tau_3}\right)C_{A3} = \frac{1}{\tau_3}C_{A2} \qquad G_{3(s)} = \frac{C_{A3(s)}}{C_{A2(s)}} = \frac{1/\tau_3}{s + k_3 + 1/\tau_3}$$

Initial Condition

$$C_{A1(0)} = C_{A2(0)} = C_{A3(0)} = 0$$

Let us assume a unit step change in the feed concentration C_{A0} and solve for the response of C_{A2}. We will take the case where all the τ_{pi}'s are the same, giving a repeaed root of order 3(a third-order pole at $s = -1/\tau_p$).

$$C_{Ao(t)} \, u_{(t)} \quad \Rightarrow \quad C_{A0(s)} = \frac{1}{s}$$

$$C_{A3(s)} = G_{(s)} \, C_{A0(s)} = \frac{K_p}{\left(\tau_p s + 1\right)^2} \frac{1}{s} = \frac{K_p / \tau_p^3}{s\left(s + 1/\tau_p\right)^3}$$

Solution for this problem in time domain is:

$$C_{A3(t)} = K_p \left[1 - \frac{1}{2}\left(\frac{t}{\tau_p}\right)^2 e^{-t/t_p} - \frac{t}{\tau_p} e^{-t/t_p} - e^{-t/t_p} \right]$$

References

Process Modeling, Simulation and Control for Chemical Engineers by W.L.Luyben, McGrawHill Publishers.

32

Dynamics of Second Order Systems

Objective: To analyse the response of second order system using SIMULINK

Procedure

- Start the Simulink environment by typing "simulink" to the matlab prompt.
- Open a new simulink mode from the File → New → Model.
- Drag and drop the blocks below from the Simulink windows into the model window.
- Double click the blocks the set up different second order transfer functions.

Define one overdamped, one critically damped and one under damped system.

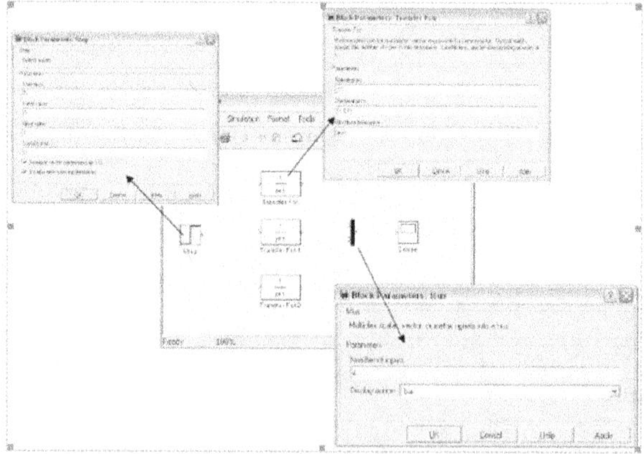

Fig. 32.1 Simulink model window

Connect the blocks together as shown below:

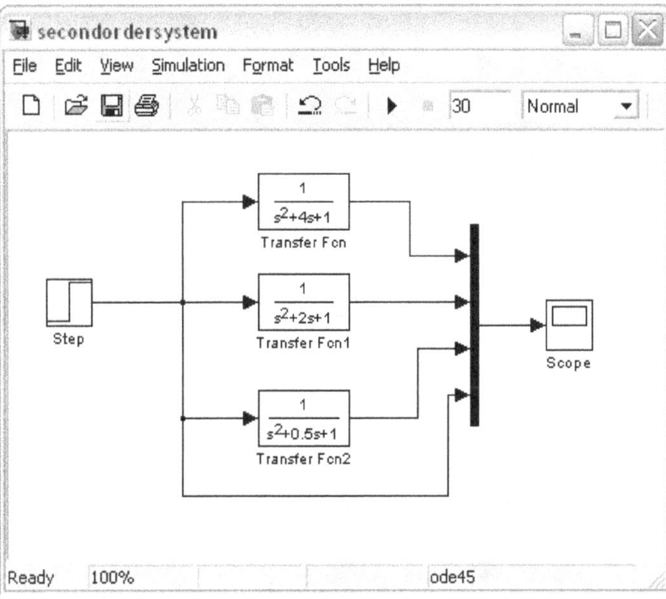

Fig. 32.2 Simulink block diagram

Set the simulation time to 30 sec from the menu "Simulation" ➔ "Configuration parameters" Simulate the process by pressing the "Run" button and then show the results by double clicking on the "Scope" block:

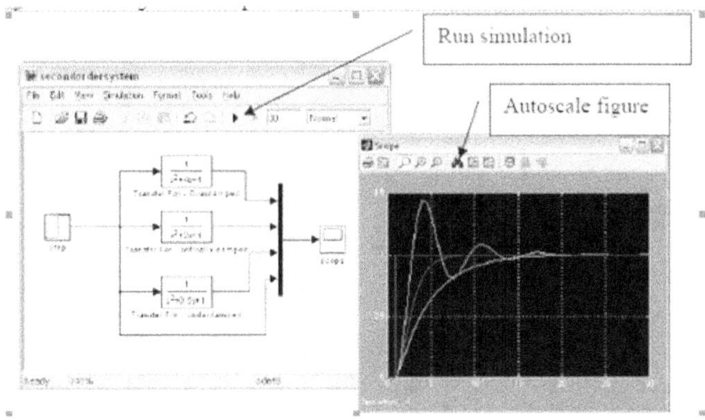

Fig. 32.3 Result plotted in scope

Process Simulation

Simulation is a process of designing an operational model of a system and conducting experiments with this model for the purpose either of understanding the behaviour of the system or of evaluating alternative strategies for the development or operation of the system. It has to be able to reproduce selected aspects of the behaviour of the system modelled to an accepted degree of accuracy.

33

Ammonia Plant Model with Recycle Loop

Aim

Simulate the Ammonia production model and calculate the recycle and purge flows to maintain Argon below 2%.

Flow sheet

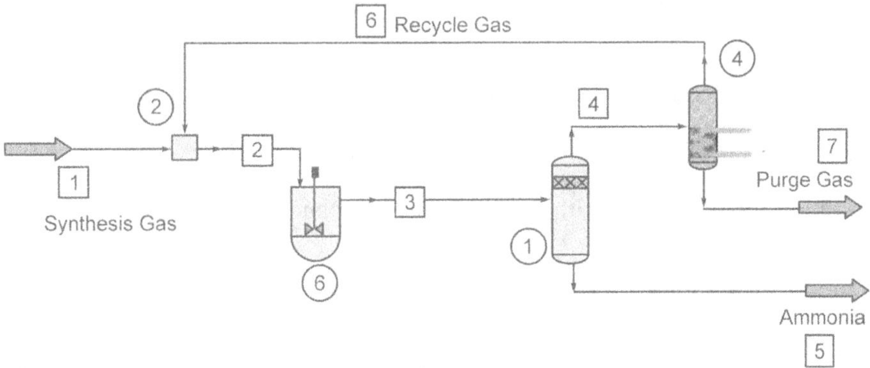

Fig. 33.1 Ammonia Synthesis loop flow diagram

Procedure

- Launch CHEMCAD

- Select Engineering units as "Alt SI" and change the pressure unit to Kg/Cm2G.

- Draw the flow sheet as shown in the diagram (1 Feed, 2 Product Streams, 1 mixer, one conversion reactor, 1 flash unit and one component separator).

- Change the mode to "Simulation" and add component H2, N2, Ar and NH3.

- Provide the feed stream information as mentioned below:

   Temperature: 229C; Pressure: 178Kg/cm2G; Total flow rate: 7855Kmol/h with composition of 74.5%Hydrogen, 24.8%Nitrogen and 0.7%Argon

- Provide Mixer pressure as 178Kg/cm2G.

- Provide the Reactor information: Pressure: 178Kg/cm2G; pressure drop=5Kg/cm2G; vapor phase reaction; Adiabatic reactor with 25% conversion with respect to Hydrogen; define ammonia reaction ; frequency factor:7; Activation energy:5345; Heat of reaction: -78000

- Flash the stream at 173Kg/cm2G and -17.8C (Condense the Ammonia for separation)

- Component separator: Vary the gas as recycle to reactor by adjusting the purge gas; pressure:170Kg/cm2G and Temperature: 400C(both top and bottom streams).

- Run the flow sheet and resolve the errors, if any. Both Recycle calculations and flow sheet should be converged at this step and the Argon content in recycle should be less than 2%.

- Generate the report.

Mass Balance with Specified Reaction and Recycle

Objective

To study mass balance with specified reaction and recycle using COMSOL multiphysics.

Problem Statement

Consider an ammonia process with nitrogen and hydrogen feed with 25% conversion per pass (in reactor) and a specified separation at the separator. The flow rates of nitrogen and hydrogen into the process are one and three moles per unit time respectively, but there is also 0.01 moles per time unit of carbon dioxide. Because of the carbon dioxide, we will add a purge stream as one percent of the recycle stream. Below is the figure.

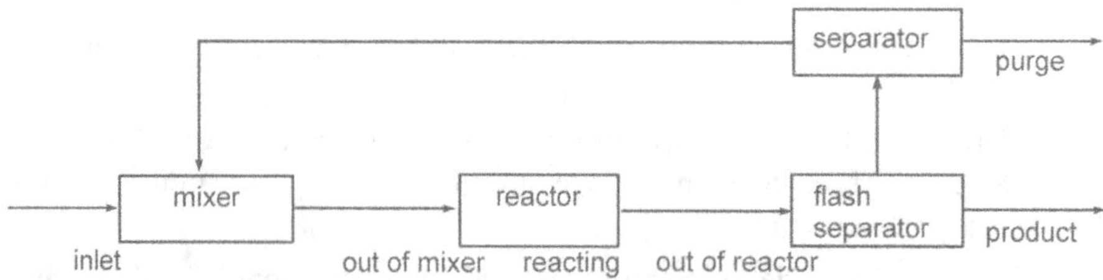

Fig. 33.2 Ammonia process with vapor-liquid equilibria and a purge stream

Solution using COMSOL:

- Follow appendix D and select 0D option – mathematics (Global ODEs and DAEs (ge)) – study (stationary) – done.

- Model 1 opens, with global ODEs and DAEs. In Model/definitions right click and choose variables. Insert the following variables.

a = variables	Model Library
Variables	
Name	Exprssion
Conv	0.25
react	conv*N2
N1	100
H1	300
A1	0
splitN	0.005
splitH	0.005
splitA	0.98

- After giving the variables, go to global ODEs/global equations and insert the equations as shown in the below figure. In COMSOL one must solve under f(u,ut,utt,t). Thus, if the equation is N2=N1+N6, the f is given as N1+N6-N2.

$$\frac{d}{dt}\ \text{Global Equations} \qquad\qquad \text{Model Library}$$

Global Equations
$$F(u, u_t, u_{tt}, t) = 0,\ u(t_0) = u_0,\ u_t(t_0) = u_{t0}$$

Name	f(u,ut,utt,t)	Initial value (u_0)	Initial value (u_t0)
N2	N1 + N6 – N2	0	0
H2	H1 + H6 – H2	0	0
A2	A1 + A6 – A2	0	0
N4	N2–react –N4	0	0
H4	H2–3*react–H4	0	0
A4	A2+2*react–A4	0	0
N5	splitH*H4–N5	0	0
H5	splitH*H4–H5	0	0
A5	splitA*A4–A5	0	0
N6	N4–N5–N6	0	0
H6	H4–H5–H6	0	0
A6	A4–A5–A6	0	0

- Now solve the problem by right clicking on study1 and choose compute. The problem converges and gives the same answers as in below figure.

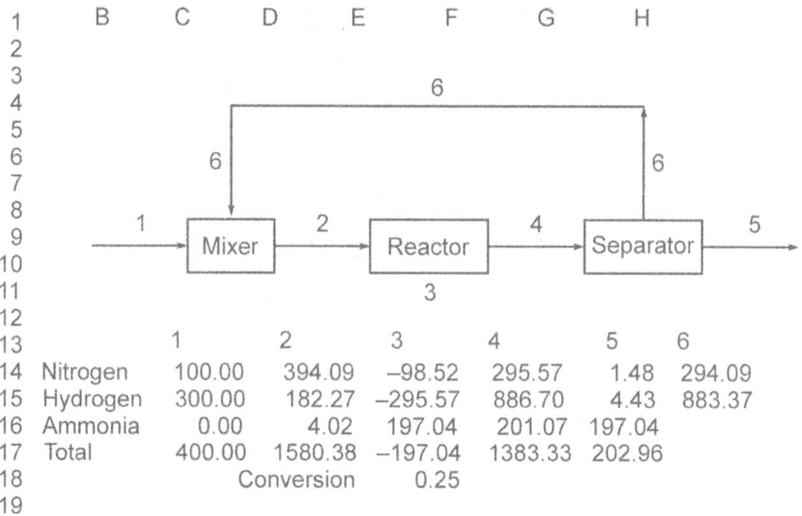

	1	2	3	4	5	6
Nitrogen	100.00	394.09	−98.52	295.57	1.48	294.09
Hydrogen	300.00	182.27	−295.57	886.70	4.43	883.37
Ammonia	0.00	4.02	197.04	201.07	197.04	
Total	400.00	1580.38	−197.04	1383.33	202.96	
		Conversion	0.25			

Fig. 33.3 Material balance for the synthesis loop

Experiment 4

Mass Balance with Equilibrium Reaction and Recycle

Objective

To study mass balance with equilibrium reaction and recycle using COMSOL multiphysics.

Problem statement

The problem is same as that of experiment 3 except for conversion which is modified so that the conversion is determined by chemical equilibrium using the below equation.

Solution using COMSOL

- Follow same as experiment 3, now include variables like equilibrium constant K_p and the pressure (pres).

Variables	
Name	**Expression**
react	conv*N2
N1	100
H1	300
A1	0
splitN	0.005
splitH	0.005
splitA	0.98
total4	N4+H4+A4
Kp	0.05
press	220

- Now global equations are modified by adding the following equation for conversion.

conv kp–total4*(A4/(N4*H4*H4)^0.5)/press

References

CHEMCAD Reference Manual

34

Condensate Stabilizer Plant

Problem Description

Condenser stabilizer plant which is existing for processing the feed of specification mentioned below:

Temperature, T = 75. F; Pressure, P = 200 psia;

Component Flow rates: N2 = 100.19 lb-mol/hr, C1 (Methane) = 4505.48, C2 (Ethane) = 514.00, C3 (propane) = 214.00, iC4(isoButane) = 19.20, nC4 (n-Butane) = 18.18, iC5 (i-Pentane) = 26.40, nC5 (n-Pentane) = 14.00, nC6 (n-Hexane) = 14.00

Our job to take this existing unit shown in the Process Flow Diagram (PFD) and determine new operating conditions and any necessary modifications to achieve:

- The cricondotherm (highest possible) dew point of the product gas must be <=20° F.

- The stabilized condensate must have maximum propane content of 1%.

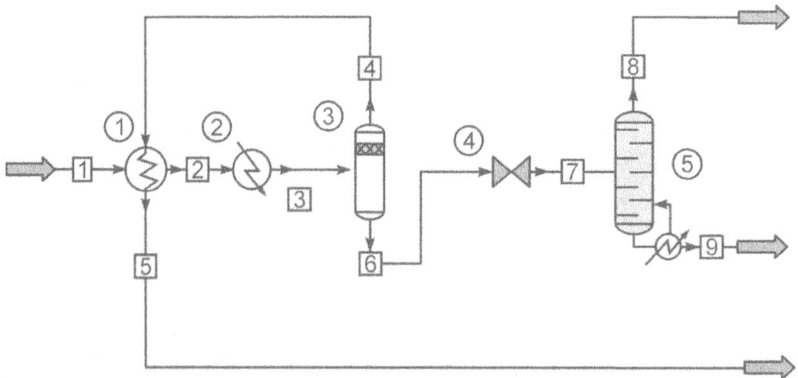

Fig. 34.1 Condensate stabilizer plant flow diagram

Solution Steps

Launch Chemcad. Observe the following:

- Menu bar with options such as File, Edit, View, Format options
- Toolbar with various small icons
- Palette on the right side of the application

1. Start a new job (*File* ➔ *New job* on the menu bar & give the model name as stabilizer).

2. Select the engineering Units (*Format* ➔ *Engineering Units* option on the menu bar)

Fig. 34.2 Engineering units menu option

The ENGLISH units option is the default and is currently highlighted. You may change the engineering units system by clicking any one of the four buttons *English, Alt SI, SI, or Metric*, or you can change any unit individually by clicking on the individual item and then selecting from the list presented.

Click [OK] and Save the file.

3. Now set up the flow sheet topology.

Creating a flow sheet is the process of placing unit operations icons on the screen, connecting them with streams, and then adding various graphical objects to enhance the drawing. All of these things may be done using the palette.

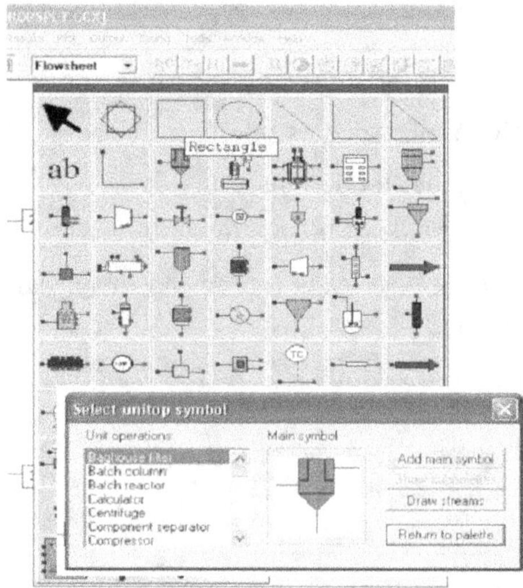

Fig. 34.3 PFD palette for selecting unit operations

By convention, CHEMCAD requires that every stream come from a unit operation and go to a unit operation. Therefore, we must place a *Feed* icon on the flow sheet to serve as the "source" unit for the feed stream. Click on the Feed icon to select the stream (typically Red color Arrow symbol in the Palette) & click again on the flow sheet to place the stream. With this, one stream is created for the flow sheet.

Drag and drop two-sided heat exchanger from the PFD Palette. Identifying the Heat exchanger icon on the PFD Pallette & right clicking on the icon can do this. Click on the heat changer icon that has the shell and tube side connections to select the heat exchanger, now click on the Flow sheet to insert one heat exchanger on the flow sheet. Similarly, Identify one-sided heat exchanger, flash tank, valve and Reboiled tower without condenser and place their icons on the flow sheet. Note that you can use the Arrow button on the palette to identify the unit operations.

Finally, put three Product stream icons on the flow sheet (Typically, purple colored Arrow symbols on the Palette), two streams for tower & one product stream for flash tank top vapors. The Product icon is located slightly to the right.

Connect the Unit operations with connections starting from the Feed stream till the product streams to make the flow sheet complete.

Click the Streams icon box on the palette. The palette will disappear and the crosshairs cursor will appear. Move the cursor close to the tip of the feed arrow. When the cursor switches to a black arrow, press the left button on the mouse. Draw the stream over to the right with your mouse. When the label for left-most inlet arrow on the first heat exchanger appears, press the left button on the mouse. CHEMCAD will draw the stream directly to that point and will place a stream ID number on the flow sheet.

NOTE: Stream and Unit ID numbers can be hidden by clicking on *View* option on menubar and select *Hide Stream IDs* and *Hide Unit IDs* options.

Click [OK] and Save the file.

4. Select Components: Identify which components are to be used in this simulation. Selecting the *ThermoPhysical* command ==> *Components List* option does this. This will show the component selection Data Entry Window (DEW).

You might observe that the *Components List* option is disabled. Components can be selected only when the flow sheet is in *Simulation* mode. Mode can be changed from *Mode* menu bar option OR the list box shown on the toolbar.

Add the following component to the model: (Select one component & click Add button)

2. Methane

3. Ethane

4. Propane

5. I-butane

6. N-butane

7. I-pentane

8. N-pentane

10. N-hexane

46. Nitrogen

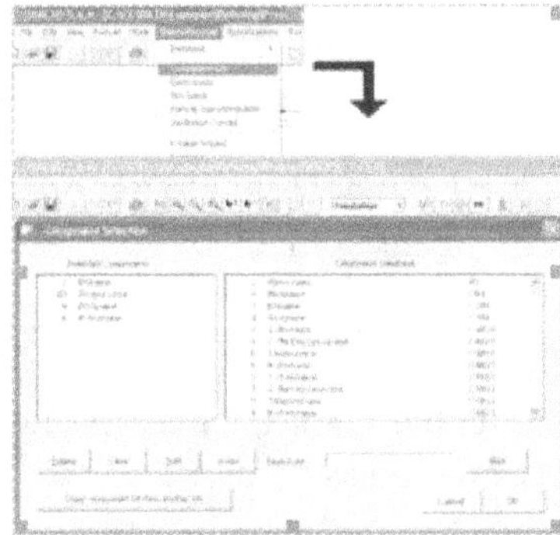

Fig. 34.4 Components selection menu option

After all the components are selected click [OK] on the DEW to go back to the flow sheet & Save the file.

5. Selecting thermodynamic options: Selecting thermodynamic options basically means selecting a model or method for calculating vapor-liquid (or vapor-liquid-liquid) phase equilibrium (called the K-value option) and selecting a method or model for calculating the heat balance (called the enthalpy option). You do this using the *ThermoPhysical* command ==> *K-value* Options OR *Enthalpy*

Options. Set Peng-Robinson as the thermodynamic method for K-Value and Enthalpy calculations.

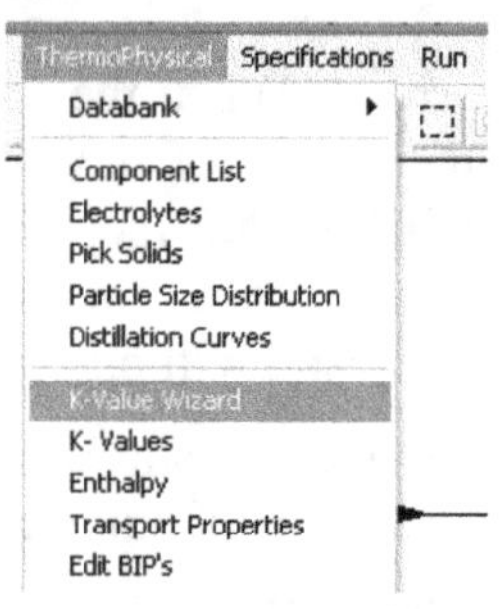

Click [OK] and Save the file.

6. Define Feed streams: Feed streams can be defined by using the Specifications command on the menu bar or by double clicking directly on the stream we wish to specify i.e., stream connecting the Feed icon and the first heat exchanger.

Fig. 34.5
Thermodynamics menu option

Specify the Stream Name as "Plant feed", thermal condition (temperature, pressure, vapor fraction, and enthalpy) as per the conditions given in the problem Description. According to the Gibbs Phase Rule, once the composition is given, specifying any two of the four thermodynamic properties of a mixture will define the other two. Thus, defining the composition, temperature, and pressure uniquely defines the vapor fraction and enthalpy as well (again this is for a mixture only). Alternately, defining the composition, pressure, and enthalpy will uniquely define temperature and vapor fraction.

Enter the data for Feed stream as mentioned in the problem description.

Click [OK] and Save the file.

7. Define Unit Operations:_Unit Operation parameters may be input using either by double clicking directly on the UnitOp OR the *Specifications* command on the menu bar (*Unit Ops* ➔ *Select Unit Ops* ➔ select the unit operation on the flow sheet ➔ [OK]).

Double click on the first heat exchanger, Enter the first heat exchanger information as below:

Pressure drop 1 = 5 psi

Pressure drop 2 = 5 psi

Vapor Fraction (1) = 1.0 click [OK] button to go back to flow sheet.

Similarly, provide the information for the second exchanger:

Pressure drop = 5 psi

Outlet Temperature = –5 F

Specifying the flash drum…

In our example, the flash drum is a vapor liquid separator and requires no specification. Therefore, we do not need to enter any input for this unit.

Specifying the valve

Enter Pressure Outlet as 125psi.

Specifying the stabilizer tower...

Enter the column details as below:

12 stages

Feed stage = 1

Column Pressure drop = 5 psi

Bottom draw = 30 lb-moles/hr [Bottom draw is not available in the main tab. Click on the *Specifications* tab in the DEW, change the "No Reboiler" specification to "Bottom mole flow rate".]

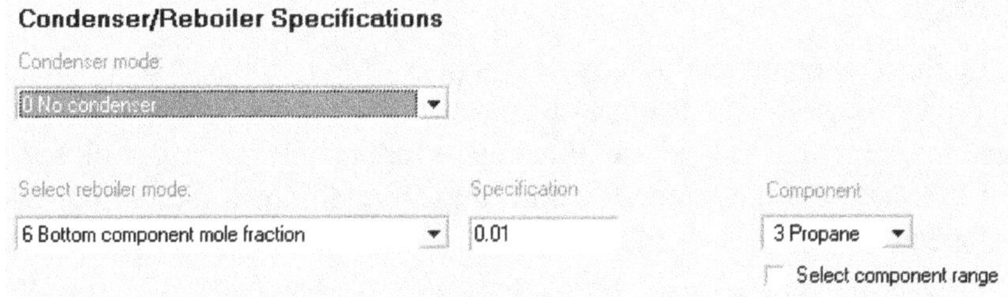

Fig. 34.6 Specifications window

Click [OK] and Save the file.

8. Running the Simulation: To run the simulation, point and click on the *RUN* command on the menu bar. This will cause the **RUN Menu** to open up & select the *Run All* option. *Run* icon is available on the toolbar, which can be used as alternative.

The program will first recheck the data and list any errors and/or warnings on the screen. In this case, we should have no errors, although we will have warnings saying we have not given certain estimates. We can ignore these warnings and proceed by clicking the [YES] button. The calculation will then proceed.

Upon finishing, the CC5 Message Box will appear with the message "Recycle calculation has converged". To close this dialog box and clear the screen, you must click on the [OK] button. After completing the calculations, status "Run finished" message will be shown on the left bottom corner of the flow sheet.

Though this step looks very easy, many users will spend most of the time to make the flow sheet running by resolving the errors. In this case study, there are no errors. Process knowledge is essential in resolving the errors. Very often, Engineers ignore the warnings as the program works even if there are few warnings, but warnings are also to be looked into very seriously to avoid unforeseen impact on the simulation results.

9. Review the results: Simulation results can be reviewed using the *Results* and *Plot* commands on the menu bar.

Checking the cricondentherm dew point...

To plot anything inside CHEMCAD, we must start with the *Plot* command on the menu bar. A list of the general categories of plots available within CHEMCAD. Please select *Envelopes* by clicking on it or by pressing the *[E]* key.

The Select streams dialog box will appear. Move the cursor to stream 5 (the product gas stream leaving the bottom of heat exchanger number 1) Select the stream. The Phase Envelope Options dialog box will appear. Enter 0.25 and the 0.5 vapor fraction lines in addition to the normal phase envelope boundaries.

CHEMCAD will perform the required flash calculations necessary to generate the phase envelope as specified above. Phase envelope results are produced in two formats (Table or Plot).

NOTE: Try to use the zoom options by selecting a region of interest on the graph to expand & right click to go back to the original graph.

Checking the bottoms stream purity...

Our second specification requires that the percent of propane in stabilizer bottom to be 1%. We can check and see if we have achieved this specification by double clicking on the product stream. Try Set Flow Units option if the results are required in mole units.

10. *Re-running the simulation*: Re-run the stabilizer specifying that we want 1% propane in the bottom. To do this, please double click on the *Tower UnitOp*, click on the *Specifications* page, change the specification from *Bottom mole flowrate* to a purity specification. Click on the field below the *Select reboiler mode* option and select *Bottom component mole fraction*. In the *Specification* field enter *.01*. Select Propane in the *component* field. Re-run the flow sheet OR Use Run Selected option in the Run menu to calculate tower only.

11. *Producing a Report*: To produce a hardcopy output of any kind, use the *Output* option on the menu bar ➜ *Report* ➜ *Calculate & give Results* button click. Try to generate output report in MS-Word and MS-Excel (using *Report Format* DEW option). Also try various options available for report printing (stream and Unit Op options).

References

CHEMCAD Reference Manual

Seader and Henley, Separation Process Principles, 2nd Edition, John Wiley & Sons Inc.

35

Refrigeration Loop

Problem Description

A refrigeration loop was originally designed to take advantage of a low temperature stream from another process, to help condense the refrigerant. Now it is proposed that some or all of this stream will be used elsewhere in the plant. You must determine the effect on the refrigeration loop of losing this auxiliary cooling duty.

Process Information

The process flow sheet is shown in *Figure*. The loop is closed and operates with a fixed quantity of refrigerant. The liquid refrigerant before being flashed through the valve is cooled in the HX2er.

The two phases in the stream leaving the cooler are separated in the flash, and the liquid portion is vaporized to provide the required refrigerant duty. The two vapor streams are then combined, recompressed, and condensed.

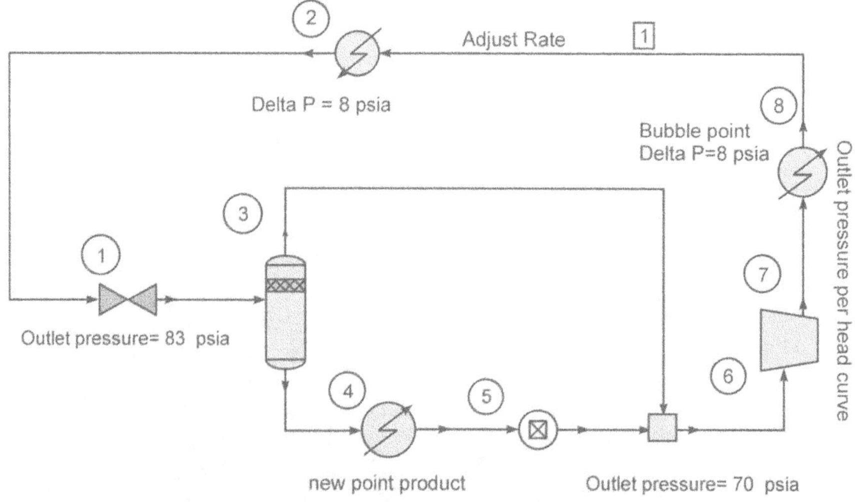

Fig. 35.1 Refrigeration Loop Flowsheet

Table 35.1 Refrigerant Composition

Component	lb moles/hr	Component	lb moles/hr
Ethane	1.04	i-Butane	1.68
Propane	96.94	n-Butane	0.34

Because the process changes will change the flowrate through the existing compressor, performance curves supplied by the manufacturer are used in the calculations. The performance curve data are shown in Table 35.2.

Table 35.2 Performance Curve Data

Volumetric Flowrate, ft3/hr	Head, ft	Adiabatic Efficiency, %
500000	39000	68
912000	38000	69
1018000	37000	71
1237000	35300	72
1356000	33100	73
1427000	27250	71

The Peng-Robinson equation of state has been selected for calculating equilibrium K-values, enthalpies, and entropies. Liquid densities are calculated using the API Lu's correlation. This combination gives accurate results for this type of light hydrocarbon system.

The independent variable in this simulation is the flow rate of the refrigerant. This flow rate is estimated for stream 1 based on the latent heat of propane, and a feedback controller is used to calculate it as shown in *Figure 1*. The vaporizer exit stream is set to its dew point, and the controller varies the rate of stream 1 in order to meet the required refrigerant duty.

We want to study the effect of reducing the duty of HX2. Using CHEMCAD, we can run the base simulation, and then double-click HX2 to lower the duty and observe its effect on other important parameters.

The existing base case is simulated. Then the duty of HX2 is reduced in steps until the program indicates that the combination of required process conditions is impossible. The refrigerant flowrate, the compressor work, and the cooling in the HX8 are inspected at each step.

Solution Steps

- Launch Chemcad.

  1. Start a new job (*File* ➜ *New job* on the menu bar & give the model name as RefrigerationLoop).

  2. Select the engineering Units (*Format* ➜ *Engineering Units* option on the menu bar)

     Change the liquid volume unit to gallons. Click [OK] and Save the file.

  3. Now set up the flow sheet topology. Creating a flow sheet is the process of placing unit operations icons on the screen, connecting them with streams, and then adding various graphical objects to enhance the drawing.

     Create the flow sheet as shown in the figure. This needs the unit operations Valve, 3 Heat exchangers; flash tank, Mixer, controller and Compressor. After placing the icons, right click on the unit and use Flip/Rotate options to change the direction of the connections to match with the figure. Connect the units. This flow sheet does not need Feed and Product icons. Save the file.

  4. Select Components: *ThermoPhysical* ==> *Components List* option will allow to add the required components to the flow sheet (Ethane, Propane,i-butane, N-butane).

     After all the components are selected click [OK] on the DEW to go back to the flow sheet & Save the file.

  5. Selecting thermodynamic options:

     ThermoPhysical ==> *K-value* Options should be used to set Peng-Robinson as the thermodynamic method for K-Value.

     ThermoPhysical ==> *Transport Properties* Option should be used to set API Lu's method for liquid density. Click [OK] and Save the file.

  6. Define Feed streams: Though this flow sheet does not have the Feed icons, Used can specify the initial estimate for HX2 feed as 140F, 315psia, and 20000 lbmol/hr flow rate with composition specified in Table1. Click [OK] and Save the file.

  7. Define Unit Operations: Unit Operation parameters may be input using either by double clicking directly on the UnitOp OR the *Specifications* command on the menu bar (*Unit Ops* ➜ *Select Unit Ops* ➜ select the unit operation on the flow sheet ➜ [OK]).

Double click on the HX2, Enter the first heat exchanger information as below:

Pressure drop = 8 psi; Duty = 12.5 MMBtu/hr. click [OK] button to go back to flow sheet.

Similarly, provide the information for the Valve:

Outlet Pressure = 83psia

Specifying the flash drum…

In our example, the flash drum is a vapor liquid separator and requires no specification. Therefore, we do not need to enter any input for this unit.

Specifying the HX4…

Pressure drop = 0; Vapor fraction=1.

Specifying the mixer…

Outlet Pressure = 70 psia

Specifying the Compressor…

Use Adiabatic Compression; Calculate outlet pressure for the specified flow rate; Supply Performance curves.

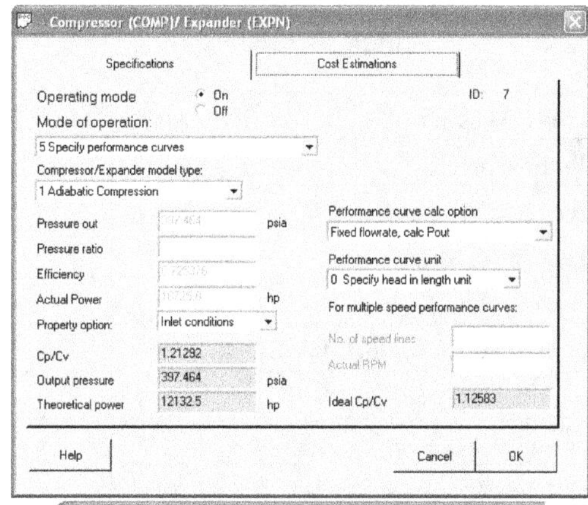

	Flow (ft3/hr)	Efficiency	Head (ft)
1	912000	0.69	38000
2	1018000	0.71	37000
3	1237000	0.72	35300
4	1356000	0.73	33100
5	1427000	0.71	27250
6	0	0	0
7	0	0	0
8	0	0	0
9	0	0	0
10	0	0	0

Specifying the HX8…

Pressure drop = 8 psi; Temperature=140F

Click [OK] and Save the file.

8. Running the Simulation: To run the simulation, point and click on the *RUN* command on the menu bar. This will cause the ***RUN Menu*** to open up & select the *Run Selected Unit Ops* option. Select the Unit operations on the flow sheet in the sequence: HX2, Valve, Flash, HX4, Mixer, Compressor, HX8 and Controller. This allows user to define own calculation sequence.

9. Review the results:_Simulation results can be reviewed using the *Results* commands on the menu bar.

List the results of the main flow sheet parameters in Table 3 for the three cases. As the duty on HX2 is reduced, more circulation is required to deliver the same amount of refrigerant to the evaporator.

The performance curve data in Table 35.3 shows that the compressor head falls with increasing flow rate. At the flow rates in these simulations, the efficiency passes through the maximum in the curve, so the compressor outlet pressure increases and then decreases.

Table 35.3 Results

HX2 duty (MM BTU/hr)	17.00	15.00	12.50
Stream 1 (lb mole/hr)			
Separator (liq. frac.)			
Valve feed pressure (psig)			
Compressor work (hp)			
Compressor actual inlet volume (ft3/hr)			
Compressor outlet temp (F)			
Compressor outlet pres (psia)			
Compressor adiabatic efficiency (%)			
HX8 duty (MM BTU/hr)			

When the refrigerant flow rate exceeds 18000 lb mole/hr, both the efficiency and head fall with increasing flow rate. When this happens, the feed pressure to the valve is reduced, and so the expansion produces less cooling. This, in turn, leads to less liquid for the vaporizer. The controller then increases the refrigerant flow rate, which again reduces the amount of liquid, and it is not possible to supply the specified refrigeration duty.

Conclusions

The plant can operate at a reduced sub-cooler duty at the cost of additional compressor power requirements. Due to increased refrigerant circulation, the line and equipment sizing will need to be validated at that rate.

This simulation shows the importance of considering the real performance characteristics of process equipment. If the compressor were specified with only a defined exit pressure, the problem with increased flow rate might not have been detected until a later and more costly stage.

10. **Producing a Report:** To produce a hardcopy output of any kind, use the *Output* option on the menu bar ➜ *Report* ➜ *Calculate & give Results* button click. Generate output report MS-Excel for HX2 duty=12.5 MMBtu/hr.

References

CHEMCAD Reference Manual

36

Predictive Control of a Wastewater Treatment Process

- Modelling of different industrial wastewater treatment processes: cellulose and paper industry, agricultural farms, spun glass industry, etc. The research team tries to model each process, de- pending on the substances involved in the process. Contrary to the case of domestic wastewater treatment which is made naturally by the microorganisms, the industrial wastewater treatment is done by cultivation of microorganisms, sometimes genetically modified, which consume a certain organic substrate.
- Conditioning the excess of active sludge in order to use it in other industrial activities, especially as a fertilizer in agriculture.

Process control: the wastewater treatment systems are complex, non-linear processes, with multiple inputs and outputs (multivariable), which determine equivocal information about the influent's charac- teristics, the model's structure and parameters. Two approaches can be distinguished in choosing the control structure for such a process: the first one is process-driven and the second one is model-based. The first approach deals with the separate control of the most important variables. Within this category, the well-known problem of controlling the dissolved oxygen level is one of the most important issues for a good operation of the wastewater treatment plants. Thus, a good level of dissolved oxygen allows the optimal growth of microorganisms used in the process. Recently, the control of nitrogen and phosphor level received also a lot of attention. The second approach has been improved a number of times. These improvements are related to the type of the mathematical model used, as it is the case for state estimators. Using simplified models allowed the application of advanced control techniques (e.g. precise linearizing or adaptive control, robust control techniques etc.). However, when using more complex models, such as the ASM1 model, the issue of automatic control became very complicated and the established results were less numerous. For the ASM1 model classic control techniques are usually used (PI, PID controllers), arranged hierarchically, in a three-level structure: at the higher level, a stable trajectory for the process is calculated for a certain period of time; the medium level deals with the trajectory optimization for the dissolved oxygen, the flow of the recycled active sludge and the re- cycled inflow for nitrogen removal; at the lower level, the control of dissolved oxygen concentration is achieved, based on the medium level reference. A well-suited approach for this type of process is the control based on artificial intelligence strategies. Thus the intelligent control exploits the knowledge and experience accumulated from managing the process and puts it across the control structures like expert, fuzzy, neuro-fuzzy systems.

The present study considers a simplified model of the biological wastewater treatment plant. The process is controlled using a model-based predictive control (MPC) strategy. The predictive controller uses a neural network as internal model of the process. This offers various possibilities for the control law adjustment by means of the following parameters: the prediction horizon, the control horizon, the weights of the error and the command. The control purpose is to maintain the substrate concentration below an admissible limit, which is indirectly achieved by controlling the dissolved oxygen concentration, considering as control input the dilution rate D. The predictive control structure has been tested during several functioning regimes, which are essential due to their frequent occurrence in current practice.

The paper is structured as follows: the second section describes the process components and the mathematical model of the plant, the third section introduces theoretical considerations about the control structure used in the paper, while the fourth section refers to the neural network used as internal model of the predictive controller. The fifth section presents the simulation results of the proposed control structure and the last section is dedicated to conclusions.

2.   The model of the wastewater treatment process

The mathematical model considered in this paper has been proposed. The model is based on the following assumptions:

- the system runs in steady-state regime ($F_{in} = F_{out} = F, D = F/V$);

- the recycled sludge is proportional to the process flow (F): $F_r = r \cdot F$, where r is the recycled sludge rate;

- the flow of the sludge removed from the bioreactor (sludge that is in excess) is considered proportional to the process flow (F): $F_\beta = \beta \cdot F$, where β is the removed sludge rate;

- there is no substrate or dissolved oxygen in the recycled sludge flow of the bioreactor;

- the output flow of the aerated tank is equal to the sum between the output flow of the clarifier tank (settler) and the recycled sludge flow.

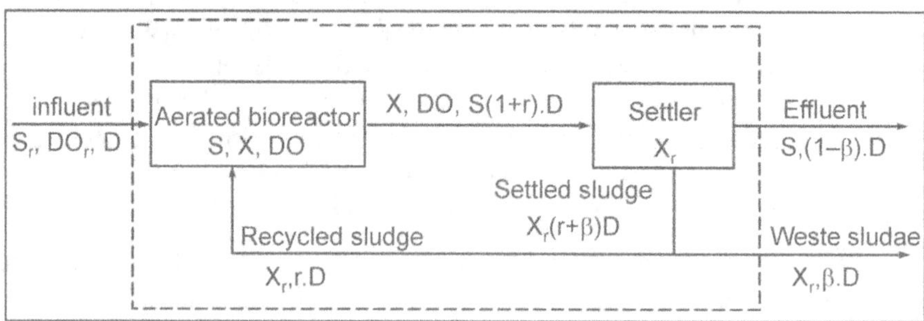

Fig. 36.1 The schematic representation of the wastewater treatment

process

Fig.36.1 presents the schematic representation of the wastewater treatment process. The *Aeration Tank* is a biological reactor containing a mixture of liquid and suspended solid, where a microorganism population is grown in order to remove the organic substrate from the mixture. The *Clarifier Tank* is a gravity settlement tank where the sludge and the clear effluent are separated. A part of the removed sludge is recycled back to the aeration tank and the other part removed.

Under these conditions the process model is given by the following mass balance equations:

$$\frac{dX}{dt} = \mu(t)X\left(1 - D(t)(1+r)X(t) + rD(t)X_r(t)\right) \qquad .....(36.1)$$

$$\frac{dS}{dt} = -\frac{\mu(t)}{Y}X(t) - D(t)(1+r)S(t) + D(t)S_{in} \qquad .....(36.2)$$

$$\frac{dDO}{dt} = K\frac{\mu(t)}{-0Y}X(t) - D(t)(1+r) + \alpha W(DO_{max} - DO(t)) + D(t)DO_m \quad ....(36.3)$$

$$\frac{Dx_r}{dt} = D(t)(1+r)X\left(t - D(t)(\beta+r)X_r(t)\right) \qquad .....(36.4)$$

$$\mu(t) = \mu\max\frac{S(t)}{k+S(t)}\frac{DO(t)}{K+DO(t)} \qquad .....(36.5)$$

where $X(t)$ - biomass, $S(t)$ - substrate, $DO(t)$ - dissolved oxygen, DO_{max} - maximum dissolved oxygen, $X_r(t)$ - recycled biomass, $D(t)$ - dilution rate, S_{in} and DO_{in} - substrate and dissolved oxygen concentra- tions in the influent, Y - biomass yield factor, μ - biomass growth rate, μ_{max} - maximum specific growth rate, k_S and K_{DO} - saturation constants, α - oxygen transfer rate, W - aeration rate, K_0 - model constant, r and β - ratio of recycled and waste flow to the influent. The model coefficients have the following values: $Y = 0.65$; $\beta = 0.2$; $\alpha = 0.018$; $K_{DO} = 2\,mg/l$; $K_0 = 0.5$ $\mu_{max} = 0.15\,mg/l$; $k_S = 100\,mg/l$; $DO_{max} = 10\,mg/l$; $r = 0.6$.

$$\frac{dx}{dt} = \mu(t)X(t) - D(t)(1+r)X(t) + rD(t)Xr(t)$$

$$\frac{dS}{dt} = -\frac{\mu(t)}{Y}X(t)(t) - D(t)(1+r)S(t) + D(t)Sin$$

$$\frac{dDO}{dt} = -kO\frac{\mu(t)}{Y}X(t) - D(t)(1+r)Do(t) + \alpha W\left(DO\,max - DO(t)\right) + D(t)DO\ in$$

$$\frac{dXr}{dt} = D(t)(1+r)S(t) - D(t)(\beta + r)Xr(t)$$

Where X(t)=biomass, S(t)=substrate, DO(t)=dissolved oxygen, DOmax = maximum dissolved oxygen, Xr(t)=recycled biomass, D(t)=dilution rate,Sin and DOin-substrate and dissolved oxygen concentrations in the influent-biomass yield factor,μ-biomass growt rate, μmax-maximum specific growth rate, ks AND Kdo-saturation constant, α-oxygen transfer rate, ko-model constant, r and β-ratio of recycled and waste flow to the influent. The model coefficient have the following values

Y=0.65,β=0.2,α=0.018,Kdo=2mg/l,ko=0.5μmax=0.15mg/l;ks=100mg/l; DOmax=10 mg/l;r=0.6

The open loop response of the system for a step input D=0.1/h(W=80/h) The initial conditions considered in this simulation are: X(0)= 200 mg/l, S(0)=88 mg/l, DO(0)= 5 mg/l, Xr(0)=320 mg/l, DOin=0.5 mg/l ,Sin=200 mg/l Use of With DEE

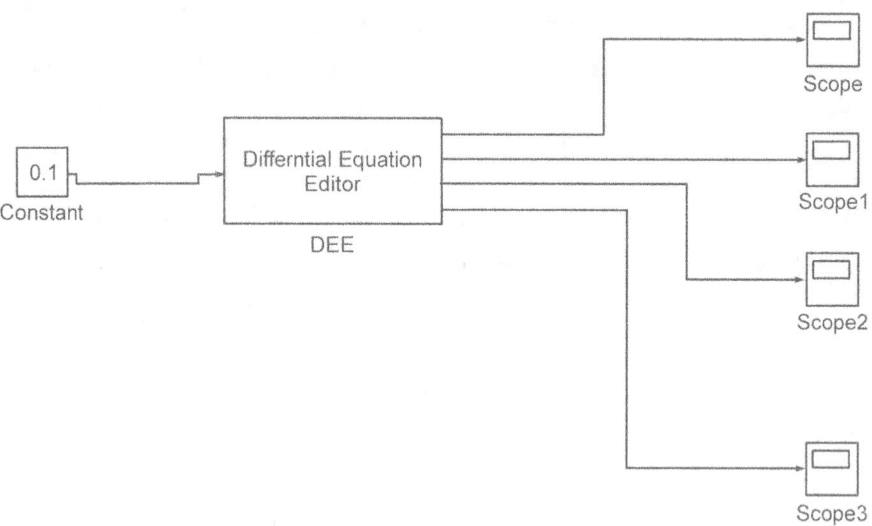

Fig. 36.2 Simulink diagram for the process

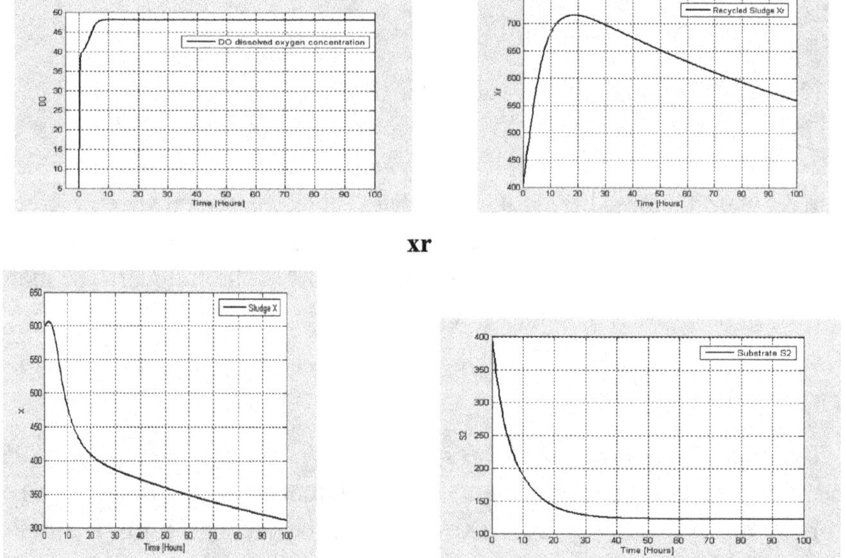

Fig. 36.3 Response curves from Simulation

```
function deriv=usabODE(t, var)
x=var(1);
s=var(2);
do=var(3);
xr=var(4);
mumax=0.15;    %mg/l
```

```
ks=100; %mg/l
kdo=2; %mg/l
mu=mumax*s*do/(ks+s*kdo+do);
d=0.1;%%%% To be verified
r=0.6;
Sin=200;%mg/l
alpha=0.018;
K0=0.5;
w=80; %h-1
D0max=10;
Doin=0.5;%mg/l
beta=0.2;
y=0.65;
deriv(1)=mu*x-d*(1+r)*x+r*d*xr;
deriv(2)=-mu*x/y-(d*(1+r)*s)+d*Sin;
deriv(3)=-K0*mu*x/y-(d*(1+r)*do)+alpha*w*(D0max-do)+d*Doin;
deriv(4)=d*(1+r)*x-d*(beta+r)*xr;
deriv=deriv';
```

Solve ODEs from MATLAB command prompt:

```
>> tspan=[0 100];
>> var0=[200 88 5 320];
>> [time variable]=ode45('usabODE',tspan,var0);
>> x=variable(:,1);
>> s=variable(:,2);
>> do=variable(:,3);
>> xr=variable(:,4);
>> plot(time,x);
>> xlabel('Time(hour)');
>> ylabel('X(mg/l)');
>> title('X vs time plot');
>>
```

References

International Journal of Computers, Communications & Control Vol. II (2007), No. 2, pp. 132-142.

37

Boiler Efficiency Theory

The estimation of the efficiency of a boiler of heater involves computation of several losses such as those due to this gases leaving the unit, unburned fuel, radiation losses, heat loss due to molten ash, and so on. Readers may refer to the ASME Power Test Code (7) for details. Two methods are widely used, one based on the measurement of input and output and the other based on heat losses. The latter is preferred, as it is easy to use.

There are two ways of stating the efficiency, one based on the HHV and the other on LHV.

$$\eta_{HHV} \times HMV = \eta_{LHv} \times LHV$$

The various losses are (1), on an HHV basis,

1. Dry gas loss, L_t:

$$Lt = 24wdg \; \frac{t_g - t_a}{HHV} \tag{37.1}$$

2. Loss due to combustion of hydrogen and moisture in fuel, L_2.

$$L_2 = (9 \times H_2 + W) \times (1080 - 0.46t_g - t_a) \times \frac{100}{HHV} \tag{37.2}$$

3. Loss due to moisture in air, L_3

$$L_3 = 46 \times Mw_{da} \times \frac{t_g - t_a}{HHV} \tag{37.3}$$

4. Radiation loss, L_4. The American Boiler Manufacturers Association (ABMA) chart may be referred to obtain this value. A quick estimate of L_4 is

$$L_4 = 10^{0.62 - 0.42 \log Q} \tag{37.4}$$

For Eqs. (37.1) to (37.4)

W_{dg} = dry flue gas produced, lb/lb fuel

W_{da} = dry air required, lb/lb fuel

H_2, W = hydrogen and moisture in fuel, fraction

  M = moisture in air, lb/lb dry air (see Q1.09b)

t_g, t_a = temperature of flue gas, air, °F

  Q = duty in MM Btu/hr

5. To losses L_1 to L_4 must be added a margin or unaccounted loss, L_5, Hence efficiency becomes

$$\eta_{HHv} = 100 - (L_1 + L_2 + L_3 + L_4 + L_5) \qquad .....(37.5)$$

Note that combustion calculations are a prerequisite to efficiency determination. If the fuel analysis is not available, plant engineers can use the MM Btu method to estimate w_{dg} rather easily and then estimate the efficiency.

The efficiency can also be estimated on an LHV basis. The various losses considered are the following.

1. Wet flue gas loss

$$W_{wg} \times C_p \times \frac{t_g - t_a}{LHV} \qquad .....(37.6)$$

(C_p, gas specific heat, will be in the range of 0.25 to 0.265 for wet flue gases.)

2. Radiation loss

3. Unaccounted loss, margin

Problem Statement

Natural gas having CH_4 = 83.4%, C_2H_6 = 15.8%, and N_2 = 0.8% by volume is fired in a boiler. Assuming 15% excess air, 70 °F ambient, and 80% relative humidity perform detailed combustion calculations and determine flue gas analysis.

Determine the efficiency of a boiler firing the fuel at 15% excess air. Assume radiation loss = 1%, exit gas temperature = 400 °F, and ambient = 70 °F. Excess air and relative humidity (15% and 80%).

Calculation Procedure

We know that air at 70 °F and 80% RH has a moisture content of 0.012 lb/lb dry air. For example, we see that CH_4 require 9.53 mol of air per mole of CH_4 and C_2H_6 requires 16.68 mol.

Let us base our calculations on 100 mol of fuel. The theoretical dry air required will be

$$83.4 \times 9.53 + 16.68 \times 15.8 = 1058.3 \text{ mol}$$

Considering 15% excess,

Actual dry air $= 1.15 \times 1058.3 = 1217$ mol

Excess air $= 0.15 \times 1058.3 = 158.7$ mol

Excess $O_2 = 158.7 \times 0.12 = 33.3$ mol

Excess $N_2 = 1217 \times 0.79 = 961$ mol

(Air contains 21% by volume O_2, and the rest is N_2.)

$$\text{Moisture in air} = 1217 \times 29 \times \frac{0.012}{18} = 23.5 \text{ mol}$$

(We multiplied moles of air by 29 to get its weight, and then the water quantity was divided by 18 to get moles of water.)

$$CO_2 = 1 \times 83.4 + 2 \times 15.8 = 115 \text{ mol}$$

$$H_2O = 2 \times 83.4 + 3 \times 15.8 + 23.5 = 237.7 \text{ mol}$$

$$O_2 = 33.3 \text{ mol}$$

$$N_2 = 961 + 0.8 = 961.8 \text{ mol}$$

The total moles of flue gas produced is $115 + 237.7 + 33.3 + 961.8 = 347.8$. Hence

$$\%CO_2 = \frac{115}{1347.8} \times 100 = 8.5$$

Similarly,

$$\%O_2 = 3.0\%, \qquad \%N_2 = 86.7\%$$

To obtain w_{da}, w_{a2}, w_{dg}, and w_{wg}, we need the density of the fuel or the molecular weight, which is

$$\frac{1}{100} \times (83.4 \times 16 + 15.8 \times 30 + 0.8 \times 28) = 18.30$$

$$W_{da} = 1217 \times \frac{29}{100 \times 18.3} = 19.29 \text{ lb dry air/lb fuel}$$

$$W_{wa} = 19.29 + \frac{23.5 \times 18}{18.3 \times 100} = 19.52 \text{ lb wet air/lb fuel}$$

$$W_{dg} = \frac{115 \times 44 + 33.3 \times 32 + 961 \times 28}{1830}$$

$$= 20.40 \text{ lb wet gas/lb fuel}$$

Dry flue gas = 18 lb/lb fuel

Moisture in air = 19.52 − 19.29 = 0.23 lb/lb fuel

Water vapor formed due to combustion of fuel = 20.4 = 18 − 0.23 = 2.17 /b/lb fuel

$$HHV = \frac{83.4 \times 1013.2 + 15.8 \times 1792}{100} = 1128 \text{ Btu/cu ft}$$

Fuel density at 60°F = 18.3/379 = 0.483 lb/cu ft so

$$HHV = \frac{1128}{0.0483} = 23{,}364 \text{ Btu/lb}$$

The losses are:

1. Dry gas loss,

$$L_t = 100 \times 18 \times 0.24 \times \frac{400 - 70}{23{,}364} = 6.1\%$$

2. Loss due to combustion of hydrogen and moisture in fuel,

$$L_2 = 100 \times 2.17 \times \frac{1080 + 0.46 \times 400 - 70}{23{,}364}$$

3. Loss due to moisture in air,

$$L_3 = 100 \times 0.23 \times 0.46 \times \frac{400 - 70}{23{,}364} = 0.15\%$$

4. Radiation loss = 1.0%
5. Unaccounted losses and margin = 0%

Total losses = 6.1 + 11.1 + 0.15 + 1.0 = 18.35%

Hence

Efficiency on HHV basis = 100 − 18.35 = 81.65%

One can convert this to LHV basis after computing the LHV

References

V. Ganapathy, Steam Plant Calculations Manual, 2nd Edition, Marcel Dekker Inc.

Appendix-I
Basics of COMSOL

Introduction to CFD

Computational fluid dynamics (CFD) is the science of predicting fluid flow, heat transfer, mass transfer, chemical reactions, and related phenomena by solving the mathematical equations, which govern these processes using a numerical process.

Applications of CFD are numerous

- Flow and heat transfer in industrial processes (boilers, heat exchangers, combustion equipment, pumps, blowers, piping, etc.).
- Aerodynamics of ground vehicles, aircraft, missiles.
- Film coating, thermoforming in material processing applications.
- Flow and heat transfer in propulsion and power generation systems.
- Ventilation, heating, and cooling flows in buildings.
- Chemical vapor deposition (CVD) for integrated circuit manufacturing.
- Heat transfer for electronics packaging applications.
- And many, many more!

Advantages

- Relatively low cost
  - Using physical experiments and tests to get essential engineering data for design can be expensive.
  - CFD simulations are relatively inexpensive, and costs are likely to decrease as computers become more powerful.
- Speed
  - CFD simulations can be executed in a short period of time.
  - Quick turnaround means engineering data can be introduced early in the design process.

- Ability to simulate real conditions.

  - Many flow and heat transfer processes cannot be (easily) tested, *e.g.* hypersonic flow.

  - CFD provides the ability to theoretically simulate any physical condition.

- Ability to simulate ideal conditions.

  - CFD allows great control over the physical process, and provides the ability to isolate specific phenomena for study.

  - Example: a heat transfer process can be idealized with adiabatic, constant heat flux, or constant temperature boundaries.

- Comprehensive information.

  - Experiments only permit data to be extracted at a limited number of locations in the system (e.g. pressure and temperature probes, heat flux gauges, LDV, etc.).

  - CFD allows the analyst to examine a large number of locations in the region of interest, and yields a comprehensive set of flow parameters for examination.

Limitations

- Physical models.

  - CFD solutions rely upon physical models of real world processes (e.g. turbulence, compressibility, chemistry, multiphase flow, etc.).

  - The CFD solutions can only be as accurate as the physical models on which they are based.

- Numerical errors.

  - Solving equations on a computer invariably introduces numerical errors.

  - Round-off error: due to finite word size available on the computer. Round-off errors will always exist (though they can be small in most cases).

  - Truncation error: due to approximations in the numerical models. Truncation errors will go to zero as the grid is refined. Mesh refinement is one way to deal with truncation error.

- Boundary conditions.

    - As with physical models, the accuracy of the CFD solution is only as good as the initial/boundary conditions provided to the numerical model.

    - *Example*: flow in a duct with sudden expansion. If flow is supplied to domain by a pipe, you should use a fully-developed profile for velocity rather than assume uniform conditions.

Important Variables: Pressure and flow velocities; Temperature for non-isothermal flows and radiation sometimes; Flow field such as shear and vorticity; Turbulent kinetic energy and dissipation rate.

Governing Equations

- The governing equations include the following conservation laws of physics:

    - Conservation of mass.

    - Newton's second law: the change of momentum equals the sum of forces on a fluid particle.

    - First law of thermodynamics (conservation of energy): rate of change of energy equals the sum of rate of heat addition to and work done on fluid particle.

Equations

- Incompressible (negligible compressibility, press. Changes before density changes, typical time scale $\ll$ typical size / speed of sound) / Compressible flow

- Steady and Unsteady flow (time derivative of any property is zero)

- Viscosity (slip / no-slip)

- Choice of variable arrangement on grid: Collocated / Staggered grid

- Convective / diffusive / source / accumulation terms (Various discretization techniques for each term)

- Lagrangian (moving fluid particle) / Eulerian form (fluid element is at fixed space and time)

- Flow type (Laminar / Turbulant) (stream line flow / vortex or eddies)

Elliptic: Poisson,

Parabolic: Diffusion equation, 1st derivative for time

Hyperbolic: Wave equation, 2nd derivative for time

Cylindrical co-ordinates:

$$x = r\cos(\theta), \qquad y = r\sin(\theta), \qquad\qquad z = z, \qquad \text{(AI.1)}$$

or

$$r = \sqrt{x^2 + y^2}, \quad \theta = \arctan\left(\frac{y}{x}\right), \qquad z = z, \qquad \text{(AI.2)}$$

Spherical co-ordinates

$$x = r\sin(\theta)\cos(\phi), \qquad y = r\sin(\theta)\sin(\phi), \qquad z = r\cos(\theta), \qquad \text{(AI.3)}$$

or

$$r = \sqrt{x^2 + y^2 + z^2}, \qquad \theta = \arctan\left(\frac{\sqrt{x^2+y^2}}{z}\right), \quad \phi = \arctan\left(\frac{y}{x}\right). \text{ (AI.4)}$$

Table 1 Equation of Continuity in Various Coordinate Systems

Rectangular coordinates (x, y, z):

$$\frac{\partial \rho}{\partial t} + \frac{\partial}{\partial x}\rho v_z + \frac{\partial}{\partial z}\rho v_y + \frac{\partial}{\partial z}\rho v_z = 0$$

Cylindrical coordinates (r, θ, z):

$$\frac{\partial \rho}{\partial t} + \frac{1}{r}\frac{\partial}{\partial r}\rho r v_r + \frac{1}{r}\frac{\partial}{\partial \theta}\rho v_\theta + \frac{\partial}{\partial z}\rho v_z = 0$$

Spherical coordinates (r, θ, ϕ):

$$\frac{\partial \rho}{\partial t} + \frac{1}{r^2}\frac{\partial}{\partial r}\rho r^2 v_r + \frac{1}{r\sin\theta}\frac{\partial}{\partial \theta}\left(\rho v_\theta \sin\theta\right) + \frac{1}{r\sin\theta}\frac{\partial}{\partial \phi}\rho v_\phi = 0$$

(Source: Bird et al., 1960)

Table 2 Equation of Motion in Rectangular Coordinates (x, y, z)

In terms of τ:

$$x \text{ component } \rho\left(\frac{\partial vx}{\partial t} + \upsilon x\frac{\partial vx}{\partial x} + vy\frac{\partial vx}{\partial y} + vz\frac{\partial vx}{\partial z}\right) =$$

$$-\frac{\partial p}{\partial x} - \left(\frac{\partial \tau_{xx}}{\partial x} + \frac{\partial \tau_{yx}}{\partial y} + \frac{\partial \tau_{zx}}{\partial x}\right) + \rho gx$$

Table 2 *contd...*

$$y \text{ component } \rho\left(\frac{\partial v_y}{\partial t} + v_x \frac{\partial v_y}{\partial x} + v_y \frac{\partial v_y}{\partial y} + v_z \frac{\partial v_y}{\partial z}\right) =$$

$$-\frac{\partial p}{\partial y} - \left(\frac{\partial \tau_{xy}}{\partial x} + \frac{\partial \tau_{yy}}{\partial y} + \frac{\partial \tau_{zy}}{\partial z}\right) + \rho gy$$

$$z \text{ component } \rho\left(\frac{\partial v_z}{\partial t} + v_x \frac{\partial v_z}{\partial x} + v_y \frac{\partial v_z}{\partial y} + vz \frac{\partial v_z}{\partial z}\right) =$$

$$-\frac{\partial p}{\partial z} - \left(\frac{\partial \tau_{xz}}{\partial x} + \frac{\partial \tau_{yz}}{\partial y} + \frac{\partial \tau_{zz}}{\partial z}\right) + \rho gz$$

In terms of velocity gradients for a Newtonian fluid and constant ρ and μ:

$$x \text{ component } \rho\left(\frac{\partial v_x}{\partial t} + v_x \frac{\partial v_x}{\partial x} + v_y \frac{\partial v_x}{\partial y} + v_z \frac{\partial v_x}{\partial z}\right) =$$

$$-\frac{\partial p}{\partial x} + \mu\left(\frac{\partial^2 v_x}{\partial x^2} + \frac{\partial^2 v_x}{\partial y^2} + \frac{\partial^2 v_x}{\partial z^2}\right) + \rho gx$$

$$y \text{ component } \rho\left(\frac{\partial v_y}{\partial t} + v_x \frac{\partial v_y}{\partial x} + v_y \frac{\partial v_y}{\partial y} + v_z \frac{\partial v_y}{\partial z}\right) =$$

$$-\frac{\partial p}{\partial y} + \mu\left(\frac{\partial^2 v_y}{\partial x^2} + \frac{\partial^2 v_y}{\partial y^2} + \frac{\partial^2 v_y}{\partial z^2}\right) + \rho gy$$

$$z \text{ component } \rho\left(\frac{\partial v_z}{\partial t} + v_z \frac{\partial v_z}{\partial x} + v_y \frac{\partial v_z}{\partial y} + v_z \frac{\partial v_z}{\partial z}\right) =$$

$$-\frac{\partial p}{\partial z} + \mu\left(\frac{\partial^2 v_z}{\partial x^2} + \frac{\partial^2 v_z}{\partial y^2} + \frac{\partial^2 v_z}{\partial z^2}\right) + \rho gz$$

(Source: Bird et al., 1960)

Introduction to Comsol Multiphysics

Many features of COMSOL Multi-physics are not used or needed until two- and three- dimensional problem. Features are explained in this section and the same should be used as reference during the COMSOL practice sessions.

For the boundary value problems, following steps generally applicable:

1. Open Comsol Multiphysics and choose the dimensions of the problem (0D, 1D, 2D or 3D) and the Physics modules based on the problem to be solved.

2. Establish preferences.

3. Create the geometry.

4. Insert the variables that define the problem and set the boundary conditions.

5. Create a mesh.

6. Solve the problem.

7. Examine the solution; (mostly limited to plots and integrals of the solution).

Following icons are worth remembering as these are frequently used:

next finish show   build plot expression range

A.1   Creating a new model

A.2   Adding Parameters

A.3   Creating Geometry, defining boundaries and wall

A.4   Adding Materials

A.5   Adding Physics

A.6   Define Boundary conditions

A.7   Define Mesh

A.8   Solve the model

A.9   Results analysis

APPENDIX 2.1: Creating a new model

1. When COMSOL starts, the model wizard will automatically open. Here define the space dimension.

2. Continue by clicking blue arrow at the top right of model wizard. Next select the applicable physics for the model.

3. The final step in the model wizard is to select the type of study you would like to perform on the model.

APPENDIX 2.2: Model builder

1. After giving the dimensions, physics and study for the model taken in the problem, build a geometry using model builder.
2. There are three ways to do it, one is to import design through CAD, second is directly selecting a shape given in the geometry icon, third is specifying it through the work plane and axis.

APPENDIX 2.3: Material

1. Open model builder – left click on materials tab – left click on material browser.
2. Right click on built in tab – select the material – click on add material to the model.
3. Your model has the properties of given material.

APPENDIX 2.4

1. Open COMSOL and choose 0D option in the model wizard. Select mathematics to expand the folder.
2. Expand 'Global ODEs and DAE interface folder' and select 'Global ODEs and DAEs'.
3. After this select the study as per the problem and click on done.

Launch Window

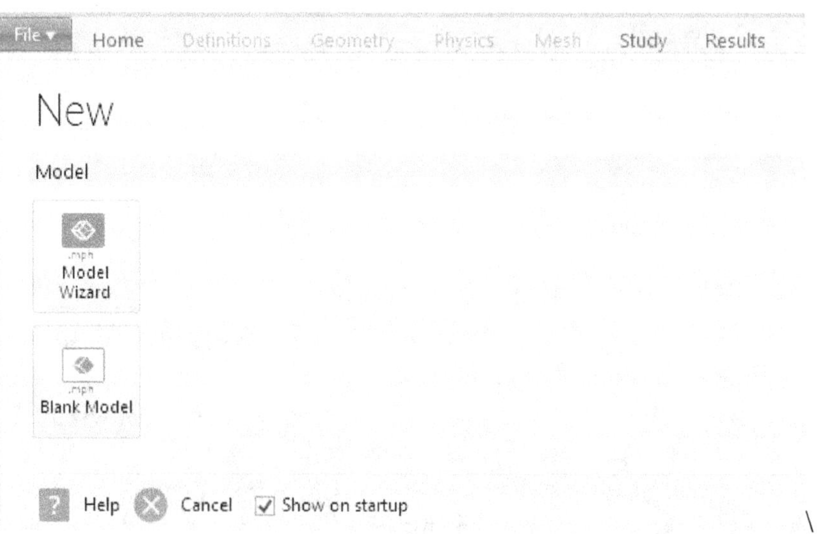

Model Wizard Window:

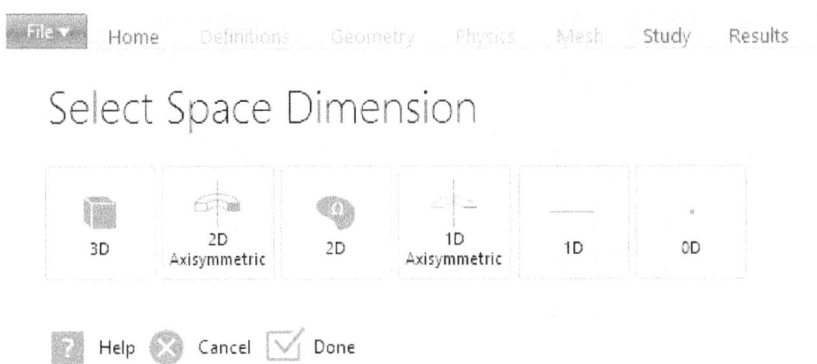

Select 0D

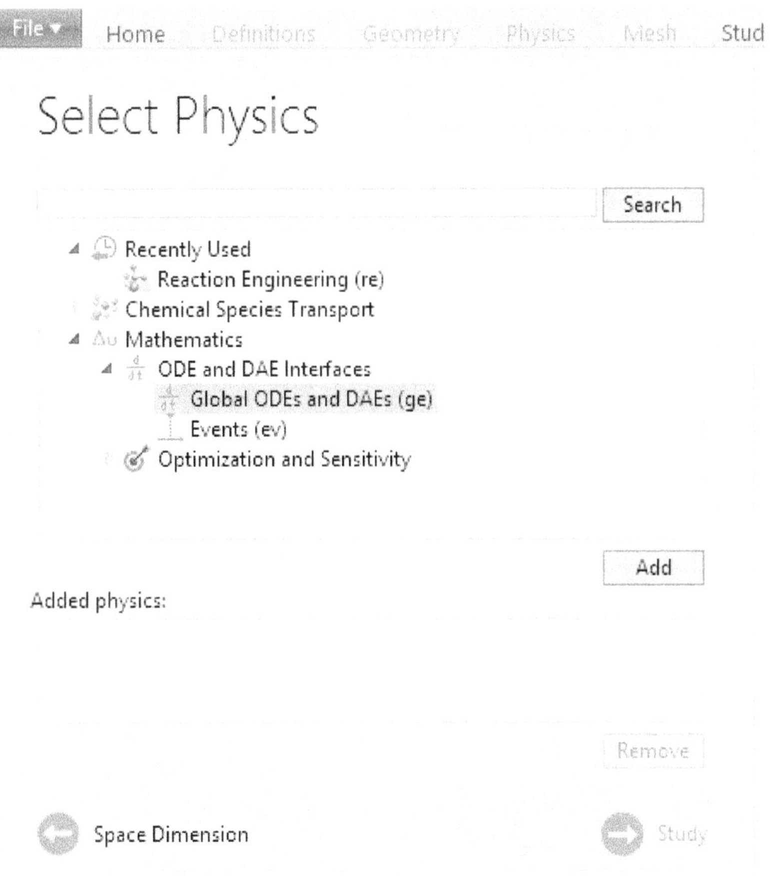

Add Physics by clicking Add button; Click to "Study" to select study type

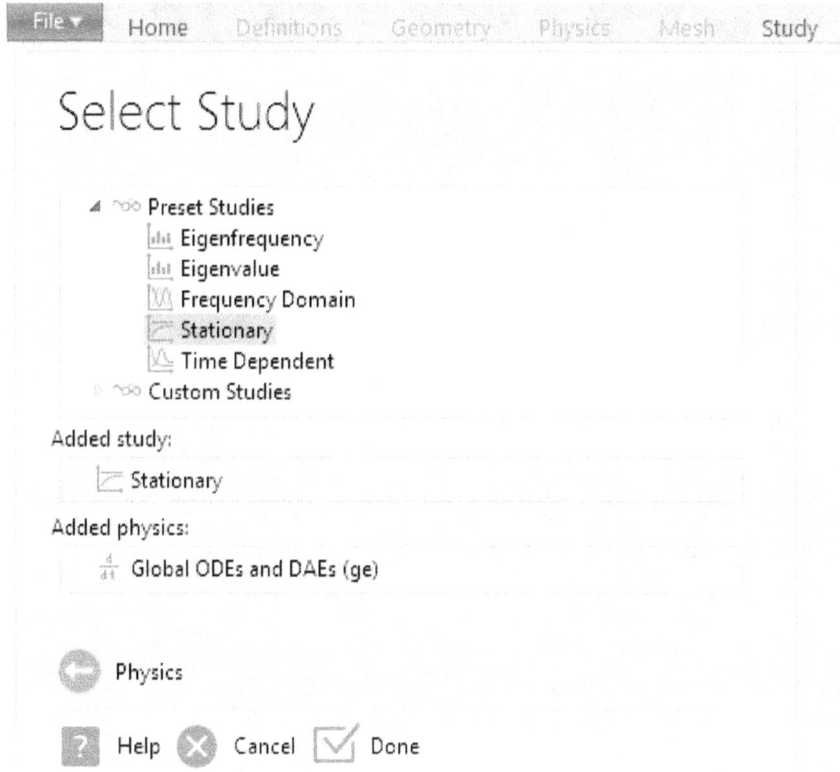

Select study type as "Stationary" and create the model by clicking "Done".

Step 2 Variables window:

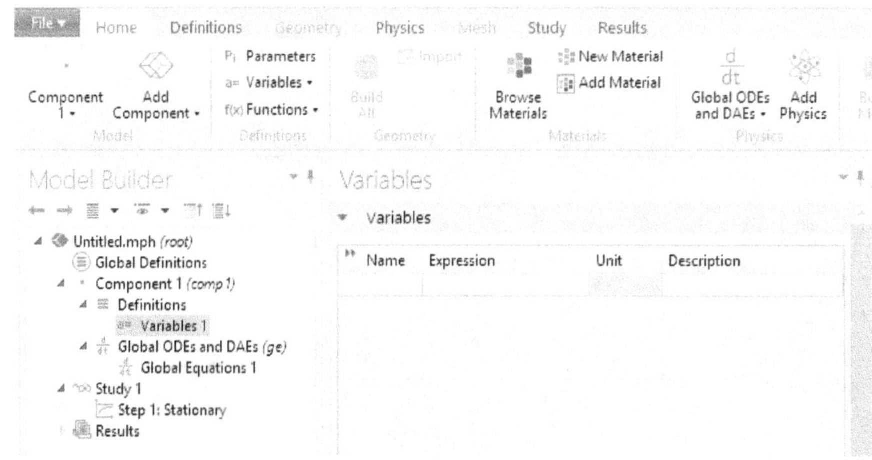

Define the variables as shown in figure.

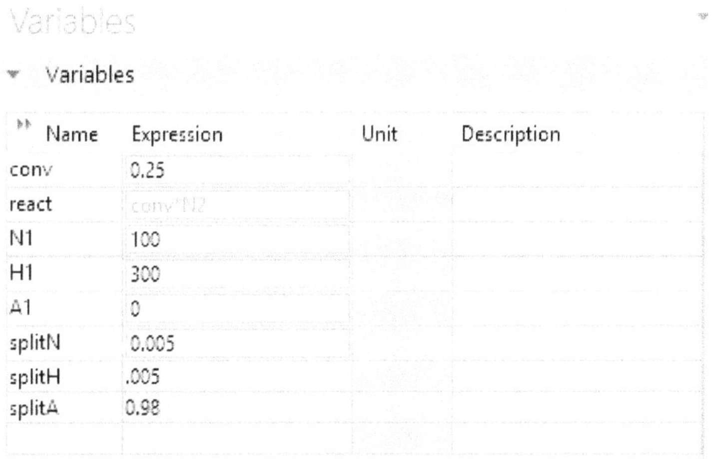

Variables

▼ Variables

Name	Expression	Unit	Description
conv	0.25		
react	conv*N2		
N1	100		
H1	300		
A1	0		
splitN	0.005		
splitH	.005		
splitA	0.98		

Define equations in Global equations section:

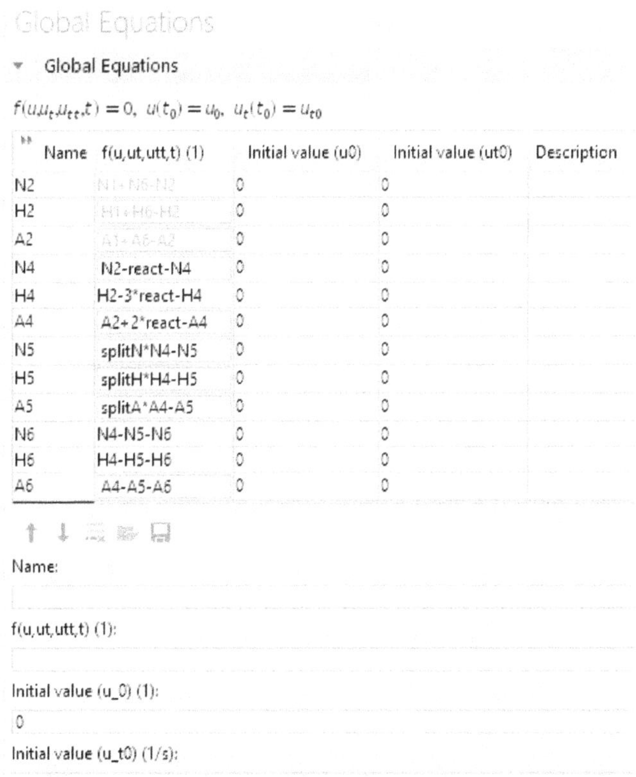

Global Equations

▼ Global Equations

$$f(u, u_t, u_{tt}, t) = 0, \quad u(t_0) = u_0, \quad u_t(t_0) = u_{t0}$$

Name	f(u,ut,utt,t) (1)	Initial value (u0)	Initial value (ut0)	Description
N2	N1+N6-N2	0	0	
H2	H1+H6-H2	0	0	
A2	A1+A6-A2	0	0	
N4	N2-react-N4	0	0	
H4	H2-3*react-H4	0	0	
A4	A2+2*react-A4	0	0	
N5	splitN*N4-N5	0	0	
H5	splitH*H4-H5	0	0	
A5	splitA*A4-A5	0	0	
N6	N4-N5-N6	0	0	
H6	H4-H5-H6	0	0	
A6	A4-A5-A6	0	0	

Name:

f(u,ut,utt,t) (1):

Initial value (u_0) (1):

0

Initial value (u_t0) (1/s):

Step3:

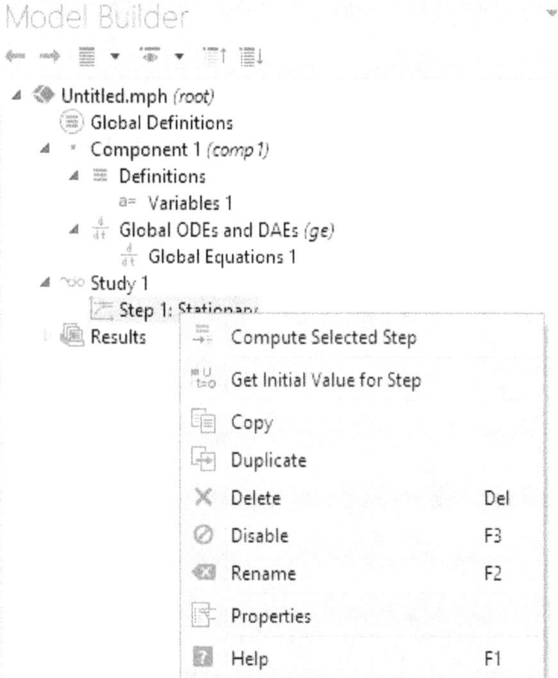

Modifying variables:

Name	Expression	Unit	Description
conv	0.25		
react	conv*N2		
N1	100		
H1	300		
A1	0		
splitN	0.005		
splitH	.005		
splitA	0.98		
total4	N4+H4+A4		
Kp	.05		
pres	220		

Global Equations

▼ Global Equations

$$f(u,u_t,u_{tt},t) = 0, \ u(t_0) = u_0, \ u_t(t_0) = u_{t0}$$

Name	f(u,ut,utt,t) (1)
A4	A2+2*react-A4
N5	splitN*N4-N5
H5	splitH*H4-H5
A5	splitA*A4-A5
N6	N4-N5-N6
H6	H4-H5-H6
A6	A4-A5-A6
conv	Kp-total4*(A4/(N4*H4^3)^0.5)/pres

NOTE: need to remove conv variable.

Global Equations

▼ Global Equations

$$f(u,u_t,u_{tt},t) = 0, \ u(t_0) = u_0, \ u_t(t_0) = u_{t0}$$

Name	f(u,ut,utt,t) (1)	Initial value (u0
N2	N1+N6-N2	300
H2	H1+H6-H2	100
A2	A1+A6-A2	1
N4	N2-react-N4	100
H4	H2-3*react-H4	300
A4	A2+2*react-A4	1
N5	splitN*N4-N5	300
H5	splitH*H4-H5	100
A5	splitA*A4-A5	1
N6	N4-N5-N6	100
H6	H4-H5-H6	300
A6	A4-A5-A6	1
conv	Kp-total4*(A4/(N4*H4^3...	.1

Geometry creation

COMSOL Multiphysics Geometry

In COMSOL Multiphysics you can use solid modeling or boundary modeling to create objects in 1D, 2D, and 3D. They can be combined in the same geometry (hybrid modeling).

During solid modeling you form a geometry as a combination of solid objects using Boolean operations like union, intersection, and difference. Objects formed by combining a collection of existing solids using Boolean operations are known as composite solid objects. Boundary modeling is the process of defining a solid in terms of its boundaries. You can combine such a solid with geometric primitives common solid modeling shapes like rectangles, circles, blocks, cones, and spheres, which are directly available in COMSOL Multiphysics.

In 3D, you can form 3D solid objects by defining 2D solids in work planes and then extrude and revolve these into 3D solids. It is also possible to embed 2D objects in the 3D geometry.

Creating Cartesian and Cylindrical Coordinate Systems

COMSOL Multiphysics uses a global Cartesian or cylindrical (axisymmetric) coordinate system. You select the geometry dimension and coordinate system when starting a new model in the Model Navigator. By default, variable names for the spatial coordinates are x, y, and z for Cartesian coordinates and r, , and z for cylindrical coordinates. These coordinate variables (together with the time for time-dependent models) make up the independent variables in COMSOL Multiphysics models.

The Coordinate Systems and the Space Dimension

The default names for coordinate systems vary with the space dimension:

- Models that you open using the space dimensions 1D, 2D, and 3D use the Cartesian coordinates x, y, and z.

- In 1D axisymmetric geometries the default coordinate is r, the radial direction. The x-axis represents r.

- In 2D axisymmetric geometries the x-axis represents r, the radial direction, and the y-axis represents z, the height coordinate.

For axisymmetric cases the geometry model must fall in the half plane $r \geq 0$.

To select the coordinate system and space dimension, select **1D, 2D, 3D, Axial symmetry (1D),** or **Axial symmetry (2D)** from the **Space dimension** list in the Model Navigator. You can do this before starting a new model or by clicking the **Add Geometry** button when creating models with multiple geometries.

Creating Composite Geometry Objects

You can form a *composite geometry object* by combining objects with Boolean operations: set union, set difference, and set intersection. You can form complex geometries using Boolean formulas that include multiple objects and operations. To perform Boolean operations, use the buttons in the Draw toolbar (2D and 3D) or work in the **Create Composite Object** dialog box.

Using Boolean Operations from the Draw Toolbar

- Click the **Union** button to create the set union of the selected solid objects.

- Click the **Intersection** button to create the set intersection of the selected objects.

- Click the **Difference** button to create the set difference; this action subtracts the union of all other selected solid objects from the largest one.

Creating an Array of Geometry Objects

To create an array of identical geometry objects, choose **Modify** in the **Draw** menu and click **Array** or click the **Array** button in the Draw menu. This opens the **Array** dialog box where you can set the displacements and array size.

Copying and Pasting Geometry Objects

Press Ctrl+C or choose **Copy** from the **Edit** menu to copy the selected geometry objects to the COMSOL Multiphysics clipboard.

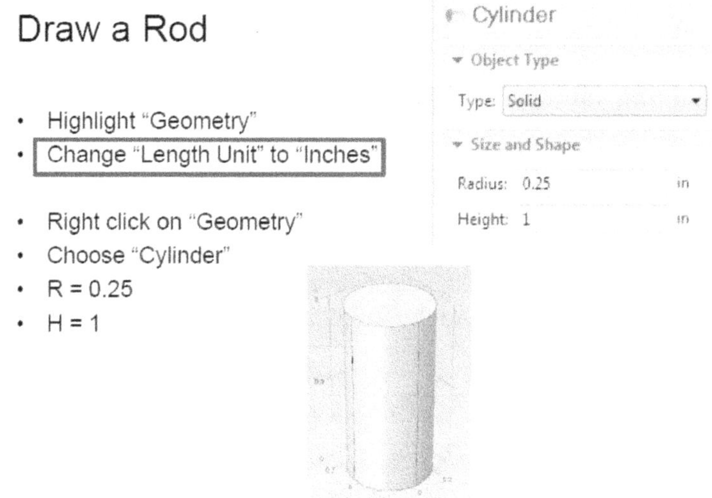

Diverging Duct

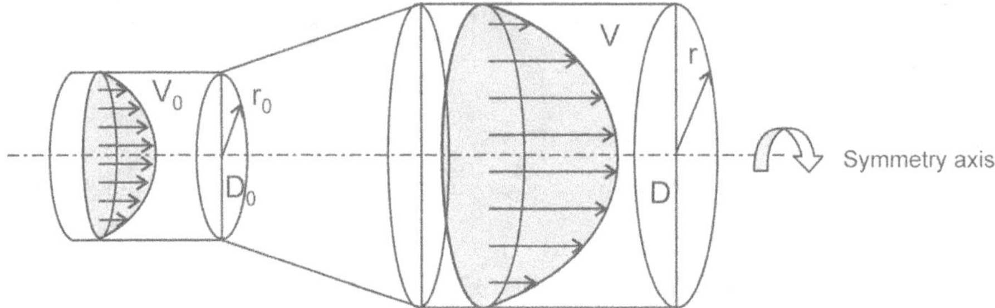

When the diameter of a pipe suddenly increases, as shown in Figure 1, the area available for flow increases. A fluid with relatively high velocity decelerates into a relatively slow moving fluid. Depending on the Reynolds number, this can cause turbulence, and much of the excess kinetic energy converts into heat and is therefore wasted. If the change of the cross section is gradual, it is possible to recover the kinetic energy as pressure energy.

The inlet condition is described by water flowing in a pipe with a diameter of 1 cm. The flow is laminar and has a fully developed laminar velocity profile, a so-called Hagen-Poiseuille profile. After 2 cm the water enters a uniformly diverging duct whose length is 5 cm and whose outlet diameter is 2 cm.

1. Start COMSOL Multiphysics.

2. In the **Model Navigator**, select **Axial symmetry (2D)** from the **Space dimension** list.

3. Choose the application mode **Chemical Engineering Module>Momentum Transport>Laminar Flow>Incompressible Navier-Stokes**.

4. Click **OK**.

Options and Settings

1. From the **Options** menu, select **Axes/Grid Settings**. Specify axis and grid settings according to the following table:

AXIS		GRID	
r min	-0.1	r spacing	0.005
r max	0.1		
z min	-0.005	z spacing	0.005
z max	0.135		

2. Be sure to clear the **Auto** check box on the **Grid** tab, then click **OK**.

3. From the **Options** menu open the **Constants** dialog box. Define the following constants for later use; when done, click **OK**.

NAME	EXPRESSION	DESCRIPTION
rho	9 98[kg/m^3]	Density
eta	1.01 e – 3[Pa*S]	Viscosity
Vmax	2 [cm/S]	Maximum flow velocity at inlet

Geometry Modeling

1. Click the Line button on the Draw toolbar and draw a sequence of lines by using the left mouse button to click at points (0, 0), (0, 0.13), (0.01, 0.13), (0.01, 0.11), (0.005, 0.02), and (0.005, 0).

2. Click the right mouse button to close the curve and create a solid object CO1.

3. Click the Line button on the Draw toolbar and draw a line between (0, 0.11) and (0.01, 0.11).

4. Click the right mouse button to create an interior boundary B1.

5. Click the Line button again and draw a line between (0, 0.02) and (0.005, 0.02).

6. Click the right mouse button to create a second interior boundary B2.

This completes the geometry modelling stage. The geometry in the drawing area on your screen should look like that in the following figure.

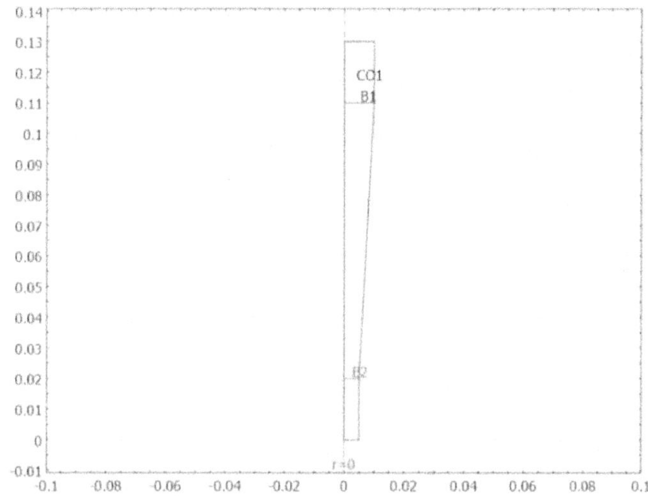

Physics Settings

Subdomain Settings

From the **Physics** menu, select **Subdomain Settings**. Select all three subdomains, then enter material properties according to the following table; when done, click **OK**.

SETTINGS	SUBDOMAINS 1-3
ρ	rho
η	eta

Boundary Conditions

From the **Physics** menu, select **Boundary Settings**. Enter boundary conditions according to the following table; when done, click **OK**.

SETTING	BOUNDARY 2	BOUNDARY 1, 3, 5	BOUNDARY 7	BOUNDARIES 8-10
Boundary type	Inlet	Symmetry boundary	Outlet	Wall
Boundary condition	Velocity	Axial symmetry	Pressure, no Viscous stress	No slip
U0	0			

SETTING	BOUNDARY 2	BOUNDARY 1, 3, 5	BOUNDARY 7	BOUNDARIES 8-10
V0	V_max*(1– S^2)			
P0			0	

Mesh Generation

1. Click the **Initialize Mesh** button on the Main toolbar.

2. Click the **Refine Mesh** button on the Main toolbar twice.

Computing the Solution

Click the **Solve** button on the Main toolbar.

Table 6.5: Equation of Motion in Cylindrical Coordinates

(r, θ, z) (Source: Bird et al., 1960)

In terms of τ:

r component

$$\rho\left(\frac{\partial v_r}{\partial t} + v_r \frac{\partial v_r}{\partial \tau} + \frac{v_\theta}{\tau}\frac{\partial v_r}{\partial \theta} - \frac{v_\theta^2}{\tau} + v_z \frac{\partial v_r}{\partial z}\right) =$$

$$-\frac{\partial p}{\partial r} - \left[\frac{1}{r}\frac{\partial}{\partial r}r\tau_{rr} + \frac{1}{\tau}\frac{\partial \tau_{r\theta}}{\partial \theta} - \frac{\tau_{\theta\theta}}{\tau} + \frac{\partial \tau_{rz}}{\partial z}\right] + \rho g r$$

θ component

$$\rho\left(\frac{\partial v_\theta}{\partial t} + v_r \frac{\partial v_\theta}{\partial \tau} + \frac{v_\theta}{\tau}\frac{\partial v_\theta}{\partial \theta} - \frac{v_r v_\theta}{\tau} + v_z \frac{\partial v_\theta}{\partial z}\right) =$$

$$-\frac{1}{\tau}\frac{\partial p}{\partial \theta} - \left[\frac{1}{\tau^2}\frac{\partial}{\partial r}r^2\tau_{r\theta} + \frac{1}{\tau}\frac{\partial \tau_{\theta\theta}}{\partial \theta} + \frac{\partial \tau_{\theta z}}{\partial z}\right] + \rho g\theta$$

z component

$$\rho\left(\frac{\partial v_z}{\partial t} + v_r \frac{\partial v_z}{\partial \tau} + \frac{v_\theta}{\tau}\frac{\partial v_z}{\partial \theta} + v_r \frac{\partial v_z}{\partial z}\right) =$$

$$-\frac{\partial p}{\partial z} - \left[\frac{1}{\tau}\frac{\partial}{\partial \tau}\tau\tau_{\tau z} + \frac{1}{\tau}\frac{\partial \tau_{\theta z}}{\partial \theta} + \frac{\partial \tau_{zz}}{\partial z}\right] + \rho g z$$

In terms of velocity gradients for a Newtonian fluid with constant ρ and μ:

r component

$$\rho\left(\frac{\partial v_r}{\partial t} + vr \frac{\partial v_r}{\partial \tau} + \frac{\partial_\theta}{\tau}\frac{\partial v_r}{\partial \theta} - \frac{v_\theta^2}{\tau} + v_z \frac{\partial v_r}{\partial z}\right)$$

$$-\frac{\partial p}{\partial \tau} + \mu\left[\frac{\partial}{\partial \tau}\left(\frac{1}{\tau}\frac{\partial}{\partial \tau}\tau v_r\right) + \frac{1}{\tau 2}\frac{\partial^2 v_r}{\partial \theta^2} - \frac{2}{\tau^2}\frac{\partial v\theta}{\partial \theta} + \frac{\partial^2 v_r}{\partial z^2}\right] + \rho g r$$

θ component

$$\rho\left(\frac{\partial v_r}{\partial t} + v_r \frac{\partial v_\theta}{\partial \tau} + \frac{v_\theta}{\tau}\frac{\partial v_\theta}{\partial \theta} + \frac{v_r v_\theta}{\tau} + v_z \frac{\partial v_\theta}{\partial z}\right)$$

$$-\frac{1}{\tau}\frac{\partial p}{\partial \theta} + \mu\left[\frac{\partial}{\partial r}\left(\frac{1}{\tau}\frac{\partial}{\partial \tau}\tau v_\theta\right) + \frac{1}{\tau^2}\frac{\partial^2 v_\theta}{\partial \theta^2} + \frac{2}{\tau^2}\frac{\partial v_r}{\partial \theta} + \frac{\partial^2 v_\theta}{\partial z^2}\right] + \rho g_\theta$$

z component

$$\rho\left(\frac{\partial v_z}{\partial t} + v_r \frac{\partial v_z}{\partial \tau} + \frac{v_\theta}{\tau}\frac{\partial v_z}{\partial \theta} + v_z \frac{\partial v_z}{\partial z}\right)$$

$$-\frac{\partial p}{\partial z} + \mu\left[\frac{1}{\tau}\left(\tau \frac{\partial v_z}{\partial \tau}\right) + \frac{1}{\tau^2}\frac{\partial^2 v_z}{\partial \theta^2} + \frac{\partial^2 v_z}{\partial z^2}\right] + \rho g_z$$

Simplified form of Navier – Stokes Equation

Conservative Form

Conservation of mass:

$$\frac{\partial \rho}{\partial t} + \frac{\partial (\rho ui)}{\partial xi} = \frac{\partial \rho}{\partial t} + \rho \frac{\partial ui}{\partial xi} + ui \frac{\partial \rho}{\partial xi} = 0$$

Conservation of momentum:

$$\frac{\partial (\rho ui)}{\partial t} + \frac{\partial (uj\rho ui)}{\partial xj} - \frac{\partial \tau ij}{\partial xj} + \frac{\partial p}{\partial xi} = 0$$

Conservation energy:

$$\frac{\partial (\rho E)}{\partial t} + \frac{\partial (uj\rho E)}{\partial xj} - \frac{\partial}{\partial xi}\left(k \frac{\partial T}{\partial xi} \right) + \frac{\partial (ujp)}{\partial xj} - \frac{\partial (\tau ij uj)}{\partial xj} = 0$$

For incompressible fluids, assuming μ as constant

$$\frac{\partial ui}{\partial xi} = 0 \quad \text{and} \quad \tau_{ij} = \mu\left(\frac{\partial ui}{\partial xj} + \frac{\partial uj}{\partial xi} \right) \quad \text{when I and j are taken as 1 and 2 for}$$

deriving one momentum equation, we get

$$\frac{\partial u1}{\partial t} + u1 \frac{\partial u1}{\partial x1} + u2 \frac{\partial u1}{\partial x2} = -\frac{1}{\rho}\frac{\partial p}{\partial x1} + v \frac{\partial^2 u1}{\partial x_1^2} + v \frac{\partial^2 u1}{\partial x_2^2}$$

Solution of the Navier-Stokes Equations

- Discretization of the convective and viscous terms
- Discretization of the pressure term
- Conservation principles
- Choice of Variable Arrangement on the Grid
- Calculation of the Pressure
- Pressure Correction Methods
  - A Simple Explicit Scheme
  - A Simple Implicit Scheme
- Nonlinear solvers, Linearized solvers and ADI solvers
- Implicit Pressure Correction Schemes for steady problems
  - Outer and Inner iterations
  - Projection Methods
    - Non-Incremental and Incremental Schemes
  - Fractional Step Methods:
    - Example using Crank-Nicholson

Appendix - II
MATLAB Basics

MATLAB, which stands for MATrix LABoratory, is a technical computing environment for high-performance numeric computation and visualization. MATLAB is both an environment and a programming language. MATLAB allows you to build your own reusable tools. You can create your own special functions and programs known as M files in MATLAB code.

- Software package for computation in engineering, science, and applied mathematics.
- Published by The MathWorks (www.mathworks.com).

Applications

- Math and computation
- Algorithm development
- Data acquisition
- Modeling, simulation, and prototyping
- Data analysis, exploration, and visualization
- Scientific and engineering graphics
- Application development, including graphical user interface building

When compared to other numerically oriented languages like C++ and FORTRAN,

- MATLAB is easy to use
- Comes with a huge standard library.
- Powerful command prompt
- An **interpreted** environment (NOT a compiled computer language)

Getting Started with MATLAB

Launch MATLAB using the shortcut icon on the desktop OR

Start -> Programs -> MATLAB xx -> MATLAB xx

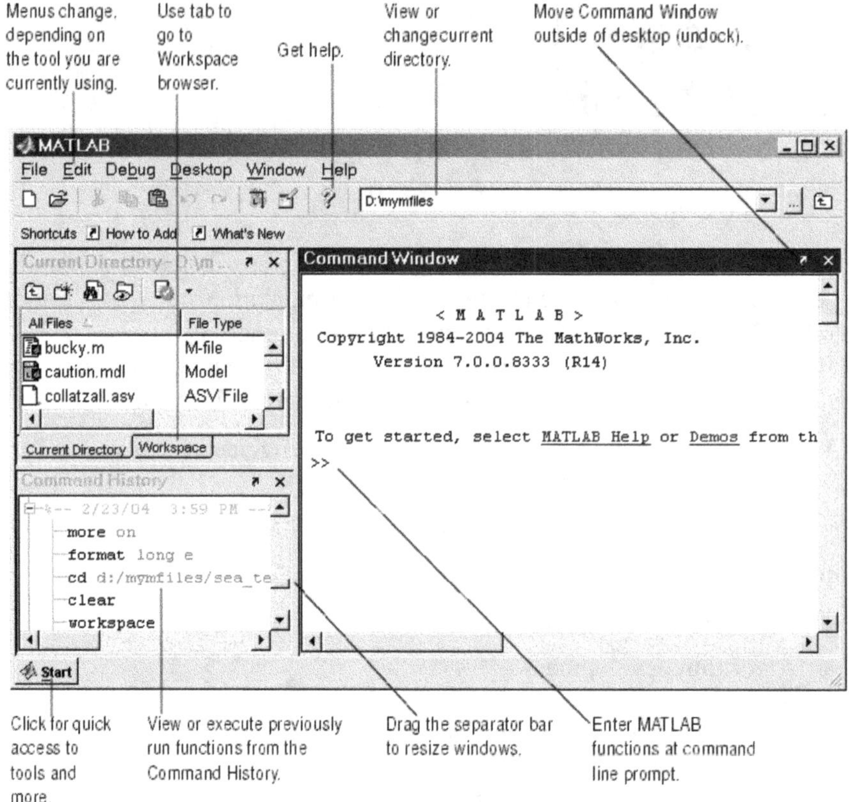

Help

- helpdesk for interactive help
- help for known commands e.g., help plot
- doc for displaying the help browser for the specific function e.g., doc plot
- lookfor when a keyword is known instead of the exact command, e.g. lookfor

Calculations at MATLAB Command line

```
>> 1+2*3

ans =7
```

Multiple commands separated by commas or semicolons can be placed on one line:

```
>> a=2, b=3; c=4;
```

Commas tell MATLAB to display results; semicolons suppress printing.

All text after a percent sign (%) is taken as a comment statement:

```
>> items=a+b+c; %calculates the total items
```

```
>> total=a*25+b*35+c*45;
```

A succession of three periods tells MATLAB that the rest appears on the next line:

```
>> average=total/ ...
```

```
items
```

```
average =37.2222
```

MATLAB offers the following basic arithmetic operations:

Operation	Symbol	Example
addition,	a+b	2+15.3
subtraction,	a-b	25-12.7
multiplication	a*b	3.14*4.56
division,	a/b / or \	56/8=8\56
exponentiation,	a^b	5^3

```
>> save mydata.mat  % saves all the data to a file with
name mydata.matTo ask MATLAB for a list of the variables
it stores currently, use the command who
```

```
>> who
```

Your variables are

a   ans   average b

c   items   total

To recall a previous command use the cursor keys ↑ → ↓ ← on your key board; ↑ recalls the most recent command, →   ← keys can be used to edit a command. Mouse & Clipboard are used to cut, paste and edit the text at the command prompt.

```
>> clear b a*;
```

```
>> who
```

Your variables are:

c   items   total

```
>> clear                    % wipes out all variables in
the workspace
```

```
>> who

>> load mydata.mat

>> who                          % displays all the variables
stored earlier
```

Variables

Variable names can contain up to 31 characters. Variable names must start with a letter, followed by any number of letters digits or underscores. Punctuation characters are not allowed. MATLAB uses the following special variables:

ans, pi, eps, flops, inf, NaN or nan, I and j, nargin, nargout, realmin, realmax

Variable names are case sensitive in MATLAB

Some Common Mathematical Functions:

abs(x)	absolute value	floor(x) round to an integer
acos(x)	inverse cosine	log(x) natural logarithm
acosh(x) yperbolic	inverse cosine	log10(x)common logarithm
		sin(x) sine
asin(x)	inverse sine	sinh(x) hyperbolic sine
asinh(x) hyperbolic sine	inverse	sqrt(x) square root
atan(x) tangent	inverse	tan(x) tangent
atanh(x) hyperbolic tangent	inverse	tanh(x) hyperbolic tangent
cos(x)	cosine	
cosh(x) cosine	hyperbolic	
exp(x)	exponential	

MATLAB works only in radians, where pi radians is equal to 180 degrees.

Mathematical Functions for working with arrays and matrices, linear algebra, data analysis, and other areas of mathematics are available in MATLAB and classified into following categories:

Arrays and Matrices

Linear Algebra

Elementary Math

Data   Analysis   and   Fourier Transforms

Polynomials

Interpolation   and   Computational Geometry

Coordinate System Conversion

Nonlinear Numerical Method

Specialized Math

Sparse Matrices

Math Constants

help documentation is available category wise and alphabetical wise also.

Trigonometric functions	sin, cos, tan, sin, acos, atan, sinh, cosh, tanh, asinh, acosh, atanh, csc, sec, cot, acsc
Exponential functions	exp, log, log10, sqrt
Complex functions	abs, angle, imag, real, conj
Rounding and Remainder functions	floor, ceil, round, mod, rem, sign

```
>> d=4.345;

>> format short

>> d

d =

    4.3450

>> format long e

>> d

d =

  4.345000000000000e+000
```

Some of the number display formatting commands:

format short	50.833	5 digits
format long	50.8333333	16 digits
format short e	5.0833e+01	5 digits
format long e	5.08333e+01	16 digits
format short g	50.833	
format long g	50.8333333	

format bank	50.83	2 decimal digits
format +	+	positive
format rat	305/6	rational

Questions

1. Convert temperature from 102F to C

2. Polonium has a half-life of 140 days. Starting with 10 grams today, how much is left after 250 days? Amount left = initial amount $(0.5)^{time/half-life}$

3. Find the value of the following expressions from MATLAB Command prompt:

2+2, 4^2, $\sin(\pi/2)$, 1/0, α/α, $x=\sqrt{2}$, $\tan^{-1}(x)$, epsilon, reminder of 123456/789, $\log_{10}(2)$

Predefined functions in MATLAB:

fzero is a one-dimensional solver

Example: Heat Capacity for CO_2

$$Cr = 1.716 - 4.257 \cdot 10^{-6} T - \frac{15.04}{\sqrt{T}}$$

For which T is $C_p = 1 (KJ/kg\ K)$?

>> syms T;

>> cpfun=0.716-4.257e-6*T-15.04/sqrt(T);

>> res=fzero('cpfun',600)

The *'fsolve'* function will probably be the most useful function for simple engineering problems. It is essentially a numeric solver, capable of solving systems of non-linear *continuous* equations.

For the system of n continuous equations (f_1--f_n) with n unknown variables (x_1-- x_n) given by

$$f(1) \equiv f_1(x_1, x_2, x_3, ..., x_n) = 0$$

$$f(2) \equiv f_2(x_1, x_2, x_3, ..., x_n) = 0$$

$$f(3) \equiv f_3(x_1, x_2, x_3, ..., x_n) = 0 ...$$

$$f(n) \equiv f_n(x_1, x_2, x_3, ..., x_n) = 0$$

Matlab's solver can be used to determine the unknown variables, x_1 through x_n. The following example will illustrate the use of the *fsolve* function

Example Solve the system of equations below using fsolve.

$$2a - b - e^{-a} = 0$$

$$2b - a - e^{-b} = 0$$

Solution: The most efficient way to solve these types of systems is to create a function m- file that contains the equations.

```
function f = example15(x)

f = [2*x(1)-x(2)-exp(-x(1)); -x(1)+2*x(2)-exp(-x(2)];
```

The equations are separated within the matrix 'f' using the semi-colon operator. Now the 'fsolve' command must be used to solve the system.

```
>>x0=[=5=5];%Initial guess, arbitrary

>>

>> fsolve('example5',xo, optimset('fsolve')) %Command calls
m-file to solve
```

The solutions to the system are $a = 0.5671$ and $b = 0.5671$. Using the 'fsolve' command, much more complicated systems can be easily solved in Matlab. NOTE: 'optimset' is used to change the default solver in Matlab to the optimization toolbox solver (fsolve, 2.0 onwards).

Arrays

- The simplest way to construct a small array is by enclosing its elements in square brackets e.g., A = [1 2 3; 4 5 6; 7 8 9], b = [0;1;0]

- Separate columns by spaces or commas, and rows by semicolons or new lines.

- Arrays can be built out of other arrays, as long as the sizes are compatible e.g., [A b]

- Bracket constructions are suitable only for very small matrices. For larger ones, there are many useful functions such as colon operator. The format is first:step:last e.g., 1:8, 0:2:10, 1:-.5:-1

- It is frequently necessary to access a block extracted from the matrix e.g., A(2,3), A(1:2,2:3), A(1,2:end), A(:,3)

- Vectors can be given a single subscript. In fact, any array can be accessed via a single subscript. Multidimensional arrays are actually stored linearly in memory, varying over the first dimension, then the second, and so on e.g., A(2), A(:),A([1;2;3;4])

- Subscript referencing can be used on either side of assignments e.g., b(1,:) = A(1,:), C = rand(2,5), C(:,4) = [],C(2,:) = 0, C(3,1) = 3

Create an array by Separating elements by spaces or commas   `>> a=[1,2,3]`

Create a column vector just separate the values with semicolons`>> c=[4;5;6]`

Create an array starts at 0, increments by 0.1*pi and ends at pi

`>> x=(0:0.1:1)*pi`

Create an array (first, last, and number of values)   `>> a=1:5,b=1:2:9`

Create an array by concatenation   `>> c=[a b], d=[a;b]`

`>> e=cat(1,a,b),f=cat(2,a,b),g=cat(3,a,b)`

Create an array (first, last, and number of values).linearly spaced

`>> x=linspace(0,pi,11)`

Create an array (first, last, and number of values) logarithmically spaced

`>> logspace(0,2,11)`

Change orientation by transpose operation   `>> d=a'`

Arrays can also be in the form of matrices: `>>g=[1 2 3 4;5 6 7 8]`

Enter matrix with Return key   `>>g=[1 2 3 4 5 6 7 8 9]`

+, -, *, and / operation by a scalar apply the operation to all elements of the array: `>> h=2*g-1`

Element wise operation for matrices of same size   `>> g.*h, g.^2`

Access an element of an array   `>> A(3)`

Access a block of elements using colon

`>>A(1:5)          >>A(1:end)          >>A(3:-1:1)`

Access all rows and 2nd column   `>> g(:,2)`

Access 3rd row and all columns   `>> g(3,:)`

Standard arrays:

Create array and initialize with zero   `>> zeros(2,3)`

Create array and initialize with one   `>> ones(2,3)`

The Identity Matrix (m-by-n)   `>> eye(m,n)`

Create diagonal matrix   `>> A = diag([12:4:32])`

Create random matrix   `>> R = rand(m,n)`

Create sparse matrix   `>> S = sparse(A)`

Create sparse identity matrix   `>> S = speye(m,n)`

Information commands:	**building commands:**
size — size in each dimension	toeplitz — constant on each diagonal
length — size of longest dimension (for vectors)	triu — upper triangle
ndims — number of dimensions	tril — lower triangle linspace evenly spaced entries
find — ndices of nonzero elements	
Linear algebra commands:	**Questions:**
rank — rank	1. Find a table that converts ^0C (from 0^0 to 100^0 with intervals of 20) to ^0F. Hint: c = 5*(f - 32)/9
det — determinant	
expm — matrix exponential	2. Solve the system of linear equations given below:
lu — LU factorization	
(Gaussian elimination)	3. 5u+2v=5
chol — Cholesky factorization	-1.5u+2.8v+1.9W=-1
eig — eigenvalue decomposition	$-2.5v+3w=2$
svd — singular value decomposition	

Graphics and Visualisation

Plots

The following reaction data has been obtained from a simple decay reaction A ➔B. Use MATLAB to plot the concentration of component A in mol/L against the reaction time, t, in minutes. Title the plot, label the axes, and obtain elementary statistics for the data

Time (Minutes)	0	1	3	6	9	12	15	18	21
Concentration (moles/lit)	100	80	65	55	49	45	42	41	38

The most fundamental plotting command is plot

```
>>C=[100 80 65 55 49 45 42 41 38]

>>time=[0 1 3 6 9 12 15 18 21]

>>plot(time,C)

>> xlabel('time (minute)')

>> ylabel('Concentration (mol/lit)')

>> title('Concentration of A in reaction')
```

Plot Syntax: *plot* (independent, dependent, options) where options (*linespec*) consists of color, marker and line style.

For formatting the following table can be used

Color (symbol)	Marker (symbol)	Line style (symbol)
blue(b)	point(.)	Solid line (-)
green(g)	circle(o)	dotted line(:)
red(r)	x-mark(x)	dash-dot line(-.)
yellow(y)	plush(+)	dashed line (–)
black(k)	star(*)	
white(w)	square(s)	
	diamond(d)	
	triangle-up(^)	
	triangle-up(v)	
	triangle-down(<)	
	triangle-right(>)	

Linear 2-D plot

```
>> plot(X,Y)
```

Plot a function between specified limits

```
>> fplot(function,limits)
```

Easy to use function plotter

```
>>ezplot
(f,[xmin,xmax,ymin,ymax])
```

Create a figure graphics object

```
>>figure
```

Hold current graph in the figure

```
>>hold on OR >>hold off
```

Log-log scale plot

```
>>loglog
(X1,Y1,X2,Y2,LineSpec,X3,Y3)
```

Semi logarithmic plot

```
>>semilogx(X,Y)
>>semilogy(X,Y)
```

Create graphs with y-axes on both left and right side

```
>>plotyy(X1,Y1,X2,Y2)
```

Plot discrete sequence data

```
>>stem(x,y)
```

Create and control multiple axes

```
>>subplot(m,n,p)
```

subplot divides the current figure into rectangular panes that are numbered rowwise. Each pane contains an axes. Subsequent plots are output to the current pane. subplot(m,n,p) creates an axes in the pth pane of a figure divided into an m-by-n matrix of rectangular panes. The new axes becomes the current axes.

```
>>  t= linspace(-2*pi,2*pi,10);

>>  stem(t,cos(t))

>> figure
```

```
>> subplot(3,3,1)
>> text(.5,.5,{'subplot(3,3,1)';'or
Subplot
  331'},'FontSize',14,'HorizontalAlignment','center')
>> subplot(3,3,2:3)
>> text(.5,.5,{'subplot(3,3,2:3)';})
>> subplot(3,3,[4 7])
>>text(.5,.5,{'subplot(3,3,[4
7])';},'HorizontalAlignment', 'center')
```

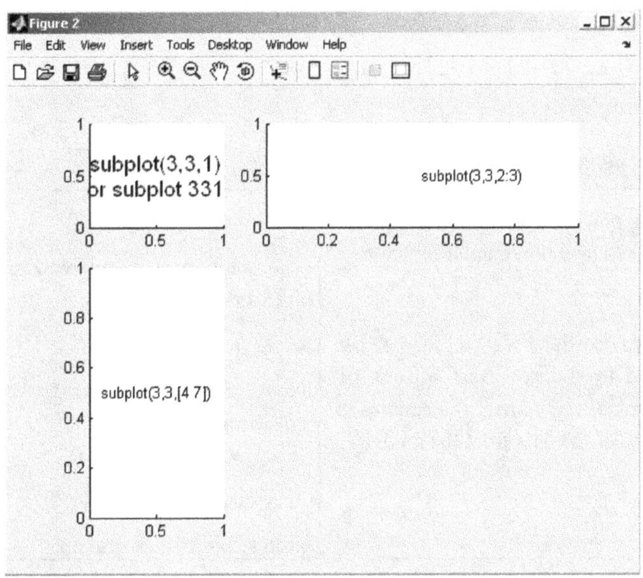

Pie charts:	Bar charts:
pie (a, b) creates a pie chart with a is a vector of values, and b is an optional logical vector describing a slice or slices to be pulled out of the pie chart. $>> x = [1\ 3\ 0.5\ 2.5\ 2]$; explode = [0 1 0 0 0]; pie (x,explode) Colormap jet	This example plots a bell-shaped curve as a bar graph and sets the colors or the bars to red. `>> x = -2.9:0.2:2.9;` bar(x,exp(x–x.*x),'r')

1. Temperature dependency of density and viscosity of water is given by the
   following equations

$$\rho = 62.122 + 0.0122T - 1.54 \cdot 10^{-4} T^2 + 2.65 \cdot 10^{-7} T^3 - 2.24 \cdot 10^{-10} T^4$$

$$\text{In } \mu = -11.0318 + \frac{1057.51}{T + 214.624}$$

where T is in $^{\circ}F$, ρ is in lb_m / ft^3, and μ is in lb_m / fts

Calculate the density ρ, viscosity μ and the kinematic viscosity (μ/ρ) between 35 and 200 $^{\circ}F$ with a 5 $^{\circ}F$ increase, and plot viscosity and kinematic viscosity against temperature.

Regression and Curve Fitting

Suppose you measure a quantity y at several values of time t.

t = [0 .3 .8 1.1 1.6 2.3]';

y = [0.5 0.82 1.14 1.25 1.35 1.40]';

plot(t,y,'o'), grid on

Polynomial regression

Based on the plot, it is possible that the data can be modeled by a polynomial function Y=a0+a1*t+a2*t^2 The unknown coefficients a0, a1, and a2 can be computed by doing a least squares fit, which minimizes the sum of the squares of the deviations of the data from the model.	X = [ones(size(t)) t t.^2] The solution is found with the backslash operator. a= X\y Now evaluate the model at regularly spaced points and overlay the original data in a plot. T = (0:0.1:2.5)'; Y = [ones(size(T)) T T.^2]*a; plot(T,Y,'-',t,y,'o'), grid on

Linear-in-the-Parameters Regression

Instead of a polynomial function, we could try using a function that is linear-in-the parameters. In this case, consider the exponential function Y=a0+a1*exp(-t)+a2*t*exp(-t)	X = [ones(size(t)) exp(-t) t.*exp(-t)]; a = X\y Now evaluate the model at regularly spaced points and overlay the original data in a plot. T = (0:0.1:2.5)'; Y = [ones(size(T)) exp(-T) T.*exp(-T)]*a; plot(T,Y,'-',t,y,'o'), grid on

Multiple Regression

Suppose we measure a quantity y for several values of parameters x1 and x2.	The observations are entered as x1 = [.2 .5 .6 .8 1.0 1.1]';

Table *Contd...*

A multivariate model of the data is	x2 = [.1 .3 .4 .9 1.1 1.4]';
y=a0+a1*x1+a2*x2	y = [.17 .26 .28 .23 .27 .24]';
Multiple regression solves for unknown coefficients a0,a1, and a2, by performing a least squares fit.	X = [ones(size(x1)) x1 x2]; a = X\y

To validate the model, find the maximum of the absolute value of the deviation of the data from the model. Y = X*a; MaxErr = max (abs(Y - y))

Demo of Curve Fitting Toolbox:

Questions

- The thermal conductivity of the metal strip at different temperatures is given in the following table. Fit the data in straight line using least square regression technique.

Temperature (T),K	300	320	340	360	380	400
Thermal conductivity (k), W/m.K	7.5	7.9	8.3	8.7	9.1	9.5

- A new microorganism has been discovered which at each cell division yields three daughter cells. The growth rate data during the batch cultivation is given below:

Time(t),h	0	.5	1	1.5	2.0
Dry wt(x),g/l	0.1	0.15	0.23	0.34	0.51

Fit the above data using least square regression in the exponential growth model $x=a*e^{bt}$ where a and b are constants

Data Structures

Cell Arrays

A cell array can store a variety of data types. A cell array is enclosed by curly brackets. Here, we might store the following data in a variable to describe the Antoine coefficients for benzene and the range they are relevant for [Tmin Tmax]

```
>> c = {'benzene' 6.9056 1211.0 220.79 [-16 104]}
>> c{5}
```

Data Structures

Structures are MATLAB arrays with named "data containers" called *fields*. MATLAB has capability to build structure arrays with any valid size or shape, including multidimensional structure arrays.

Building Structure Arrays Using Assignment Statements:

```
>> patient.name = 'John';

patient.billing = 127.00;

patient.test

= [79 75 73; 180 178 177.5; 220 210 205];

>> patient(3).name = 'Johnson'

>> patient(1).test(2,2)

%Accessing fields
```

Adding a field to structure:

```
>> patient(2).PAN = 'ADQPJ2354A';
```

Remove a field from structure:

```
>> patient = rmfield(patient, 'name');
```

Build using *struct* function:

```
>> s = struct

('name','benzene','A',6.9056,'B',1211.0')

>> s.C = 220.79, s.Trange = [-16 104]
```

Accessing field names:

```
>> fieldnames(s)
```

Build using *repmat* function:

```
>>repmat(struct('name','benzene','A',6.9056,'B',1211.0
'),1,4);
```

struct with cell array syntax:

```
>> weather = struct('temp', {68, 80, 72}, 'rainfall',
{0.2, 0.4, 0.0});

>> weather(2).rainfall
```

Building nested structures:

```
>> A = struct('data', [3 4 7; 8 0 1], 'mystruct',...

struct('testnum', 'Test 1', 'xdata', [4 2 8], 'ydata',
[7 1 6]));
```

List of Functions

struct	Create structure array
fieldnames	Field names of structure, or public fields of object
getfield	Field of structure array
isfield	Determine whether input is structure array field
isstruct	Determine whether input is structure array
orderfields	Order fields of structure array
rmfield	Remove fields from structure
setfield	Assign values to structure array field
arrayfun	Apply function to each element of array
structfun	Apply function to each field of scalar structure
cell2struct	Convert cell array to structure array
struct2cell	Convert structure to cell array

Differential Equations

Symbolic differentiation and integration can be performed using *diff*, *int* commands e.g., diff(a*x^2+b*x), int(2*a*x+b) after defining a,b,x as symbols using *syms* command	```>> syms a b x```
	
	```>> diff(a*x^2+b*x)```
	
	```ans = 2*a*x+b```
	
	```>> int(ans)```
	
	``` ans = a*x^2+b*x```

- *dsolve* can solve ordinary differential equations symbolically with or without boundary conditions or initial value parameters

- dsolve('D2y=6*y-Dy','x') to solve the ordinary differential equation $\dfrac{d^2y}{dx} = 6y - \dfrac{dy}{dx}$ dsolve ('D2m+m=0','m(0)=2','Dm(0)=3','x') to solve the ode $\dfrac{d^2m}{dx^2} + m = 0, m(0) = 2$ and $\dfrac{dm}{dx}(0) = 3$ where m(x).

- **Numerical solution for ODEs:** *ode45, ode15* etc. can be used to numerically differentiate the ODEs

Solving Single ODE

A fluid of constant density starts to flow into an empty and infinitely large tank at 8 L/s. A valve regulates the outlet flow to a constant 4L/s. Derive and solve the differential equation describing this process over a 100 second interval.

Create a function ODE1.m with 2 line code:

```
Function dvdt=ODE1(t,v)

dvdt=4

>>timespan=[0 100];

>>v0=0       %Initial condition

>>[t,v]=ode45('ODE1',timespan,v0)

[independent, dependent]=ode45('file_name',tspan,initial_condition_vector)

>>plot(t,v(:,1))

>>xlabel('Time(s)')

>>ylabel('Volume in tank(L)')

>>title('ODE describing the tank volume')
```

Solver	Solves These Kinds of Problems	Method
ode45	Nonstiff differential equations	Runge-Kutta
ode23	Nonstiff differential equations	Runge-Kutta
ode113	Nonstiff differential equations	Adams
ode15s	Stiff differential equations and DAEs	NDFs (BDFs)
ode23s	Stiff differential equations	Rosenbrock
ode23t	Moderately stiff differential equations and DAEs	Trapezoidal rule
ode23tb	Stiff differential equations	TR-BDF2

Solving Multiple ODEs

The following set of differential equations describes the change in concentration three species in a tank. The reactions A ➜B ➜C occur within the tank. The constants k1, and k2 describe the reaction rate for A➜ B and B➜ C respectively. The following ode's are obtained:

$$\frac{dCa}{dt} = -k1Ca$$

$$\frac{dCb}{dt} = k1Ca - k2Cb$$

$$\frac{dCc}{dt} = k2Cb$$

Where k1= 1 hr-1 and k2 = 2 hr-1 and at time t=0, Ca=5 mol and Cb=Cc=0mol. Solve the system of equations and plot the change in concentration of each species over time.

Select an appropriate time interval for the integration.

Solution:

Data:

```
K1 = 1 hr⁻¹        k2 = 2 hr⁻¹
At t=0             Ca = 5mol          Cb = Cc = 0 mol
      function dcdt=batch(t,c)
      %c(1)=ca, c(2)=cb, c(3)=cc
      global k1 k2
      dcdt=zeros(3,1);
      dcdt=[(-k1*c(1));((k1*c(1))-
      (k2*c(2)));(k2*c(2))];
```

Main Program

```
clc
clf
clear
global k1 k2
k1=1;
k2=2;
tspan[08];
c0=[500];
[t,c]=ode45('batch',tspan,c0);
plot(t,c(:,1),'+',t,c(:,2),'*',t,c(:,3),'x')
legend('ca','cb','cc')
xlabel(Time(hr)')
ylabel('Conc.(mol/hr)')
title('Batch Reactor')
```

Solve Higher order ODE:

$$Y'' + Y' + Y = 0, Y(0) = 1 \text{ and } Y'(0) = 0$$

```
function derive=higherode(t,y)
dYdt=y(2);
dY2dt=-(y(2)+y(1));
derive=[d Y dt dY2dt];
>> tspan=[0 10]
```

```
>> y0=[10]
>> plot(t,Y]=ode45('higherode',tspan,y0);
>> legend('Y','dYdt');
>> xlabel('t');
>> ylabel('Y and dY/dt');
```

Differential Equation Editor Demo is available in demos by typing dee at MATLAB command prompt

Files

MATLAB provides a number of file-system functions and commands to list file names, view and delete M-files, show and change the current directory of folder, etc. Some of these commands are:

```
cd                      show  present directory or folder

cd path                 change to directory given by path

delete test.m           delete the M-file test.m

dir                     list all files

edit test               open test.m for editing

exist ('test','file')   check existence of file test.m

Opening file           fid = fopen

                        ('filename','permission')

Reading file           A  =  fread(fid)  %reads  data  in
                        binary format from fid into matrix
                        A.

Read formatted
data from file         A = fscanf(fid,format)

Check for
end-of-file            eofstat = feof(fid)

Writing to a file count    =    fwrite(fid,A,precision)
                       %file  should  be  opened  in  write
                       mode

Write formatted
data to file           count = fprintf(fid,format,A)

Get file position indicator

                       position = ftell(fid)

Seek file position
indicator              status = fseek(fid,offset,origin)

Query about errors
```

in file I/O   message = ferror(fid)

Move the file position
to the bof   frewind(fid)

Close one or more
open files   status = fclose(fid) or status =
                     fclose('all')

Read line from file, discard newline charactertline =
fgetl(fid)

Read line from file, keep newline character   tline =
fgets(fid,nchar)

Find a string within another, longer string   k   =
findstr(str1,str2)

File name construction:
Return parts of filename
                [pathstr,name,ext,versn]
                = fileparts('filename')

Return directory separator for this platform
                f = filesep

full filename from parts
                f = fullfile('dir1','dir2',...,
                'filename')

name of system's
temporary directory tmp_dir = tempdir

unique string for use as

temporary filename tmp_nam = tempname

Microsoft Excel Functions

Determine if file contains Microsoft Excel (.xls)
spreadsheet
                [type, sheets] = xlsfinfo
                ('filename')

Read Microsoft Excel spreadsheet file (.xls)
                N = xlsread('filename',

                sheet, 'range')

Write Microsoft Excel spreadsheet file (.xls)
                xlswrite('filename', M,

                sheet, 'range')

Lotus123 Functions

Read Lotus123 WK1 spreadsheet file into matrix

```
M = wk1read(filename,r,c)
```

Write matrix to Lotus123 WK1 spreadsheet file

```
wk1write(filename,M,r,c)
```

Text Files

Read numeric data from text file, using comma delimiter

```
M = csvread('filename', row, col)
```

Write numeric data to text file, using comma delimiter

```
csvwrite('filename',M,row,col
```

Read numeric data from text file, specifying your own delimiter   M = dlmread('filename',

```
delimiter, R, C)
```

Write numeric data to text file, specifying your own delimiter

```
dlmwrite('filename', M, 'D', R, C)
```

Read data from text file, write to multiple outputs
[A,B,C,...] =

```
textread('filename','format')
```

Read data from text file, convert and write to cell array   C = textscan(fid, 'format')

Parse XML document   DOMnode = xmlread(filename)

Serialize XML Document Object Model node

```
xmlwrite(filename, DOMnode)
```

Transform XML document using XSLT engine result   =
xslt(source, style, dest)

Images

Read image from graphics file I = imread(filename)

Convert image to instance of Java image object

```
jimage = im2java(I) % I is handle
```
       from *imread*

Return information about graphics file

```
info = imfinfo(filename)
```

Write image to graphics file

```
imwrite(A,filename)
```

Thermo database access example is available by typing thermos at the MATLAB command prompt

Programming

Script M-file is the MATLAB commands in a simple text file with an extension ".m". To create a M-file choose New from the File menu and select M-file. This brings up a text editor window where you enter MATLAB commands.

After entering the required commands, file can be saved as the M-file frictionfactor.m on your disk by choosing Save from the File menu. Then MATLAB will execute the commands in the file when you type the file name at the MATLAB prompt:

```
>> frictionfactor
```

```
% frictionfacor.m
% Done on 16-Dec-2001
D = input('Dia in meter = ');
v = input('Velocity in m/s = ');
rho = 1000; % Density of water in kg/m3
mu = 0.001; % Viscoity of water in kg/m.s
NRe = D*v*rho/mu
f = 0.079*NRe^(-0.25);
disp(f);
```

```
» frictionfactor

Dia in meter = 2

Velocity in m/s = 3
```

There are two types of m-files, *functions* and *scripts*. A MATLAB function has variables that can be passed into and out of the function. Any other variables used inside the function are not saved in memory when the function is finished. Scripts, on the other hand, save all their variables in the MATLAB workspace. Functions and scripts have names like *myfunction.m*. the first line of a function must contain a function declaration, using the following format:

```
function [output1, output2,
output3]=myfunction(input1, input2,input3)
```

Command lines immediately following the function declaration comprise the help file for the function. To obtain information on any function, simply type help function.

```
Function m-File flow.m
function [NRe, f] = flow(D,v,rho,mu)
% flow(Diamter, Velocity, Density, Viscosity)
% Outputs NRe and f
% Done on 16-Aug-2012
NRe = D*V*rho/mu;
f = 0.079*NRe^(-0.25);
```

Calling a function

```
» [Re1, f1] = flow(3,1.5,900,0.01)
```

Some useful functions with M-files:

```
disp(ans)               display results without variable names
echo                    control the Command window echoing of
script file commands
input                   prompt user for input
pause                   pause  until user  presses any key
pause(n)                pause for n seconds
waitforbuttonpress      pause until user presses mouse button
```

Constructing a Function Handle   fhandle = @functionnam

```
» trigFun = {@sin, @cos, @tan};
» plot(trigFun{2}(-pi:0.01:pi))
```

Calling a Function Using Its Handle   fhandle(arg1, arg2, ..., argn)

Return information about a function handle   S = functions(funhandle)

Construct a function name string from a function handle   s = func2str(fhandle)

Construct a function handle from a function name string   fhandle = str2func('str')

Determine if a variable contains a function handle.   K = isa(fhandle, 'function_handle')

Determine if two function handles are handles to the same function.
tf = isequal(A,B)

Return maximum filename length   len = namelengthmax

Determine if input is a valid variable name   tf = isvarname('str')

Construct valid variable name from string   varname = genvarname(str)

e.g genvarname(['reading'
datestr(clock, 'HHMMSS')])

- The *nargin* and *nargout* functions enable you to determine how many input and output arguments a function is called with.

- The *varargin* and *varargout* functions let you pass any number of inputs or return any number of outputs to a function.

Global Variables

Use global variables sparingly. The global workspace is shared by all of your functions and also by your interactive MATLAB session. The more global variables you use, the greater the chances of unintentionally reusing a variable name, thus leaving yourself open to having those variables change in value unexpectedly. This can be a difficult bug to track down.

If several functions, and possibly the base workspace, all declare a particular name as global, then they all share a single copy of that variable. Any assignment to that variable, in any function, is available to all the other functions declaring it global.

```
>> global MAXLEN
```

Alternatives to Using Global Variables:

- Pass the variable to other functions as an additional argument.
- Use a persistent variable (You can declare and use them within M-file functions only; Only the function in which the variables are declared is allowed access to it.)
- persistent SUM_X % At beginning of M-file

Debugging and Profiling

Use the Editor/Debugger to create and debug M-files, which are programs you write to run MATLAB functions. The Editor/Debugger provides a graphical user interface for text editing, as well as for M-file debugging.

- Standard Break point: To set a standard breakpoint using the Editor/Debugger, click in the breakpoint alley at the line where you want to set the breakpoint. OR click the Set/Clear Breakpoint button on the toolbar after keeping the cursor at required line.

- Conditional Break point: Select Set/Modify Conditional Breakpoint from Debug or Breakpoints or context menu and then specify the condition in pop-up window.

- Error breakpoints: Select Debug -> Stop if Errors/Warnings. In the resulting Stop if Errors/Warnings for All Files dialog box, specify error breakpoints on all appropriate tabs and click OK.

- Disable a breakpoint: right-click the breakpoint icon and select Disable Breakpoint from the context menu, or click anywhere in a line and select Enable/Disable Breakpoint from the Breakpoints or context menu.

- Icon bar options:

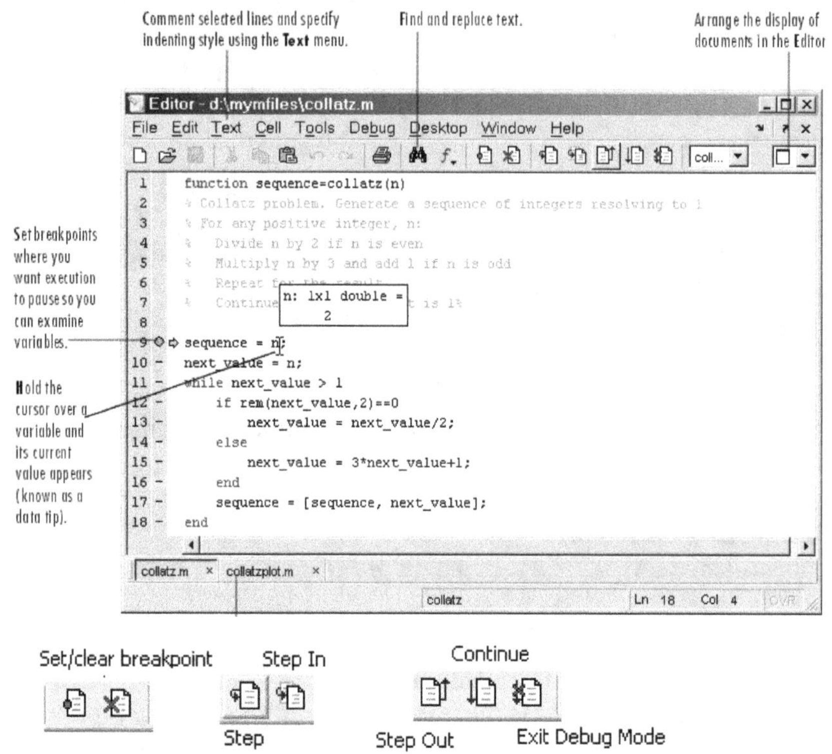

Profiling

ne way to improve the performance of your M-files is using profiling tools. MATLAB provides the M-file Profiler, a graphical user interface that is based on the results returned by the profile function.

The Profiler is a useful tool for isolating problems in your M-files. For example, if a particular part of the file did not run, you can look at the detail reports to see what lines did run, which might point you to the problem. You can also view the lines that did not run to help you develop test cases that exercise that code. If you get an error in the M-file when profiling, the

Profiler provides partial results in the reports. You can see what ran and what did not to help you isolate the problem. Similarly, you can do this if you stop the execution using Ctrl+C, which might be useful when a file is taking much more time to run than expected.

- Select Desktop-> Profiler from the MATLAB desktop.
- Select Tools->Open Profiler from the menu in the MATLAB Editor.

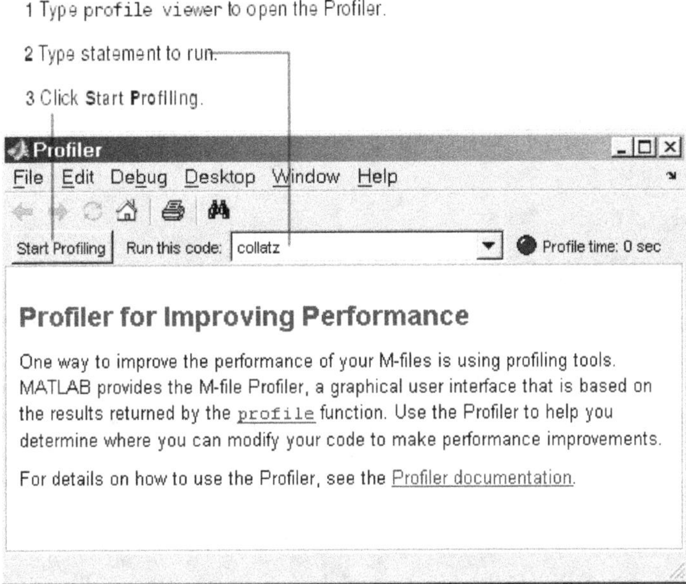

When profiling is complete, the Profile Summary report appears in the Profiler window.

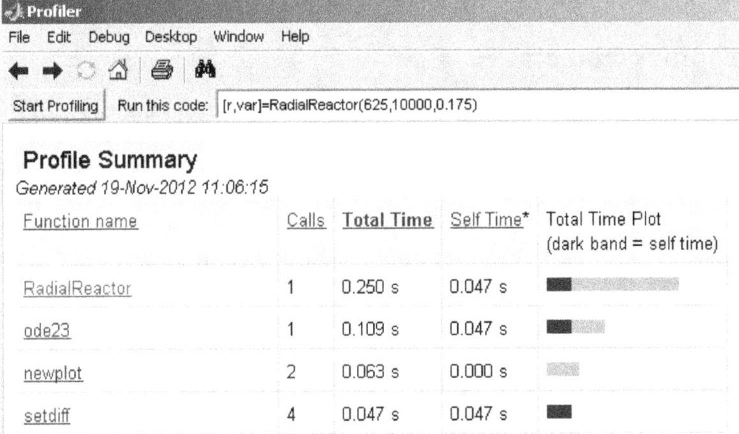

Function name	Calls	Total Time	Self Time*	Total Time Plot (dark band = self time)
RadialReactor	1	0.250 s	0.047 s	
ode23	1	0.109 s	0.047 s	
newplot	2	0.063 s	0.000 s	
setdiff	4	0.047 s	0.047 s	

flowtempder	49	0 s	0.000 s

Self time is the time spent in a function excluding the time spent in its child functions. Self time also includes overhead resulting from the process of profiling.

To get more detailed information about a particular function, click its name in the Function Name column to get the following details:

- Parent function
- Lines where the most time was spent
- Children (called functions)
- M-Lint results
- Coverage results
- File listing

Graphics

Organization of Graphics Objects

Graphics objects are the basic drawing elements used by MATLAB to display data. Each instance of an object is associated with a unique identifier called a handle. Using this handle, you can manipulate the characteristics (called object properties) of an existing graphics object. You can also specify values for properties when you create a graphics object. These objects are organized into a hierarchy, as shown by the following diagram.

There are two basic types of graphics objects:

- Core graphics objects -- Used by high-level plotting functions and by composite objects to create plot objects

- Composite objects -- Composed of core graphics objects that have been grouped together to provide a more convenient interface

Core Graphics Objects

h = axes('Position',position_rectangle) where position_rectangle = [left, bottom, width, height];

```
>> axes('position',[.1   .1   .8   .6])
```

```
>> mesh(peaks(20)); % 20x20 matrix obtained by
translating and scaling Gaussian distributions
```

```
>> axes('position',[.1   .7   .8   .2])
```

```
>> pcolor([1:10;1:10]); %rectangular array of cells with
colors determined by Cell argument
```

```
>>
set(gca,'CameraViewAngle',get(gca,'CameraViewAngle')+5)
%gca gets the current axes
>> str_path  = '';
>> str_fname = '04112009071.jpg';
>> img_orig = imread(strcat(str_path, str_fname));
>> image(img_orig)
>> image(img_orig)
>> image(1000,1000, img_orig)
>> img_gray = rgb2gray(img_orig);
>> level    = graythresh(img_gray);
img_bw    = im2bw(img_gray, level); % Here particles are
black
img_bw    = 1 - img_bw; % Here particles are white and
background is black
>> img_blob    = bwlabeln(img_bw);
img_blobrgb = label2rgb(img_blob);
>> figure;
subplot(2, 2, 1); imshow(img_orig); title('Original')
subplot(2, 2, 2); imshow(img_gray); title('Gray')
subplot(2, 2, 3); imshow(img_bw); title('Thresholded')
subplot(2, 2, 4); imshow(img_blobrgb);
title('Particles')
>> figure, h = surf(peaks);
>> figure, h = surf(peaks);
set(h,'FaceLighting','phong','FaceColor','interp',...
     'AmbientStrength',0.5)
light('Position',[1 0 0],'Style','infinite');
>> t = 0:pi/20:2*pi;
hline1 = plot(t,sin(t),'k');
>> hline2 = line(t+.06,sin(t),'LineWidth',4,'Color',[.8
.8 .8]); %add a shadow
>> set(gca,'Children',[hline1 hline2]) %pop the first
line to the front
```

```
x = [0 0;0 1;1 1];

y = [1 1;2 2;2 1];

z = [1 1;1 1;1 1];

tcolor(1,1,1:3) = [1 1 1];

tcolor(1,2,1:3) = [.7 .7 .7];

patch(x,y,z,tcolor)

>> rectangle('Position',[0.59,0.35,3.75,1.37])

>>
rectangle('Position',[0.59,0.35,3.75,1.37],'Curvature',[
0.8,0.4])

>> rectangle('Position',[0.59,0.35,3.75,1.37],...

          'Curvature',[0.8,0.4],...

          'LineWidth',2,'LineStyle','--')
```

This example creates a surface using the peaks M-file to generate the data, and colors it using the clown image. The ZData is a 49-by-49 element matrix, while the CData is a 200-by-320 matrix. You must set the surface's FaceColor to texturemap to use ZData and CData of different dimensions.
load clown

```
surface(peaks,flipud(X),...

'FaceColor','texturemap',...

 'EdgeColor','none',...

 'CDataMapping','direct')

colormap(map)

view(-35,45)
```

plot(0:pi/20:2*pi,sin(0:pi/20:2*pi))

text(pi,0,' \leftarrow sin(\pi)','FontSize',18)

Function	Related Functions
axes	
image	
light	
line	
patch	
rectangle	
surface	
Text	

Polynomials

MATLAB represents polynomials as row vectors containing coefficients ordered by descending powers. For example, consider the equation: $x^4 - 12x^3 + 25x + 116$

Enter this polynomial   » P = [1 -12 0 25 116];

Polynomial roots   » r=roots(P)

Characteristic Polynomial   » PP = poly(r)

Polynomial Evaluation at a specified value   >> polyval(PP,0)

Polynomial Multiplication / Convolution   >> a = [1 2 3 4]; b = [4 9 16]; c
= conv(a,b)

Polynomial division / Deconvolution   >> aa=deconv(c,b)

Derivative of the polynomial   >> polyder(aa)

Polynomial curvefitting

```
>> x = [1 3 7 21];
>> y = [2 9 20 55];
>> p=polyfit(x,y,3)
>> yCalc = polyval(p,x);
>> y-yCalc
```

Partial Fraction Expansion

$$\frac{-4s+8}{s^2+6s+8}$$

Consider the transfer function

```
>> b = [-4 8]; a = [1 6 8]; [r,p,k] = residue(b,a)
```

Given three input arguments (r, p, and k), residue converts back to polynomial form.

```
>> [b2,a2] = residue(r,p,k)
```

Mathematics using symbols:

```
» syms x
» int('x^3')
» eval(int('x^3',0,2))
» syms a b
» [a, b] = solve('a+b=7', 'a-b=1')
```

Solving nonlinear equations:

Write fcn1.m file as shown below

function y = fcn1(x)

```
% To solve f(x) = x^2 - 2x - 3 = 0

y = x^2 - 2*x - 3;

» y1 = fzero('fcn1', 0)
```

Zero found in the interval: [-1.28, 0.9051].

Another example nle.m

function f = nle(x)

```
% To solve

% f1(x1,x2) = x1^2 - 4x1^2 - x1x2 = 0

% f2(x1,x2) = 2x^2 - x2^2 + 3x1x2 = 0

f(1) = x(1) - 4*x(1)*x(1) - x(1)*x(2);

f(2) = 2*x(2) - x(2)*x(2) + 3*x(1)*x(2);

» x0 = [1 1]';

» x = fsolve('nle', x0)
```

The derivative can be found by using the polynomial derivativefunction polyder:

for function: $p??? -9.818x^2 ?? 201293x? -0.0317$

```
>>p=[-9.818 20.1293 -0.0317]

>> pd=polyder(p)
```

In MATLAB, the function polyfit solves the least square fitting problem. The data to be fitted:

```
>> x=[0 0.1 0.2 0.3 0.4 0.5]

>> y=[-0.447 1.978 3.28 6.16 7.08 7.34]
```

We should supply the preceding data and the order of the polynomial; n=1 for linear regression; n=2 for a quadratic polynomial:

```
>> n=2;

>> p=polyfit(x,y,n)

p=

- 22.0589 27.5497 -0.5835
```

The solution is: $y\ □22.0589x^2.\ 27.5497\ \tilde{x}\ 0.5835$

To compare the curve fit solution to the data points, let's plot both:

```
>>xi=linspace(0,0.5,100);
```

creates x-axis

```
>> z=polyval(p,xi);
```

calculates the polynomial p at points xi

```
>>plot(x,y,'-o',xi,z, ':')
```

plots the original data, marking the data points with 'o' and connecting them with straight lines, together with the polynomial data xi and z using a dotted line':'.

```
>>xlabel('x'),ylabel('y(x)'),title('second order curve
fitting')
```

Questions

1. A function of volume, f(V), is defined by the equation and setting it to zero.

$$pV^3 - bV^2 - RTV^2 + aV - ab = 0$$

Solve either by using the fzero command or by using the roots command to find all the roots of the polynomial.

```
Pcrit=111.3; % in atm

Tcrit=405.5; % in Kelvin

R=0.08206; % in atm.liter/g-mol.K

T=450; % K

a=27/64*R^2*Tcrit^2/Pcrit;

b=R*Tcrit/(8*Pcrit);

vols=roots([press, -(press*b+R*T), a, -a*b]);  %
Finds al roots

vol=max(vols(find(imag(vols) == 0)));
% finds largest real root
```

2. Fit 3rd and 4th order polynomial for vapor pressure:

```
vp = [    1      5     10     20     40     60    100
200    400    760]

T = [-36.7   -19.6   -11.5    -2.6    7.6    15.4    26.1
42.2    60.6    80.1]

m = 4
```

```
%fit the polynomial

p =polyfit(T,vp,m)

%p =    3.9631e-06    4.1312e-04    3.6044e-02
1.6062e+00    2.4679e+01

%evaluate the polynomial for every T (if desired)

z=polyval(p,T)

%z =    1.0477e+00    4.5184e+00    1.0415e+01
2.0739e+01    3.9162e+01

%    5.9694e+01    1.0034e+02    2.0026e+02
3.9977e+02    7.6005e+02
```

Numerical Analysis

Interpolation:

MATLAB provides a number of interpolation functions. Interp1 interpolates one dimensional data; Interp2 interpolates two dimensional data.

Typical measurements data are:

```
>>T=[50 60 70 80]

>> % temperatures in C

>>H=[2592.2 2609.7 2626.9 2643.8]

>> % vapor enthalpies in kJ/kg
```

Let's estimate the enthalpy at 65C in different ways:

```
>> h= interp1(T, H, 65)

>> % linear interpolation

>> h= interp1(T, H, 65,'linear')

>> % linear interpolation

>> h= interp1(T, H, 65,'cubic')

>> % cubic interpolation

>> h= interp1(T, H, 65,'spline')

>> % spline interpolation

>> h= interp1(T, H, 65,'nearest')

>> % nearest-neighbor
```

For a two dimensional interpolation of x, y, z data at x1 and y1 use:

```
>>zi=interp2(x,y,z,x1,y1,'linear')
```

```
>> % linear interpolation
>>zi=interp2(x,y,z,x1,x2,'cubic')
>> % cubic interpolation
```

Numerical and Symbolic Integration

Whenever it is difficult to integrate, differentiate, or determine some specific value of a function analytically, a computer may be called upon to numerically approximate the desired solution

Numerical Integration-Quadrature

MATLAB can perform in-depth numerical integrations or quadrature effortlessly and accurately. The first part of this section will look at some of the simple methods of numerical integration within MATLAB.

The Simpson's Rule and Lobotto Quadrature

In MATLAB, the functions 'quad' or in more recent versions, 'quadl' perform numerical integration based on the Simpson's rule and the adaptive Lobatto quadrature respectively. The syntax for the Simpson's based approximation and the Lo botto quadrature are the same. The difference is in each function's name. The syntax below is for the Lobotto quadrature (quadl), and it is the same for the Simpson's quadrature (quad).

```
q = quadl(fun,a,b)
q = quadl(fun,a,b,tol)
q = quadl(fun,a,b,tol,trace)
q = quadl(fun,a,b,tol,trace,p1,p2,...)
[q,fcnt] = quadl(fun,a,b,...)
```

The functions can be defined in function m-files or as inline functions (those functions entered directly into the command). MATLAB also has a function called 'trapz' which enables numerical integration using the trapezoid rule. More information about 'trapz' can be found in the MATLAB help files.

Differential Equations

- Symbolic differentiation and integration can be performed using *diff*, *int* commands e.g., diff(a*x^2+b*x), int(2*a*x+b) after defining a,b,x as symbols using *syms* command

```
>> syms a b x
>> diff(a*x^2+b*x)
ans =
```

```
2*a*x+b
>> int(ans)
ans =
a*x^2+b*x
```

- *dsolve* can solve ordinary differential equations symbolically with or without boundary conditions or initial value parameters

  - dsolve('D2y=6*y-Dy','x') to solve the ordinary differential equation

  $$\frac{d^2y}{dx} = 6y - \frac{dy}{dx}$$

  - dsolve('D2m+m=0','m(0)=2','Dm(0)=3','x') to solve the ode

  $$\frac{d^2m}{dx^2} + m = 0, m(0) = 2 \text{ and } \frac{dm}{dx}(0) = 3 \text{ where m(x).}$$

- Numerical solution for ODEs: *ode45, ode15* etc. can be used to numerically differentiate the ODEs

- A fluid of constant density starts to flow into an empty and infinitely largetank at 8 L/s. A valve regulates the outlet flow to a constant 4L/s. Derive and solve the differential equation describing this process over a 100 second interval.

  Create a function ODE1.m with 2 line code:

  Function dvdt=ODE1(t,v)

```
dvdt=4
>>timespan=[0 100];
>>v0=0        %Initial condition
>>[t,v]=ode45('ODE1',timespan,v0)
[independent, dependent]=ode45('file_name',tspan,initial_condition_vector)
>>plot(t,v(:,1))
>>xlabel('Time(s)')
>>ylabel('Volume in tank(L)')
>>title('ODE describing the tank volume')
diff((exp1,independent)
dsolve('ode1,ode2,oden', 'bc1,bc2,bcn', 'IV')
[Indepent, dependentbase] = ode#('filename', tspan, y0)
```

Solver	Solves These Kinds of Problems	Method
ode45	Nonstiff differential equations	Runge-Kutta
ode23	Nonstiff differential equations	Runge-Kutta
ode113	Nonstiff differential equations	Adams
ode15s	Stiff differential equations and DAEs	NDFs (BDFs)
ode23s	Stiff differential equations	Rosenbrock
ode23t	Moderately stiff differential equations and DAEs	Trapezoidal rule
ode23tb	Stiff differential equations	TR-BDF2